Veterinary Virology: An Evidence-Based Approach

Veterinary Virology: An Evidence-Based Approach

Edited by Travis Schroeder

AMERICAN
MEDICAL PUBLISHERS
www.americanmedicalpublishers.com

American Medical Publishers,
41 Flatbush Avenue,
1st Floor, New York,
NY 11217, USA

Visit us on the World Wide Web at:
www.americanmedicalpublishers.com

ISBN: 978-1-63927-531-1

Cataloging-in-Publication Data

Veterinary virology : an evidence-based approach / edited by Travis Schroeder.
p. cm.
Includes bibliographical references and index.
ISBN 978-1-63927-531-1
1. Veterinary virology. 2. Evidence-based medicine. 3. Veterinary medicine. I. Schroeder, Travis.
SF780.4 .V48 2022
636.089 601 94--dc23

Table of Contents

Preface

This book was inspired by the evolution of our times; to answer the curiosity of inquisitive minds. Many developments have occurred across the globe in the recent past which has transformed the progress in the field.

The study related to the viruses in animals is known as veterinary virology. It is a sub-field of veterinary medicine. Each animal species is affected by different types of viruses. However, there are also some viruses which can affect different species as well as both vertebrates and invertebrates. Some of the viruses which affect vertebrates are bluetongue virus, rabies virus and rhabdovirus. Bluetongue virus generally infects livestock while rabies virus can infect a large variety of animals such as dogs, monkeys, foxes and bats. Rhabdovirus is single stranded, negative sense RNA virus inheriting six genera that infect a wide variety of animals such as cattle, fish, horse, bovine, etc. Invertebrates such as honey bees are infected by deformed wing virus. This book contains some path-breaking studies in the field of veterinary virology. Different approaches, evaluations, methodologies and advanced studies on veterinary virology have been included herein. Researchers and students in this field will be assisted by the content of this book.

This book was developed from a mere concept to drafts to chapters and finally compiled together as a complete text to benefit the readers across all nations. To ensure the quality of the content we instilled two significant steps in our procedure. The first was to appoint an editorial team that would verify the data and statistics provided in the book and also select the most appropriate and valuable contributions from the plentiful contributions we received from authors worldwide. The next step was to appoint an expert of the topic as the Editor-in-Chief, who would head the project and finally make the necessary amendments and modifications to make the text reader-friendly. I was then commissioned to examine all the material to present the topics in the most comprehensible and productive format.

I would like to take this opportunity to thank all the contributing authors who were supportive enough to contribute their time and knowledge to this project. I also wish to convey my regards to my family who have been extremely supportive during the entire project.

Editor

Genetic characterization of hantaviruses isolated from rodents in the port cities of Heilongjiang, China, in 2014

Suya Cao[1†], Jian Ma[2†], Cheng Cheng[3], Wendong Ju[3] and Yulong Wang[1*]

Abstract

Background: Hantavirus is a tripartite negative-sense RNA virus. It can infect humans through contaminated rodent excreta and causes two types of fatal human diseases: hemorrhagic fever with renal syndrome (HFRS) and hantavirus pulmonary syndrome (HPS). China exhibits the highest HFRS occurrence rate in the world, and the Heilongjiang area is one of the most severely infected regions.

Results: To obtain additional insights into the genetic characteristics of hantaviruses in the port cities of the Heilongjiang area in China, a molecular epidemiological investigation of hantaviruses isolated from rodents was performed in 2014. A total of 649 rodents (11 murine species and 1 shrew species) were caught in 12 port cities in Heilongjiang. Among these rodents, the most common species was A. agrarius, and the second-most common was R. norvegicus. A viral gene PCR assay revealed the presence of two specific genotypes of hantavirus, referred to as Hantaan virus (HTNV) and Seoul virus (SEOV), and the positive SEOV infection rate was higher than that for HTNV. A genetic analysis based on partial M segment sequences indicated that all of the isolates belonging to SEOV could be assigned to two genetic lineages, whereas the isolate belonging to HTNV could be assigned to only one genetic lineage.

Conclusions: These results suggested that HTNV and SEOV are circulating in A. agrarius and R. norvegicus in the port cities in the area of Heilongjiang, but SEOV may be the dominant common hantavirus.

Keywords: Hantavirus, SEOV, HTNV, Epidemiology, Heilongjiang

Background

Hantaviruses, which belong to the Hantavirus genus in the *Bunyaviridae* family, are tripartite negative-sense RNA viruses. Hantaviruses possess a tripartite negative-sense RNA genome consisting of the following three segments: the large (L) segment encodes a viral RNA-dependent RNA polymerase; the medium-sized (M) segment encodes two viral glycoproteins (GPs, Gn and Gc); and the small (S) segment encodes the viral nucleocapsid protein (NP) [1–5]. The GPs along with the NP determine the virulence and pathogenicity of the hantavirus. Unlike other viruses of the Bunyaviridae family, hantaviruses are not transmitted by arthropods; rather, they infect people though the

urine, saliva and feces excreted by rodent hosts, especially muroids [6, 7]. Hantaviruses only generate transient pathology in rodents, and they do not affect the life span and reproduction of their hosts. In contrast, they can cause two severe clinical manifestations in humans: hemorrhagic fever with renal syndrome (HFRS) in the old world and hantavirus cardiopulmonary syndrome (HPS) in the new world [8–12]. Previous studies have indicated that at least 40 species and 30 genotypes belonging to the hantavirus genus have been isolated worldwide [13].

Etiological studies have shown that HFRS that has spread around the world, resulting in the production of variant hantaviruses. In Asia and Europe, five types of hantaviruses can cause HFRS: Hantaan virus (HTNV), Seoul virus (SEOV), Dobrava virus (DOBV), Saaremaa virus (SAAV), and Puumala virus (PUUV) [14]. In the USA, the Sin Nombre virus (SNV) and the Andes virus (ANDV) are stable viruses that can cause HPS. Most HFRS cases occur in Europe and East Asia (Korea,

* Correspondence: wangyl@nefu.edu.com
Suya Cao and Jian Ma are co-first authors.
[†]Equal contributors
[1]Department of Wildlife Medicine, Wildlife Resources Faculty, Northeast Forestry University, Harbin 150040, China
Full list of author information is available at the end of the article

China and the eastern part of Russia) [11, 15]. China is the country that is most seriously affected by hantavirus infection worldwide. Previous investigations have shown that HFRS-infected patients in East Asian countries, including China, Russia and Korea, account for at least 90 % of HFRS patients around the world. At least 100,000 cases of HFRS are reported annually in China, and more than 900 cases are reported in Korea and the eastern part of Russia [16, 17]. HFRS is caused by different types of hantaviruses in different countries, and phylogenetic analysis suggests that hantaviruses and rodent hosts have coevolved [3]. In China, there are two common types of hantavirus, HTNV and SEOV, which are reportedly carried by *A. agrarius* and *R. norvegicus,* respectively, and some hantaviruses isolated from *R. norvegicus* seemed to be HTNV [16, 18, 19].

HTNV and SEOV could cause serious public health problems in China, especially in the Heilongjiang area of China. Heilongjiang is located in northeastern China, and it is adjacent to northern Russia and the nearby Jilin area to the south. It was the first area in which the etiological agent of HFRS was isolated in China, [11] and the Heilongjiang area has remained a high-incidence region [20]. Previous studies have shown that SEOV and HTNV are circulating in the Heilongjiang area [21, 22].

The port of Heilongjiang serves as bridge between countries. A single commercial port is used for economic trade, the exchange of technology and culture, tourism, immigration and so on. Given that many anthropozoonoses are transmitted by vectors (material, people and animals), and these vectors might be introduced into China through ports, it is essential to investigate the molecular epidemiology of hantaviruses by monitoring the rodents that are the natural hosts of hantavirus in the port cities of the Heilongjiang area. In this study, 12 port cities have been selected as trapping sites for our hantavirus investigation. We captured 649 rodents and characterized 29 hantaviruses isolated from various rodent species in the port cities. A phylogenetic analysis of the partial M segments indicated that 19 of 29 viruses belonged to SEOV, and 10 of 29 viruses belong to HTNV. Entire M segment of HTNV (TJ strain) which isolated from *R. norvegicus* have been sequenced and compared with the reference HTNVs. The results indicated that the spillover of HTNV has occurred under natural condition in port cities of Heilongjiang area.

Methods
Rodent collection
In 2014, rodents were trapped with rat traps or baited cages in the 12 port cities of Luobei, Jiayin, Jiamusi, Dongning, Hulin, Mudanjiang, Mishan, Harbin, Fujin, Tongjiang, Suifenhe, and Raohe (Fig. 1). The trap sites included tussocks, brushwood, residential areas, canals and fields, among other areas. All of the trapped rodents were identified by zootaxy experts, and they were dissected as soon as they were authenticated. Lung tissues were collected from the captured rodents and transported to the HEILONGJIANG ENTRY-EXIT INSPECTION AND QUARANTINE BUREAU at −80 °C and stored in a liquid nitrogen tank until further processing. All procedures used in this study were approved by the Institutional Animal Care and Use Committee of Heilongjiang International Travel Healthcare Center (Harbin, China) (XD-25; 12 April 2014).

RNA extraction and RT-PCR
The RNAprep Pure Tissue Kit (TIANGEN, Beijing, China) was used to extract total RNA from lung tissues, after the lung tissues were ground into a homogeneous mixture with a Tissue Lyser (Qiagen, Hilton, Germany). Purified total RNA was reverse-transcribed using the Superscript First-Strand Synthesis System (Invitrogen, Beijing, China) and random primers (Invitrogen, Beijing, China) to obtain cDNA. The HTNV and SEOV genotypes were amplified via nested PCR. The initial round of PCR product amplification employed the primer pairs HTm1F (5′-AAA GTA GGT GTT AYA TCY TTA CAA TGT GG-3′) and HTm1R (5′-GTA CAT CCT GTR CCT ACC CC-3′), and the second round of PCR production used the primer pairs HTm2F (5′-GAA TCG ATA CTG TGG GCT GCA AGT GC-3′)/HTm2R (5′-GGA TTA GAA CCC CAG CTC GTC TC-3′) and SEm2F (5′-GTG GAC TCT TCT TCT CAT TAT T-3′)/SEm2R (5′-TGG GCA ATC TGG GGG GTT GCA TG-3′). The initial round of PCR cycling consisted of an initial denaturation at 95 °C for 5 min, followed by 35 cycles of denaturation at 94 °C for 30 s, annealing at 52 °C for 30 s, and elongation at 72 °C for 30 s, with a final extension at 72 °C for 10 min. The second round of PCR cycling consisted of an initial denaturation at 95 °C for 5 min, followed by 30 cycles of denaturation at 94 °C for 30 s, annealing at 55 °C for 30 s, and elongation at 72 °C for 30 s, with a final extension at 72 °C for 10 min. The amplified products were approximately 330 bps.

For avoiding possible false positive results caused by RT-PCR assays, a commercial hantavirus detection kit (Hantavirus qPCR assay kit, Huiruibio, China) was also used to confirm the positive results determined by RT-PCR.

The whole M gene of HTNV isolate (TJ) was further amplified by RT-PCR method as following: The 5′ fragment of M gene was amplified by the primer pairs as : HM1F (5′- CAA CAT TAT ATA TGA TTG TAC CGA T -3′)/HM1R (5′- TGA ACC TGT GAG TTA CCT GGC ATA C -3′), and the 3′ fragment of M gene was amplified by the primer pairs as : HM2F (5′- AGA TGT TAT ATC

Fig. 1 Map showing the trapping sites for rodents in the Heilongjiang area, China. Almost all of the port cities are located along the border with Russia, except for Harbin and Jiamusi. 48 rodents were captured in Luobei, 79 in Jiayin, 60 in Jiamusi, 22 in Dongning, 36 in Hulin, 27 in Mudanjiang, 102 in Mishan, 95 in Harbin, 29 in Fujin, 60 in Tongjiang, 22 in Suifenhe, and 69 in Raohe

TTT ACA ATG TGG G -3′)/HM2R (5′- CAC TCT CTG CAC CAT AAC AGA TAG C -3′). The PCR cycling consisted of an initial denaturation at 95 °C for 5 min, followed by 35 cycles of denaturation at 94 °C for 30 s, annealing at 55 °C for 30 s, and elongation at 72 °C for 90 s, with a final extension at 72 °C for 10 min. The sequence of entire M gene with approximately 1500 bps was obtained by assembling the above two fragments sequences by using DNAStar software.

Phylogenetic analysis of M gene sequences

The PCR products were purified with the EZNA TM Gel Extraction Kit (OMEGA, USA) according to the manufacturer's instructions and sequenced with the same primers used for PCR amplification. The nucleic acid sequences of the hantaviruses and those downloaded from GenBank were edited and analyzed with the DNASTAR program (DNASTAR, Madison, WI, USA). The identities of the isolated and downloaded hantavirus sequence were also calculated with the DNASTAR program.

The Seqman program was used to edit the nucleic acid sequences. The MEGA 4.0 program was employed to generate a viral phylogenetic tree via the neighbor-joining (N J) method. The alignment program of DNAStar software and Gendoc software were used to complete the alignment analyses of amino acid sequence of M segment

Table 1 Detection of hantavirus in rodent species captured at trap sites in various locations

Location	A. agrarius		R. norvegicus		Clethrionomys rufocanus Sundevall		A. peninsulae		Clethrionomys rutilus Pallas		Mus musculus		Microtus fortis Buchner		R. rattus		E. sibiricus		C. triton		S. dauricus		Sorex araneus Linnaeus	
	n/N	%	n/N	%	n/N	%	n/N	%	n/N	%	n/N	%	n/N	%	n/N	%	n/N	%	n/N	%	n/N	%	n/N	%
Luobei	0/5[a]	0[b]	0/3	0	–[c]	0	0/2	0	0/11	0	–	–	0/25	0	–	–	0/2	0	–	0	–	–	–	–
Jiayin	1/43	2.3	0/14	0	0/5	0	–	–	0/1	0	0/5	0	0/2	0	0/1	0	0/6	0	0/2	0	–	0	–	0
Jiamusi	0/15	0	4/43	9.3	–	–	–	–	–	–	0/1	0	–	–	–	–	–	–	–	–	–	–	0/1	0
Dongning	0/5	0	5/14	35.7	–	0	–	–	0/3	0	–	–	–	–	–	–	0/3	0	–	–	–	–	–	–
Hulin	1/21	4.8	0/4	0	0/7	0	0/1	0	–	–	–	–	–	–	–	–	–	–	–	–	–	–	–	–
Mudanjiang	0/21	0	2/6	33.3	–	0	–	–	–	–	–	–	–	–	–	–	–	–	–	–	–	–	–	–
Mishan	1/51	1.9	0/25	0	0/3	0	0/17	0	–	–	0/1	0	–	–	–	–	0/2	0	0/3	0	–	–	–	–
Harbin	4/36	11	0/18	0	–	–	–	–	–	–	–	–	–	–	0/2	0	–	–	0/36	0	0/3	0	–	–
Fujin	1/11	9.1	0/15	0	–	–	–	–	–	–	–	–	0/3	0	–	–	–	–	–	–	–	–	–	–
Tongjiang	3/46	6.5	1/11	9.1	–	–	–	0	–	–	–	–	0/1	0	–	–	–	–	–	–	–	–	0/2	0
Suifenhe	0/5	0	0/5	0	–	–	0/7	0	–	–	–	–	–	–	–	–	0/1	0	0/3	0	–	–	0/1	0
Raohe	3/23	13	3/22	13.6	–	–	0/5	0	0/5	0	0/2	0	0/12	0	–	–	–	–	–	–	–	–	–	–
Total	14/282	4.9	15/180	8.3	0/15	0	0/32	0	0/20	0	0/9	0	0/43	0	0/3	0	0/14	0	0/44	0	0/3	0	0/4	0

[a] Number of positive samples/Number of tested samples
[b] Positive rate (%)
[c] No Samples

Table 2 The information of the Hantavirus islalated in this Study

Genotype	Isolate	Geographic Location	Accession No.
HTNV	CJAp93	Jilin	EF208930
HTNV	CGHu1	Guizhou	EU092222
HTNV	Lee	South Korea	D00377
HTNV	76-118	South Korea	Y00386
HTNV	LR1	China	AF288293
HTNV	S85-46	China	AF288658
HTNV	84FLi	Sanxi	AF366569
HTNV	SN7	China	AF288656
HTNV	N8	-	EF077656
HTNV	Z10	Zhejiang	NC_006437
HTNV	ZLS6-11	Zhejiang	FJ753397
HTNV	A6	China	AF288645
HTNV	HV114	Hubei	L08753
HTNV	h5	Helongjiang	L08753
HTNV	a16	Sanxi	AF288645
HTNV	AH09	China	AF285265
HTNV	jilinap06	Jilin	EF371454
HTNV	Q32	Guizhou	DQ371905
HTNV	CGRni1	Guizhou	EU363815
HTNV	TJJ16	Tianjin	EU074672
HTNV	CGAa4MP9	Guizhou	EF990929
HTNV	CGAa1015	Guizhou	EF990926
HTNV	NC167	Anhui	DQ989237
HTNV	Bao14	Helongjiang	AB127995
HTNV	TJF3	Tongjiang	KT885159
HTNV	TJF2	Tongjiang	KT885160
HTNV	TJF1	Tongjiang	KT885161
HTNV	HRH14	Raohe	KT885162
HTNV	HRH13	Raohe	KT885163
HTNV	HMS1	Mishan	KT885164
HTNV	HMDJ6	Mudanjiang	KT885165
HTNV	HJY10	Jiayin	KT885166
HTNV	HJI14	Harbin	KT885167
HTNV	HHL11	Hulin	KT885168
SEOV	hebei4	Hebei	AB027089
SEOV	Z37	Zhejiang	AF190119
SEOV	ZT71	Zhejiang	EF117248
SEOV	Sapporo	Sapporo	M34882
SEOV	80-39	South Korea	S47716
SEOV	hubei-1	Hubei	S72343
SEOV	Tchoupitoulas	-	U00473
SEOV	Vietnam5CSG	Vietnam	AB618130
SEOV	guang199	-	AB027086
SEOV	Houston	UK	U00465

Table 2 The information of the Hantavirus islalated in this Study
(Continued)

SEOV	HN71-L	Hainan	AB027084
SEOV	K24-V2	-	AF288654
SEOV	L99	Jiangxi	AF288298
SEOV	HB55	Henan	AF035832
SEOV	R22	Henan	AF035834
SEOV	Brazil	Brazil	U00460
SEOV	humber	UK	JX879768
SEOV	IR461	UK	AF458104
SEOV	Gou3	zhejiang	AF145977
SEOV	SDN2	Dongning	KT885169
SEOV	SDN8	Dongning	KT885170
SEOV	SDN10	Dongning	KT885171
SEOV	SDN12	Dongning	KT885172
SEOV	SDN13	Dongning	KT885173
SEOV	SDN19	Dongning	KT885174
SEOV	SJC12	Harbin	KT885175
SEOV	SJC20	Harbin	KT885176
SEOV	SJI1	Harbin	KT885177
SEOV	SJMS16	Jiamusi	KT885178
SEOV	SJMS19	Jiamusi	KT885179
SEOV	SJMS21	Jiamusi	KT885180
SEOV	SJMS30	Jiamusi	KT885181
SEOV	SMDJ24	Mudanjiang	KT885182
SEOV	SRH17	Raohe	KT885183
SEOV	SRH19	Raohe	KT885184
SEOV	SRH27	Raohe	KT885185
SEOV	SRHA1	Raohe	KT885186
SEOV	TJF11	Tongjiang	KT885187

Results

Descriptions of the sampled rodents

In 2014, 649 rodents were captured from the 12 port cities in the Heilongjiang area of China. The rodents included 282 specimens of *A. agrarius*, 180 *R. norvegicus*, 15 *Clethrionomys rufocanus Sundevall*, 32 *A. peninsulae*, 20 *Clethrionomys rutilus Pallas*, 9 *Mus musculus*, 43 *Microtus fortis Buchner*, 3 *R. rattus*, 14 *E. sibiricus*, 44 *C. triton*, 3 *S. dauricus*, and 4 *Sorex araneus Linnaeus*. The Heilongjiang area harbors a diversity of species, with *A. agrarius* and *R. norvegicus* representing the dominant species. Nested PCR targeting partial M segment sequences could be applied to screen the infected rodents in all samples. The results showed that the infection rate of *A. agrarius* was 4.9 %, and the infection rate of *R. norvegicus* was 8.3 % (Table 1).

Genetic diversity of SEOV isolates

Phylogenetic tree was constructed with the partial M segment sequences of the viruses (Fig. 2). In the phylogenetic tree analyses of the SEOV isolates, the calculations showed that all of isolates formed six lineages. Lineages 1 to 4 were quite closely related to one another. All of the isolates in lineages 1, 2, and 3 came from China, while the viruses in lineage 4 were isolated from China and neighboring countries (Japan, Korea, Singapore and Vietnam) as well as the United States. In lineage 4, the isolates came from Jiamusi, Mudanjiang, Harbin and Raohe. The isolates show large sequence distances from strains Z37 and CP211, which were isolated from Zhejiang and Beijing. Lineage 3 consisted of the viruses isolated from Dongning and Raohe, and their virus sequences were very similar. All of the isolates were closely related to the Vietnam5CSG strain previously isolated from small mammals in Vietnam, which is far from the Heilongjiang area. Within lineage 1, strain L99 was

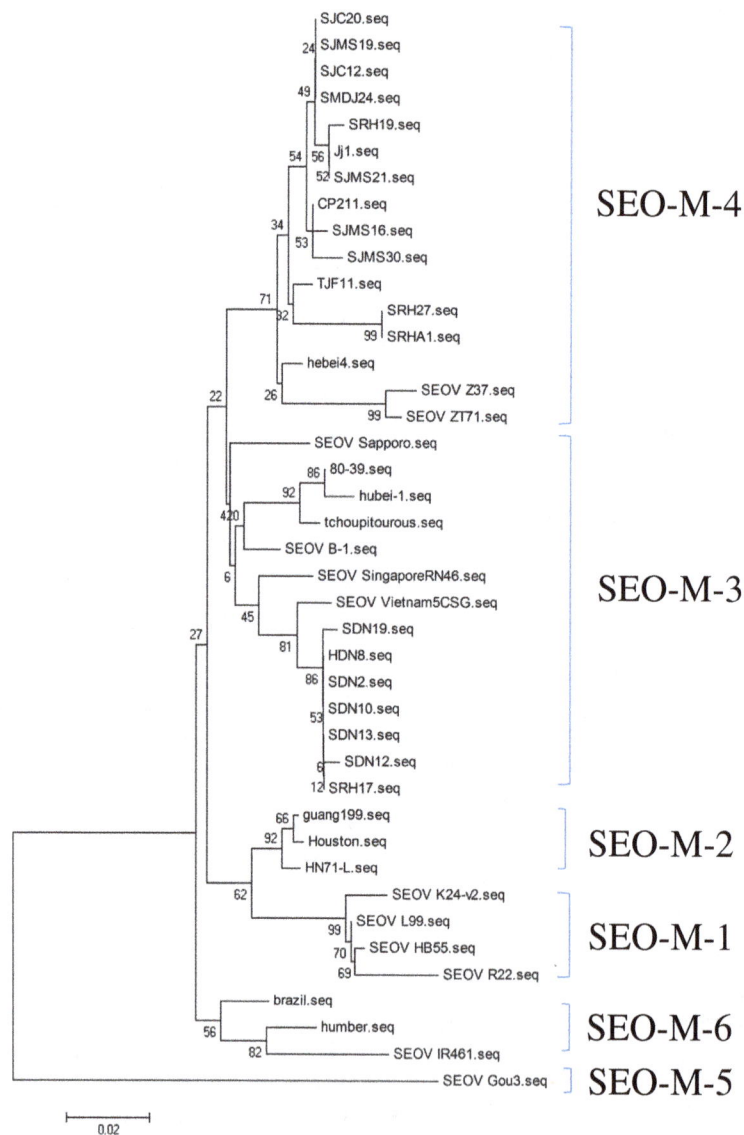

Fig. 2 Phylogenetic tree of Seoul virus isolates based on the partial sequences of M segments. The tree was constructed by the neighbour-joining method (NJ) of Mega software. The HTNVs isolated from *R. norvegicus* or *A. agrarius* that were trapped in Harbin were designated SJC20, ji1 and SJC12; those iaolated from Jiamusi, SJMS19, SJMS21, SJMS16, and SJMS30; those isolated from Suifenhe, SRH19, SRH27, SRH17 and SRHA1; that isolated from Mudanjiang, SMDJ24; that iaolated from Tongjiang, TJF11; and those isolated from Dongning, HDN8, SDN2, SDN10, SDN12 and SDN13. The GenBank accession numbers of the viruses isolated in this study and the reference HTNV stains were listed in Table 2 in detailedly

isolated from the Jiangxi area, while strain K24-v2 came from the Zhejiang area, and these two areas share borders. The district from which strains R22 and HB55 were isolated, in the Henan area, is located far from the Zhejiang area. Lineage 5 consisted of Gou3, which was isolated from the Zhejiang area, and it appeared to be distinct from the other SEO viruses. Lineage 6 consisted of UK viruses, with the exception of one Brazil strain.

Genetic diversity of HTNV

The genetic diversity of the HTN isolates was lower than that of the SEO isolates based on the partial M segment sequences. The analysis of the obtained HTNV phylogenetic tree revealed nine lineages from all of the isolated genes (Fig. 3). The partial M segment sequences of HTNV that were isolated in this study belonged to only one lineage, lineage 6. This lineage has been designated the Far East (FE) lineage,

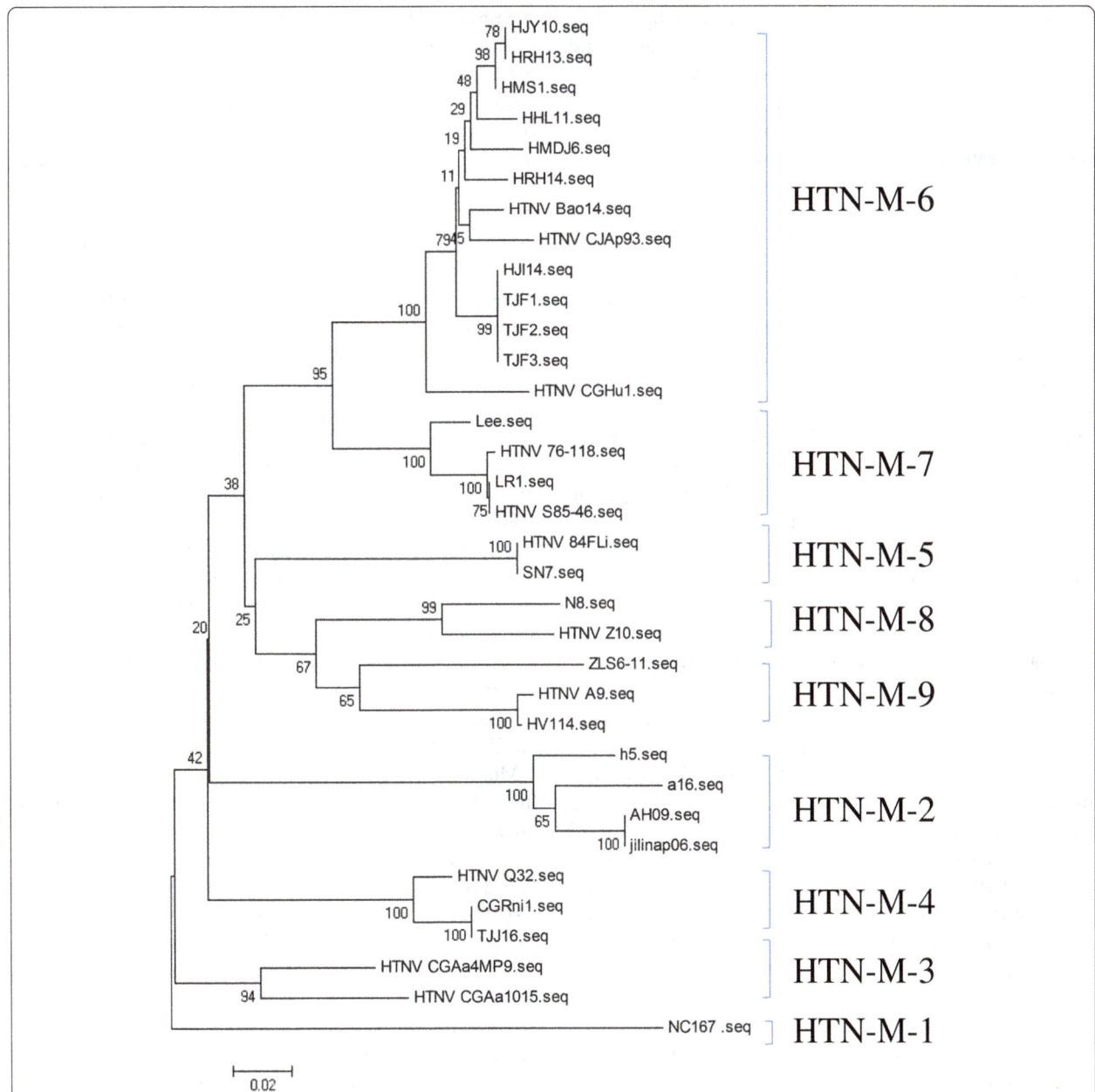

Fig. 3 Phylogenetic tree (NJ) of hantaviruses based on partial M segment sequences of HTNVs from China. The HTNV that isolated from *R. norvegicus* or *A. agrarius* that were trapped in Harbin was designated HJI14; those isolated from Tongjiang, TJF1, TJF2 and TJF3; that isolated from Hulin, HHL11; that isolated from Mishan, HMS1; that isolated from Jiayin, HJY10; that isolated from Mudanjiang, HMDJ6; and those isolated from Raohe were designated HRH14 and HRH13. The GenBank accession numbers of the viruses isolated in this study and the referencial SEOV stains were showed in Table 2 in detailedly

which was identified in patients with serious HFRS. The viruses isolated from the port cities (Jiayin, Raohe, Mishan, Hulin, Mudanjiang and Harbin) in the Heilongjiang area appeared to be very closely related to one another. The Bao14 strain showed a similar sequence to the viruses isolated from the Heilongjiang area and the CJAp93 strain isolated from the Jilin area, which adjoins Heilongjiang. Some of the partial M segment sequences of the isolates from port cities were also closely related to HTNV strains that were isolated far from Heilongjiang, such as the CGHu1 strain originating from the Guizhou area. Lineage 7 consisted of several Korean viruses and shared a common evolutionary source with lineage 6. Lineage 1 consisted of the novel Hantaan virus type NC167, which was isolated from Anhui. It forms a separate branch in the phylogenetic tree.

Phylogenetic analysis of HTN, SEO and prototype hantaviruses

To understand more about the HTNV and SEOV genetic variants circulating in rodent hosts in the port cities of the Heilongjiang area, we employed 19 partial M segment sequences of SEOV (KT885169-KY885187) and 10 partial M segment sequences of HTNV (KT885159-KT885168) to construct a phylogenetic tree with the reference sequences Hantaan 76-118 and Seoul 80-39 (Fig. 4). The phylogenetic analysis showed that the mutation rates of HTNV isolates were higher than the mutation rates of SEOV isolates, based on the reference hantavirus sequences.

Comparison of amino acid sequences of M segment among HTNVs

The nucleotide sequence for M segment of TJ strain which isolated from R. norvegicus is 3408 nt, and a 1135 amino acid sequence could be translated by the nucleotide sequence. The alignment result in Fig. 5 showed that there was no deletion or insertion found in the deduced amino acid sequence based on M gene sequence of TJ strain. The identity of the deduced amino acid sequence between TJ strain and other HTNVs (Bao14, 84Li, N8 and Q32 strain) were more than 95 %. Among all the HTNVs, TJ strain showed a highest identity with Bao 14 strain (99.2 %), which was isolated from Heilongjiang too. The previous studies indicated that Five N- glycosylation sites in Gn (position 134, 235, 347, 399 and 609) and one N- glycosylation sites in Gc (position 928) were related with the function of M gene coding proteins [23]. However, all of the N- glycosylation sites were conserved among these different HTNVs (Fig. 5).

Discussion

The Heilongjiang area is the most seriously affected endemic area of HFRS in China [13]. To gain a better understanding of the genetic characteristics of the hantaviruses present in the port cities in the Heilongjiang area, we collected 649 rodent specimens, and we detected hantaviruses in A. agrarius and R. norvegicus. The primary rodent hosts in the Heilongjiang area were A. agrarius and R. norvegicus for all of 2014. Among all of the evaluated rodents, 4.47 % were positive for hantavirus infection. In addition, we analyzed the genetic evolution of hantaviruses based on all of the positive isolates obtained during this study.

The rodent hosts in a specific area and the epidemic genotypes of hantaviruses exhibit a close relationship to epidemics of HFRS. This study showed that two hantaviruses (HTNV and SEOV) are circulating in the Heilongjiang frontier area of northeastern China at present. The epidemiological investigation showed that A. agrarius and R. norvegicus are the dominant rodent species and carriers of hantavirus in the suburban district and residential areas of the port cities in the Heilongjiang area (Table 1). Although the number of R. norvegicus specimens captured in this study was smaller than the number of A. agrarius, the R. norvegicus infection rate was higher than that of A. agrarius. R. norvegicus has become an advantageous rodent host for hantavirus, and the SEOV genotype carried by R. norvegicus has become the primary genotype of the virus. These results showed that HFRS occurring in the Heilongjiang frontier area might be caused primarily by SEOV. However, the possibility that other hantaviruses also play important roles in causing HFRS in this area could not be eliminated. Therefore, further studies will be necessary to determine whether other viruses are present in the Heilongjiang frontier area, and clinical samples will be needed to clarify the true etiological agents of HFRS in these regions.

Molecular epidemiological analysis has indicated that the SEOVs obtained in this study can be divided into six lineages in phylogenetic trees based on partial M segment sequences [24, 25]. Previous studies showed that lineages 1, 3 and 4 were present in northeastern China [26, 27]. However, we only isolated the viruses belonging to lineages 3 and 4 in the Heilongjiang frontier area in this study. Notably, these two virus lineages are distributed widely in China, especially in northeastern China and the middle and lower reaches of the Yangtze river [26, 28, 29]. Lineage 4 includes strains from Jilin, Zhejiang and Jiangxi, while lineage 3 includes strains from Japan, Vietnam and South Korea. We surmised that hantaviruses have spread around the world along with their primary natural reservoir, harbored by R. norvegicus. Moreover, according to Fig. 2, we observed that lineages 1— 4 exhibit a

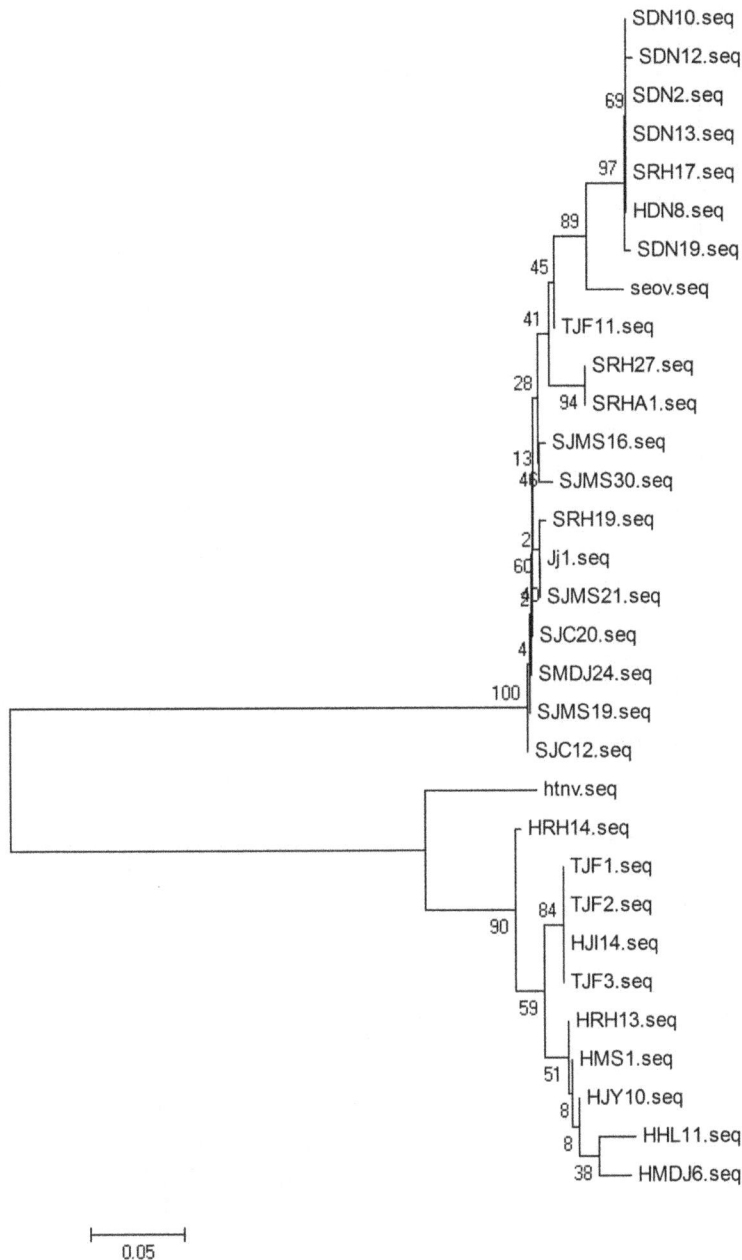

Fig. 4 Phylogenetic tree (NJ) of hantaviruses based on partial M segment sequences of all Hantaviruses isolated in this study. The analysis was performed with MEGA software. The numbers at the nodes are bootstrap confidence levels for 1000 replicates. 19 partial M segment sequences of SEOV isolates and 10 partial M segment sequences of HTNV isolates were used to construct this phylogenetic tree with the reference sequences Hantaan 76-118 and Seoul 80-39. The GenBank accession numbers of the reference SEOV and HTNV genes were S47716 and Y00386 as showed in Table 2

very close evolutionary relationship to lineage 6. These results indicated that most of the SEOVs isolated from around the world might have evolved from a single ancestor. However, a few variants, such as Gou3, which was isolated from Zhejiang and belongs to lineage 5, show a high degree of variability from most of the other SEOVs, which may suggest the evolution of an ancient hantavirus.

The first time that HTNV was reported was in Korea, and it was carried by *A. agrarius* [30]. Phylogenetic analysis showed that HTNV can be divided into nine lineages [25]. Previous work demonstrated the presence of lineage 6 in Heilongjiang Province, and lineage 7, which originated from Korea, is found widely throughout China [29, 31]. Several of the

Gn

```
                    *        20         *        40         *        60         *        80         *       100        *
HTN-TJ.pro : MGMWKWLVIASLVWPALALRNVYDMKIECPHTVSFGENSVIGYVELPPMPLADTAQMVPESSCSMDNHQSINTITKYTQVTWRGKADPGQSSQNSFETVSAEVDLKGTCVLKHK : 114
HTN-bao14. : ..................................................................................I.......................T........... : 114
HTN-84Li.p : ..............CT.............................T..............................K...QA............V.......... : 114
HTN-N8.pro : ..............YT.........................................N.....................QA.A......A.............. : 114
HTN-Q32.pr : ..............F.............................T.............................I.......QS.........T........... : 114

                   120        *       140         *       160         *       180         *       200        *       220
HTN-TJ.pro : MVEESYRSRKSITCYDLSCNSTVCKPTLYMIVPIHACNMMKSCLIALGPYRVQVVYERTYCMTGVLIEGKCFVPDQSVVSIIKHGIFDIASVHIVCFFVAVKGNTYKLFEQVKK : 228
HTN-bao14. : .................................................................................................. : 228
HTN-84Li.p : ............................................................................K..................... : 228
HTN-N8.pro : ...............N................................................................................. : 228
HTN-Q32.pr : .................................................................................................. : 228

                  *           240         *       260         *       280         *       300         *       320         *       340
HTN-TJ.pro : SFESTQNDTENKVQGYYICIVGGNSAPIYVPTLDDFRSMEAFTGIFKSPHGEDHDLAGEEIASYSIVGPANAKVPHSASSDTLSLIAYSGIPSYSSLSILTSSTDAKHVFSPGF : 342
HTN-bao14. : .......................................................................................................L : 342
HTN-84Li.p : ..D...........................................................................................E........ : 342
HTN-N8.pro : ....N.......................................................................G..E.................... : 342
HTN-Q32.pr : ....N.......................S.............................................G..F...........E........L : 342

             FSKLNHTNCDKSAIPLTWTGMIDLPGYYEAIHPCTVFCVLSGPGASCEAFSEGGIFNITSPMCLVSKQNRFRLTEQQVNFVCQRVDMDIVVYCNGQRKVILTKTLVIGQCIYTI : 456
HTN-bao14. : .P...................................................................................................... : 456
HTN-84Li.p : .PQ...K............................L........................Q...K........................................ : 456
HTN-N8.pro : .PQ...K.................................................................................................. : 456
HTN-Q32.pr : .PQ...K.................................................................................................. : 456

                  460         *       480         *       500         *       520         *       540         *       560         *
HTN-TJ.pro : TSLFSLLPGVAHSIAVELCVPGFHGWATAALLVTFCFGWVLIPAVTFIILTVLKFIANIFHTSNQENRLKSVLRKIKEEFERTKGPMVCDICKYECETYKELKAHGVSCPQSQC : 570
HTN-bao14. : .........................................................................S.............................. : 570
HTN-84Li.p : .........................................................................S.............................. : 570
HTN-N8.pro : ............................................................A............S.............................. : 570
HTN-Q32.pr : .........................................................................S.............................. : 570
```

Gc

```
                  580         *       600         *       620         *       640         *       660         *       680
HTN-TJ.pro : PYCFTHCEPTEAAFQAHYKVCQVTHRFRDDLKKTVTEQNFTPGCYRTLNLFRYKSRCYIFTMWVFLLILESILWAASASETPLTPVWNDNAHGVGSVPMHTDLELDFSLTSSSK : 684
HTN-bao14. : .................................................................................................. : 684
HTN-84Li.p : .........................................................................V.N................... : 684
HTN-N8.pro : ..........................................................C..............V.N................... : 684
HTN-Q32.pr : .........................................................................I.N................... : 684

             *           700         *       720         *       740         *       760         *       780         *       8
HTN-TJ.pro : YTYRRKLTNPLEEAQSVDLHIEIEEQTIGVDVHALGHWFDGRLNLKTSFHCYGACTKYEYPWHTAKCHYERDYQYETSWGCNPSDCPGVGTGCTACGLYLDQLKPVGSAYKIIT : 798
HTN-bao14. : .................................................................................................. : 798
HTN-84Li.p : ............T.A................................................................................... : 798
HTN-N8.pro : ............T.A..........................................................A........................ : 798
HTN-Q32.pr : ............S.A................................................................................... : 798

                  00          *       820         *       840         *       860         *       880         *       900         *
HTN-TJ.pro : IRYSRRVCVQFGEENLCKIIDMNDCFVSRHVKVCIIGTVSKFSQGDTLLFFGPLEGGGLIFKHWCTSTCQFGDPGDIMSPRDKGFLCPEFPGSFRKKCNFATTPICEYDGNMVS : 912
HTN-bao14. : .................................................................................................. : 912
HTN-84Li.p : .................................................................................................. : 912
HTN-N8.pro : .................................................................................................. : 912
HTN-Q32.pr : .................................................................................................. : 912

                  920         *       940         *       960         *       980         *      1000         *      1020
HTN-TJ.pro : GYKKVMATIDSFQSINTSTMHFTDERIEWKDPDGMLRDHINILVTKDIDFDNLGENPCKIGLQTSSIEGAWGSGVGFTLTCLVSLTECPTFLTSINACDKAICYGAESVTLTRG : 1026
HTN-bao14. : ....................................................................................K............. : 1026
HTN-84Li.p : ....................................................................................K............. : 1026
HTN-N8.pro : ....................................................................................K.......A.... : 1026
HTN-Q32.pr : .................................................................................S...K..........I.. : 1026

                  *          1040         *      1060         *      1080         *      1100         *      1120         *
HTN-TJ.pro : QNTVRVSGKGGHSGSTFKCCHGEDCSKIGLHAAAPHLDKVNGISEIENSKVYDDGAPQCGIKCWFVKSGEWISGIFSGNWIVLIVLCVFLLFSLVLLSILCPVRKHKKS- : 1135
HTN-bao14. : .............................................................................................- : 1135
HTN-84Li.p : ......................QT.................................T..................................- : 1135
HTN-N8.pro : ...........E..Q.................................N...........................................- : 1135
HTN-Q32.pr : ...........E..Q............................................................................- : 1135
```

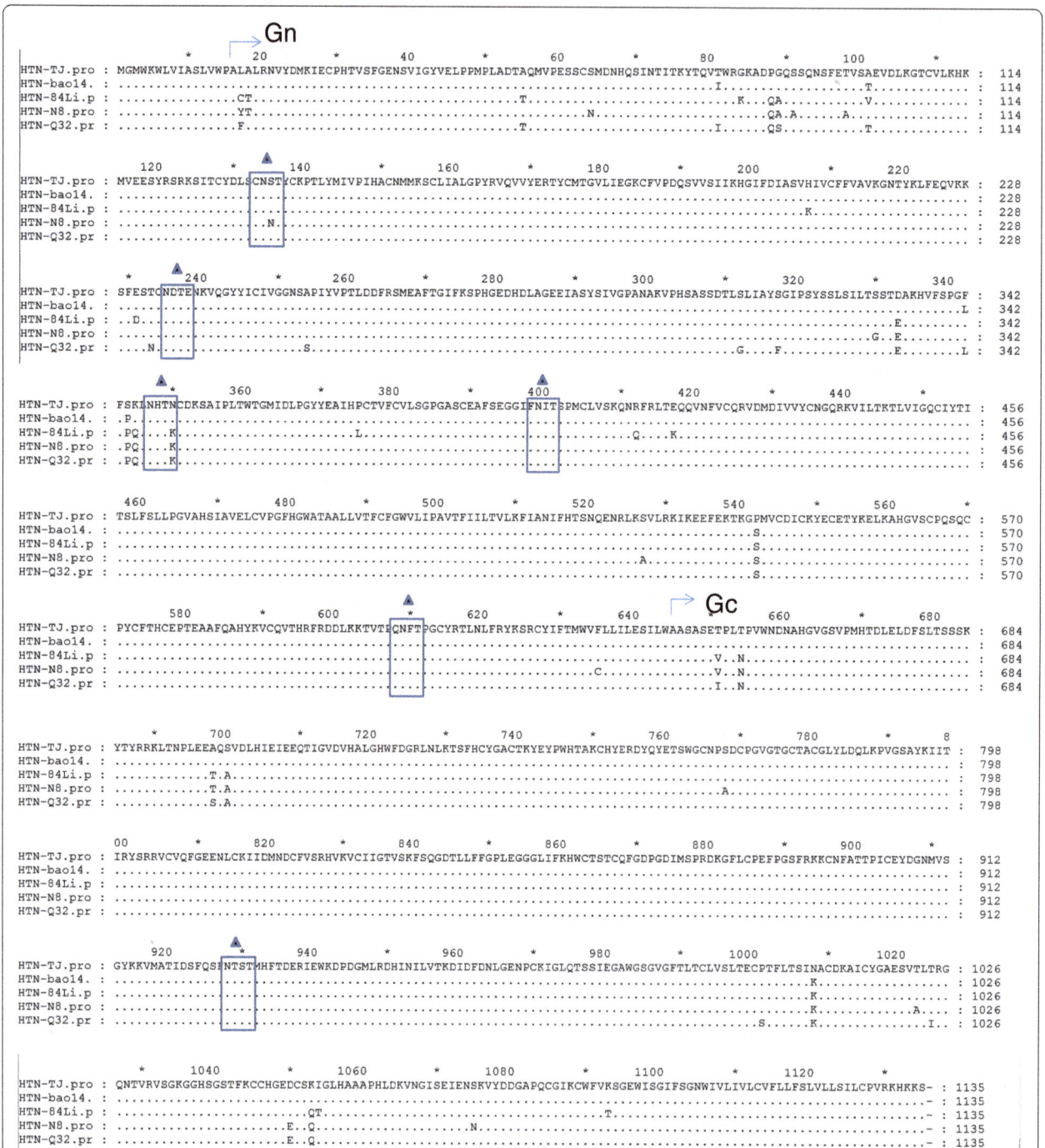

Fig. 5 Multiple alignment of the deduced amino acid sequences based on M gene of hantaviruses isolate TJ and other reference stains. The amino acid differences between TJ isolate and other strains were shown. The function related N- glycosylation sites were framed and labled with filled triangles

HTNVs that we detected are similar to the representative strain Bao14. Strains Bao14 and CJAp93, which belong to lineage 6 (lineage FE) and were isolated from Heilongjiang and Jilin, respectively, are the main pathogens responsible for HFRS in the northeast region of China. This finding may indicate that this HTNV variant is one of the pathogens associated with HFRS in northeastern China at present. However, we have not found any isolates that are similar to strain 76-118, which belongs to lineage 7 [31]. Previous studies have indicated that each dominant hantavirus genotype is associated with a specific rodent host, as HTNV is carried by *A. agrarius*, whereas the virus isolated from *R. norvegicus* has been identified as

SEOV [32]. However, we isolated HTNVs from *R. norvegicus* in this study. This finding may be explained as a spillover infection, though we have not identified spillover of SEOV from *R. norvegicus* to *A. agrarius*. Furthermore, The entire M segment sequence of HTNV isolate (TJ strain) shared high identity with Bao 14 strain (99.2 %). It indicated that both TJ and Bao14 iaolates may evolve from the same ancestry. And we predicted that the transfer of HTNV infection host might be the approach employed by hantavirus to adapt to environmental change. In addition, all of the isolates (HTNVs and SEOVs) obtained in the present study belong to known lineages, and no new lineage of isolates has emerged in the Heilongjiang area.

Conclusions

In this study, we found that HTNV and SEOV are circulating in the Heilongjiang frontier area and that SEOVs are the dominant common strains detected in rodent hosts. The HTNV isolates belong to lineage 6, and the SEOV isolates belong to lineages 3 and 4. All of the lineages are common in the northeast region of China. Although we have not identified a new lineage of hantavirus, regular epidemiological surveillance of local murine colonies is necessary and should be performed intensively.

Abbreviations
ANDV: Andes virus; DOBV: Dobrava virus; HFRS: hemorrhagic fever with renal syndrome; HPS: hantavirus cardiopulmonary syndrome; HTNV: Hantaan virus; PUUV: Puumala virus; SAAV: Saaremaa virus; SEOV: Seoul virus; SNV: Sin Nombre virus.

Competing interests
The authors declare that they have no competing interests.

Authors' contributions
YLW and JM proposed the conception of this study and supervised the project. SYC, CC and WDJ performed the experiments, SYC and JM analyzed the results and wrote this manuscript. All authors read and approved the final manuscript.

Acknowledgments
This study was supported by the grant from the Fundamental Research Funds for the Central Universities (Grant no. DL13CA04), the grant from the Heilongjiang provincial natural science foundation (Grant no. C201116), and a grant from the General Administration of Quality Supervision, Inspection and Quarantine of the People's Republic of China (AQSIQ) for scientific research subjects (Grant no. 2014IK407).

Author details
[1]Department of Wildlife Medicine, Wildlife Resources Faculty, Northeast Forestry University, Harbin 150040, China. [2]State Key Laboratory of Veterinary Biotechnology, Harbin Veterinary Research Institute, the Chinese Academy of Agriculture Sciences, Harbin 150001, China. [3]Heilongjiang International Travel Healthcare Center, Harbin 150001, China.

References
1. Antic D, Lim BU, Kang CY. Nucleotide sequence and coding capacity of the large (L) genomic RNA segment of Seoul 80-39 virus, a member of the hantavirus genus. Virus Res. 1991;19(1):59–65.
2. Hung T, Xia SM, Zhao TX, Zhou JY, Song G, Liao GX, Ye WW, Chu YL, Hang CS. Morphological evidence for identifying the viruses of hemorrhagic fever with renal syndrome as candidate members of the Bunyaviridae family. Brief report. Arch Virol. 1983;78(1-2):137–44.
3. Plyusnin A, Vapalahti O, Vaheri A. Hantaviruses: genome structure, expression and evolution. J Gen Virol. 1996;77(Pt 11):2677–87.
4. Schmaljohn CS, Dalrymple JM. Analysis of Hantaan virus RNA: evidence for a new genus of bunyaviridae. Virology. 1983;131(2):482–91.
5. Luan VD, Yoshimatsu K, Endo R, Taruishi M, Huong VT, Dat DT, Tien PC, Shimizu K, Koma T, Yasuda SP, et al. Studies on hantavirus infection in small mammals captured in southern and central highland area of Vietnam. J Vet Med Sci. 2012;74(9):1155–62.
6. Douron E, Moriniere B, Matheron S, Girard PM, Gonzalez JP, Hirsch F, McCormick JB. HFRS after a wild rodent bite in the Haute-Savoie–and risk of exposure to Hantaan-like virus in a Paris laboratory. Lancet (London, England). 1984;1(8378):676–7.
7. Tsai TF. Hemorrhagic fever with renal syndrome: mode of transmission to humans. Lab Anim Sci. 1987;37(4):428–30.
8. CfDCaP CDC. Update: hantavirus pulmonary syndrome–United States, 1993. MMWR Morb Mortal Wkly Rep. 1993;42(42):816–20.
9. Clement J, Heyman P, McKenna P, Colson P, Avsic-Zupanc T. The hantaviruses of Europe: from the bedside to the bench. Emerg Infect Dis. 1997;3(2):205–11.
10. Hughes JM, Peters CJ, Cohen ML, Mahy BW. Hantavirus pulmonary syndrome: an emerging infectious disease. Science (New York, NY). 1993; 262(5135):850–1.
11. Schmaljohn C, Hjelle B. Hantaviruses: a global disease problem. Emerg Infect Dis. 1997;3(2):95–104.
12. Mackow ER, Dalrymple NA, Cimica V, Matthys V, Gorbunova E, Gavrilovskaya I. Hantavirus interferon regulation and virulence determinants. Virus Res. 2014;187:65–71.
13. Jonsson CB, Figueiredo LT, Vapalahti O. A global perspective on hantavirus ecology, epidemiology, and disease. Clin Microbiol Rev. 2010;23(2):412–41.
14. Sane J, Reimerink J, Harms M, Bakker J, Mughini-Gras L, Schimmer B, van Pelt W. Human hantavirus infections in the Netherlands. Emerg Infect Dis. 2014;20(12):2107–10.
15. Vapalahti O, Mustonen J, Lundkvist A, Henttonen H, Plyusnin A, Vaheri A. Hantavirus infections in Europe. Lancet Infect Dis. 2003;3(10):653–61.
16. Song G. Epidemiological progresses of hemorrhagic fever with renal syndrome in China. Chin Med J (Engl). 1999;112(5):472–7.
17. Tkachenko EA, Lee HW. Etiology and epidemiology of hemorrhagic fever with renal syndrome. Kidney Int Suppl. 1991;35:S54–61.
18. Zhang YZ, Xiao DL, Wang Y, Wang HX, Sun L, Tao XX, Qu YG. The epidemic characteristics and preventive measures of hemorrhagic fever with syndromes in China. Zhonghua Liu Xing Bing Xue Za Zhi. 2004;25(6):466–9.
19. Zou Y, Hu J, Wang ZX, Wang DM, Yu C, Zhou JZ, Fu ZF, Zhang YZ. Genetic characterization of hantaviruses isolated from Guizhou, China: evidence for spillover and reassortment in nature. J Med Virol. 2008;80(6):1033–41.
20. Li CP, Cui Z, Li SL, Magalhaes RJ, Wang BL, Zhang C, Sun HL, Li CY, Huang LY, Ma JJ, et al. Association between hemorrhagic fever with renal syndrome epidemic and climate factors in Heilongjiang Province, China. Am J Trop Med Hyg. 2013;89(5):1006–12.
21. Chen HX, Qiu FX, Dong BJ, Ji SZ, Li YT, Wang Y, Wang HM, Zuo GF, Tao XX, Gao SY. Epidemiological studies on hemorrhagic fever with renal syndrome in China. J Infect Dis. 1986;154(3):394–8.
22. Chen LF XJ, Liu ZW, Liu YC, Sun CQ, Li JH, et al.: First molecular identification of Seoul hantavirus isolated from Rattus norvegicus in Heilongjiang, China. China Public Health 2001(17):343.
23. Kariwa H, Isegawa Y, Arikawa J, Takashima I, Ueda S, Yamanishi K, Hashimoto N. Comparison of nucleotide sequences of M genome segments among Seoul virus strains isolated from eastern Asia. Virus Res. 1994;33(1):27–38.
24. Cosgriff TM. Mechanisms of disease in Hantavirus infection: pathophysiology of hemorrhagic fever with renal syndrome. Rev Infect Dis. 1991;13(1):97–107.
25. Kanerva M, Mustonen J, Vaheri A. Pathogenesis of puumala and other hantavirus infections. Rev Med Virol. 1998;8(2):67–86.
26. Zhang YZ, Dong X, Li X, Ma C, Xiong HP, Yan GJ, Gao N, Jiang DM, Li MH, Li LP, et al. Seoul virus and hantavirus disease, Shenyang, People's Republic of China. Emerg Infect Dis. 2009;15(2):200–6.

27. Zhang YZ, Zhang FX, Wang JB, Zhao ZW, Li MH, Chen HX, Zou Y, Plyusnin A. Hantaviruses in rodents and humans, Inner Mongolia Autonomous Region, China. Emerg Infect Dis. 2009;15(6):885–91.

28. Shi X, McCaughey C, Elliott RM. Genetic characterisation of a Hantavirus isolated from a laboratory-acquired infection. J Med Virol. 2003;71(1):105–9.

29. Wang H, Yoshimatsu K, Ebihara H, Ogino M, Araki K, Kariwa H, Wang Z, Luo Z, Li D, Hang C, et al. Genetic diversity of hantaviruses isolated in china and characterization of novel hantaviruses isolated from Niviventer confucianus and Rattus rattus. Virology. 2000;278(2):332–45.

30. Lee HW, Lee PW, Johnson KM. Isolation of the etiologic agent of Korean hemorrhagic fever. 1978. J Infect Dis. 2004;190(9):1711–21.

31. Jiang JF, Zhang WY, Yao K, Wu XM, Zuo SQ, Zhan L, Zhang PH, Cao WC. A new Hantaan-like virus in rodents (Apodemus peninsulae) from Northeastern China. Virus Res. 2007;130(1-2):292–5.

32. Plyusnin A, Morzunov SP. Virus evolution and genetic diversity of hantaviruses and their rodent hosts. Curr Top Microbiol Immunol. 2001;256:47–75.

2

Caprine herpesvirus 2-associated malignant catarrhal fever of captive sika deer (*Cervus nippon*) in an intensive management system

Hongwei Zhu[1], Qingrong Huang[1], Xiaoliang Hu[2], Wenhui Chu[3], Jianlong Zhang[1], Linlin Jiang[1], Xin Yu[1], Xingxiao Zhang[1]* (iD) and Shipeng Cheng[3*]

Abstract

Background: Caprine herpesvirus 2 (CpHV-2) infection usually induces chronic malignant catarrhal fever (MCF) in sika deer (*Cervus nippon*), with the primary signs of weight loss, dermatitis and alopecia.

Case presentation: Here, we report a case of CpHV-2-associated acute MCF in a sika deer herd raised in an intensive management system distant to the reservoir goats. Affected deer developed clinical signs of high fever (41 °C) followed by nasal discharge and lameness. Severe lesions of hemorrhage, necrosis and infiltration of lymphoid cells could readily be observed in the lung, kidney, heart valves and subcutaneous tissue surrounding a tendon. Etiologically, identical CpHV-2 specific DNA sequences were detected in peripheral blood lymphocyte (PBL) from the affected deer and reservoir goats.

Conclusion: In summary, domestic goats were the reservoir of the CpHV-2, which is the causative agent of the outbreak of MCF in the three hinds. The disease was probably transmitted via aerosol infection. In addition, necrosis and inflammation in subcutaneous tissue surrounding a tendon was the reason for lameness. Therefore, MCF should be put into a differential diagnostic list when similar disease occurs in sika deer herds.

Keywords: Sika deer, *Cervus nippon*, Malignant catarrhal fever, CpHV-2, Lameness, Aerosol route transmission

Background

Malignant catarrhal fever (MCF) is a lymphoproliferative disease, caused by a group of closely related ruminant viruses within the genus *Macavirus*, in the subfamily *Gammaherpesvirinae* [1]. The disease is characterized as sporadic with low incidence and high mortality in bison, cattle, buffalo, deer and antelope [2, 3]. Etiologically, alcelaphine herpesvirus 1 (AlHV-1) and ovine herpesvirus 2 (OvHV-2) are the most common causative agents. Subclinically infected wildebeest and domestic sheep act as natural reservoirs, causing wildebeest-associated MCF (WA-MCF) and

sheep-associated MCF (SA-MCF), respectively. In general, AIHV-1 causes MCF syndromes in ruminant wild animals in Africa and zoological animals worldwide, whereas OvHV-2 is more prevalent in domestic sheep, causing MCF in most regions of the world. In addition to AlHV-1 and OvHV-2, caprine herpesvirus 2 (CpHV-2), ibex-MCFV, MCFV–white-tailed deer (WTD), and alcelaphine herpesvirus 2 (AlHV-2) have been reported to be pathogenic under natural conditions [4].

Upon infection, different clinical manifestations could be observed in the host depending on the affected organs. In most cases, early clinical signs of the disease usually begin with high fever and depression, followed by catarrhal nasal discharges, mucosal ulcers and/or cloudy corneas. Dysentery or bloody diarrhea is common in most cases [2, 5]. Oculonasal discharge could be an indication of the so-called head-and-eye form of MCF, whereas the intestinal

* Correspondence: zhangxingxiao@163.com; tcscsp@126.com
[1]School of Life Sciences, Ludong University, No. 186 Hongqi Middle Rd., Zhifu District, Yantai 264025, China
[3]Institute of Special Economic Animal and Plant Sciences, Chinese Academy of Agricultural Sciences, No. 4899 Juye St., Jingyue District, Changchun 130112, China
Full list of author information is available at the end of the article

tract form, the central nervous system (CNS) form and the cutaneous form of the disease are easily confused with other diseases.

Pathologically, heart, brain, lung and kidney are the most easily affected organs, although lesions may be found in any organ with different severities and frequencies. Characteristic lymphoid cell accumulation and infiltration in various tissues can be readily observed during microscopic investigation. Lymphocytic vasculitis is a typical histological lesion of the disease [5, 6].

A wide range of animals in the family *Cervidae* are susceptible to malignant catarrhal fever virus (MCFV) infections. These deer species include sika deer (*Cervus nippon*), white-tailed deer (*Odocoileus virginianus*), Père David's deer (*Elaphurus davidiansus*), sambar deer (*Cervus unicolor*), mule deer (*Odocoileus hemionus*), moose (*Alces alces*), red deer (*Cervus elaphus*), red brocket deer (*Mazama americana*), elk (*Cervus canadensis*), rusa deer (*Cervus timorensis*), Chinese water deer (*Hydropotes inermis*) and axis deer (*Axis axis*) [4, 7–13]. In general, sika deer seem highly susceptible to OvHV-2 [14–16]. Peracute cases with sudden death and acute cases with hemorrhagic diarrhea are well documented and common in our experience of clinical investigation. CpHV-2 infection in deer, on the contrary, is rarely encountered and usually produces mild and chronic clinical signs such as weight loss, dermal inflammation and alopecia. The pathogenic significance of CpHV-2 was reported shortly after its recognition: four outbreaks of CpHV-2-associated MCF in sika deer and white-tailed deer in Florida, Arizona, Minnesota and Texas in the US were described with the primary signs of weight loss and alopecia [17–19].

Before the causative agent was identified, a case of white-tail deer infection was reported in a North American zoo as well [20, 21]. Later, in Europe, CpHV-2-induced MCFs were also documented in deer species in Norway and UK [12, 22]. In this report, we described, for the first time in China, the outbreak of an acute MCF caused by CpHV-2 from domestic goats 1.0 km in distance. Significant differences were found with respect to clinical signs and gross and histological lesions, compared with the previous CpHV-2-induced MCFs in sika deer.

Case presentation

In December 2016, an outbreak of a deadly disease in a captive sika deer herd was reported by a local veterinarian in Siping City, Jilin Province, China. The deer herd consisted of 186 sika deer raised in group-housing sheds. Deer of different ages and sex were grouped separately in different sheds. Each shed had a roof and its own approximately 200 m^2 yard. Adjacent sheds were separated with bars and brickwork. In this outbreak, 3 deer hind were affected, and they were housed in two adjacent sheds. Clinical signs included high fever (> 41 °C) followed by nasal discharge. Oral vesicles developed on the tongue for each hind (Fig. 1a). Mild scabs appeared on nose of the third affected deer (Fig. 1b). In addition, all three hinds showed moderate lameness, however, no interdigital cleft vesicle, foot rot or hoof deformities could be observed. No dermal inflammation or alopecia was found. Two hinds died on the 5th and 7th day of clinical signs and were disposed of by the farm owner with only one blood sample taken. The third hind was then euthanized for postmortem examination on the 6th day when the condition deteriorated.

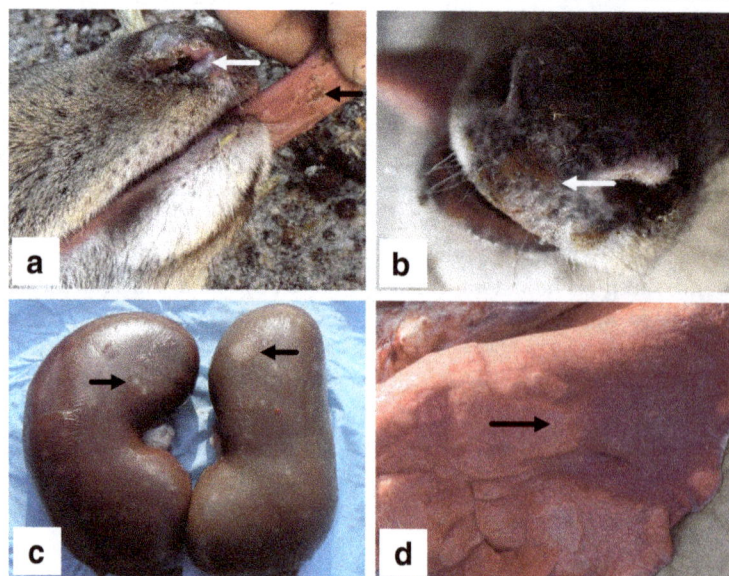

Fig. 1 Main clinical signs and gross lesion of the third affected hind. Nasal discharge (white arrow), oral vesicles on tongue (black arrow) and mild scabs on the nose (white arrow) of the affected deer are indicated in **a** and **b**. Gross lesions on the kidney and lung are indicated in **c** and **d** respectively

Main gross lesions included slight swelling of the extensor tendon and flexor tendon near the joint of hoofs, and enlargement of the kidneys and mesenteric lymph nodes. Notably, multiple pale necrotic foci approximately 0.3 cm in diameter were present on both kidneys (Fig. 1d). In addition, hemorrhagic areas appeared on the surface of the lungs (Fig. 1c).

Histopathological examination also revealed more severe lesions of the lung, kidney, heart valves and tendon. More specifically, in the lung, multifocal hemorrhages and obvious interstitial fibrosis with a small number of infiltrating lymphoid cells appeared in pulmonary tissue (Fig. 2a, c and Additional file 1: Figure S2F). In the affected kidney, large numbers of lymphoid cells, mainly lymphocytes and neutrophils, accumulated and infiltrated in the renal parenchyma, resulting in the disappearance of the inherent structure of the tissue, which was replaced by the infiltrating lymphoid cells (Fig. 2b and d). Cardiac muscle was not affected; however, extensive valvular tissue degeneration and necrosis, inflammatory infiltration and severe hemorrhage were noticed in the heart valves (Additional file 1: Figure S2E). No lesion was found in the tendon (Fig. 3c). However, in the subcutaneous tissue surrounding the tendon, a wide range of necrosis, lymphoid cell infiltration and fibroblast proliferation occurred (Fig. 3a). In addition, arteritis caused by infiltration of lymphoid cells in arterial walls in the subcutaneous tissue could be observed, and thrombosis was occasionally seen in the arteries (Fig. 3a–d). Other mild lesions included hemorrhage, lymphoid cell infiltration and/or necrosis in the liver (Additional file 1: Figure S2B), adrenal glands (Additional file 1: Figure S2D), tongue (Additional file 1: Figure S2C) and salivary glands (not shown). Lymphocytes were decreased in the lymph nodule of mesenteric lymph node (Additional file 1: Figure S2A). No lesions were found in the brain, spleen, derma and muscle. The lesions described above were in most cases suggestive of MCF.

For etiological agent identification, initial diagnostic attempts including virus isolation and viral RNA detection for foot and mouth disease virus (FMDV), peste des petits ruminants virus, epizootic hemorrhagic disease virus, bluetongue virus and vesicular stomatitis virus failed to identify any virus. Pathogenic bacterial infection was also ruled out by 16S rDNA detection. For MCFVs detection, EDTA-anticoagulated blood samples were collected from the two hinds (SP-1 and SP-2) followed by peripheral blood lymphocyte (PBL) isolation. DNA polymerase (DPOL) fragments were amplified from PBLs using a pan-herpesviral nested PCR with five degenerate primers as described previously (Additional file 2: Figure S1A). Presence of CpHV-2 in both samples was subsequently confirmed by amplicons sequencing and phylogenetic analysis (Additional file 2: Figure S1B). The resultant sequences were then submitted to the GenBank database under the accession numbers of KY475595 and MF318872. In a retrospective questionnaire for reservoir hosts, the deer farm owner claimed a well-fenced feedlot with approximately 30 domestic goats was approximately 1.0 km away in the north. Eight goats were then tested for the presence of CpHV-2 DNA in blood samples. Three were determined to be positive for the virus. Sequence alignment suggested that CpHV-2 from these goats shared 100% DPOL sequence identity with those from sika deer (data not shown).ac.

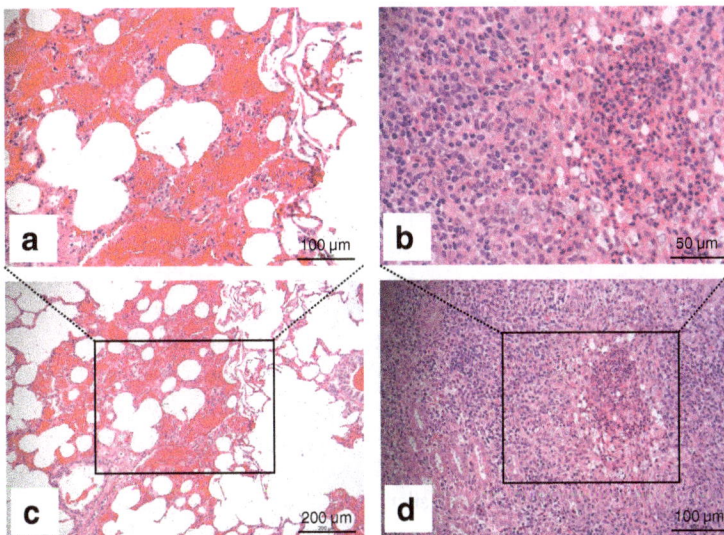

Fig. 2 Histopathological findings. Sika deer. Lung and kidney. Lung: multifocal hemorrhage with a small amount of lymphoid cell infiltration (**a** and **c**). Kidney: large quantity of lymphoid cells, mainly lymphocytes and neutrophils, accumulated and infiltrated in the renal parenchyma. Inherent structures were replaced by the infiltrating lymphoid cells (**b** and **d**). H&E. Scales are indicated over the bars

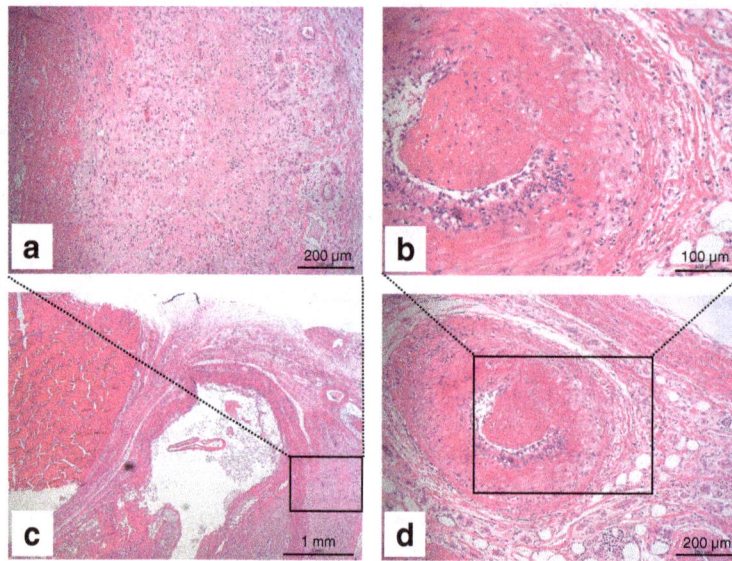

Fig. 3 Histopathological findings. Sika deer. Tendon and surrounding subcutaneous tissue. A wide range of necrosis, lymphoid cell infiltration and fibroblast proliferation in subcutaneous tissue (**a** and **c**). Arteritis and thrombosis in subcutaneous tissue (**b** and **d**). H&E. Scales are indicated over the bars

Discussion and conclusion

To the best of our knowledge, this is the first confirmed case of CpHV-2-induced MCF, although the disease has probably long existed in China. MCF is sporadic, easily neglected and easily confused with other diseases with similar clinical signs such as foot and mouth disease or enterotoxemia. In this outbreak of MCF, the clinical signs of oral vesicles and lameness could be confused with FMD, which, however, could be ruled out by evaluating the following facts. First, epidemiologically, FMDV is a rapidly spread virus with high infectivity and transmissibility [23]. More affected deer would be expected with an FMD outbreak. Additionally, clinically, lameness in the affected deer was not due to vesicles on the interdigital cleft, but to inflammatory lesions of subcutaneous tissue. Since MCF is a systemic disease, any organ could be affected, therefore, it should be included in diagnostic considerations when an infectious disease occurs in sika deer herds.

The CpHV-2-associated MCF described in this outbreak is significantly different from MCFs in sika deer from the US and UK with respect to disease course, clinical signs and histopathological lesions. First, the hinds in Florida and in Arizona underwent chronic disease courses of ten days, one month and six months respectively, except for one deer euthanized due to seizures [17, 18]. In addition, the sika deer in the UK also experienced a two-month history of weight loss and crusting [22], while hinds in this outbreak underwent acute courses of five to seven days. Second, previously described sika deer infected with CpHV-2-

associated MCF usually developed mild clinical signs of dermatitis, crusting and generalized alopecia and weight loss in most cases. However, in our case description, hinds showed no signs of weight loss or dermatitis. They displayed not only the typical MCF signs of high fever and nasal discharge but also the rarely occurring sign of lameness. Third, gross and histopathological lesions suggest the virus in this report caused more extensive inflammation and more severe hemorrhage in the lung, kidney, heart valve and liver. Whether these apparent differences were due to pathogenic characteristics of the viruses or to host aspects is yet to be elucidated.

There are many factors that affect clinical duration, clinical severity and even tissue tropism in a single MCF outbreak. Clinical signs in the mucosa and subcutaneous are most common. Lameness is rarely seen in both SA-MCF and WA-MCF. In contrast, in this MCF outbreak, we noticed the clinical sign of lameness in all affected deer while unaffected deer showed no signs of this kind. Further histological lesions in the subcutaneous tissue around a tendon correlated with clinical manifestations. In our observation, the lameness of the affected deer was most likely caused by necrosis, fibroblast proliferation and arteritis of the subcutaneous tissue. These lesions are all related to lymphoid cell infiltration, which is typical in MCF infection. No observable lesions in the brain were recorded for the euthanized deer, thus, CNS dysfunction did not contribute to the lameness in this MCF outbreak. Accordingly, we noticed that the affected deer appeared to suffer from pain in the feet rather than clinical ataxia.

It is convincing that goats near the deer farm were the natural reservoir responsible for the outbreak of MCF in sika deer, due to the identical sequence polymorphism of the detected viral DNAs. The infection route, however, was quite intriguing. The affected deer were raised in an intensive management system. They had no chance of direct contact or feed cross contamination with the reservoir goats. We thus tend to advocate an airborne infection route here. Previous studies have demonstrated that sheep shed large quantities of cell-free OvHV-2 virions via nasal secretions. Experimental infection of sheep and rabbits with ovine herpesvirus 2 via aerosolization of nasal secretions over a short distance have also been established [24–26]. In another report, Li et al. described a long distance (> 5.0 km) spread of OvHV-2 from feedlot lambs to ranch bison, causing MCF [27]. In this report, we observed that transmission of CpHV-2 from goats via aerosolization could reach and establish efficient infection at a distance of at least 1.0 km. In this scenario, 1.0 km is far from a safe distance for sika deer to prevent CpHV-2 associated MCF. However, we could not rule out the possibilities of CpHV-2 transmission through harmful fauna, given the proximity between the farms, as has been described for other viruses [28]. Wild rodents, ticks or bats could serve as vectors in this situation.

This case reminds us again of the potential risk of cross-species transmission of MCFVs, as different species demonstrate various susceptibilities to the viruses. One species carries its own sets of viruses, which may be well adapted to the original host, causing asymptomatic infection, but may bear unknown risks for another species. Therefore, keeping a safe distance from reservoirs is an effective way to avoid MCF. In addition, different deer species should be raised separately as a precaution, as MCF-like disease has been reported in sika deer co-raised with fallow deer in Inner Mongolia (personal communication with a veterinarian in a red deer farm).

In conclusion, the CpHV-2-associated MCF in this outbreak presented differently than previously described, with a more chronic, rather than acute presentation regarding disease course and severity. The domestic goat was the reservoir of the CpHV-2 which caused outbreak of the disease. The disease was transmitted at least 1.0 km in distance probably via an aerosol route. Necrosis, fibroblast proliferation and arteritis in subcutaneous tissue surrounding a tendon were the cause of lameness in this outbreak. The causative agent, CpHV-2, should be put into a differential diagnostic list when similar disease occurs in sika deer herds. Since there is no cure or vaccine for the disease, prophylactic measures should be taken accordingly. Keeping a safe distance from reservoirs is an effective approach to avoid MCF.

Additional files

Additional file 1: Figure S2. Histopathological findings. Sika deer. Mesenteric lymph node, liver, tongue, adrenal glands, heart valve and lung. Lymphocytes decreased in the lymph nodule of a mesenteric lymph node (**Figure S2A.**) Hemorrhage, necrosis and slight lymphoid cell infiltration in the liver (**Figure S2B.**) Slight lymphoid cell infiltration in the lamina propria of the tongue (**Figure S2C.**) Focal hemorrhage in the cortex of adrenal glands (**Figure S2D**). Necrosis, inflammatory infiltration and severe hemorrhage in a heart valve (**Figure S2E.**) Interstitial fibrosis with slight lymphoid cell infiltration in the lung (**Figure S2F.**) H&E. Bar = 100 μm. (PDF 667 kb)

Additional file 2: Figure S1. Pan-herpesvirus detection and DPOL phylogeny of the isolate from affected hinds. Ethidium bromide-stained agarose gel of two amplicons from SP-1 and SP-2 respectively, using the consensus PCR assay targeting herpesviral DNA polymerase (DPOL) (**Figure S1A.**) Based on the resultant DPOL sequences, a phylogenic tree was constructed using the PhyML software (version 3.0) with LG substitution model (**Figure S1B.**) Approximate likelihood ratio test (aLRT) was performed and indicated in the node. aLRT values less than 0.50 were collapsed. Scale bar indicates 0.1 amino acid substitutions per site. DPOL GenBank accession number is AAC59454 for outgroup Bovine gammaherpesvirus 4 (BoHV-4); NC_002531 for Alcelaphine herpesvirus 1(AlHV-1); AAO88177 for MCFV-WTD; ADY17131 for Ovine gammaherpesvirus 2 (OvHV-2); APG30119 for Muskox rhadinovirus 1 (Muskox-LHV); ADY17115 for Caprine gammaherpesvirus 2 (CpHV-2); KY475595 for isolate SP-1. (PDF 129 kb)

Abbreviations

AlHV-2: Alcelaphine herpesvirus 2; CNS: Central nervous system; CpHV-2: Caprine herpesvirus 2; DPOL: DNA polymerase; FMDV: Foot and mouth disease virus; MCF: Malignant catarrhal fever; MCFVs: Malignant catarrhal fever viruses; OvHV-2: Ovine herpesvirus 2; PBL: Peripheral blood lymphocyte; SA-MCF: Sheep-associated MCF; WA-MCF: Wildebeest-associated MCF; WTD: White-tailed deer

Acknowledgements

The authors wish to thank Jun Fu from the Meihekou Agricultural Bureau in Jilin Province, for his assists in postmortem examination.

Funding

This work was financially supported by the National Key R&D Program of China (2016YFD0501010 and 2016YFD0501001–3) and the Innovation Team Project for Modern Agricultural Industrious Technology System of Shandong Province (SDAIT-11-10).

Authors' contributions

HZ, XZ and SC followed the clinical case and designed the experiment. QH and HZ carried out the necropsy. XH and WC performed the H&E staining and histopathological examination. JZ, LJ and XY collected the samples and performed bacterial and viral detection. HZ drafted the manuscript. All authors read and approved the final manuscript.

Consent for publication

The deer owner was informed and gave his verbal consent to publish this case report.

Competing interests

The authors declare that they have no competing interest.

Author details
[1]School of Life Sciences, Ludong University, No. 186 Hongqi Middle Rd., Zhifu District, Yantai 264025, China. [2]Harbin Veterinary Research Institute, Chinese Academy of Agricultural Sciences, Harbin 150069, China. [3]Institute of Special Economic Animal and Plant Sciences, Chinese Academy of Agricultural Sciences, No. 4899 Juye St., Jingyue District, Changchun 130112, China.

References

1. DJ MG, Gatherer D, Dolan A. On phylogenetic relationships among major lineages of the Gammaherpesvirinae. J Gen Virol. 2005;86(2):307–16.

2. Russell GC, Stewart JP, Haig DM. Malignant catarrhal fever: a review. Vet J. 2009;179(3):324–35.

3. Taus NS, O'Toole D, Herndon DR, Cunha CW, Warg JV, Seal BS, Brooking A, Li H. Malignant catarrhal fever in American bison (Bison Bison) experimentally infected with alcelaphine herpesvirus 2. Vet Microbiol. 2014; 172(1–2):318–22.

4. Li H, Cunha CW, Taus NS, Knowles DP. Malignant catarrhal fever: inching toward understanding. Annu Rev Anim Biosci. 2014;2:209–33.

5. O'Toole D, Li H. The pathology of malignant catarrhal fever, with an emphasis on ovine herpesvirus 2. Vet Pathol. 2014;51(2):437–52.

6. Vikoren T, Klevar S, Li H, Hauge AG. Malignant catarrhal fever virus identified in free-ranging musk ox (Ovibos Moschatus) in Norway. J Wildl Dis. 2013; 49(2):447–50.

7. Tomkins NW, Jonsson NN, Young MP, Gordon AN, McColl KA. An outbreak of malignant catarrhal fever in young rusa deer (Cervus Timorensis). Aust Vet J. 1997;75(10):722–3.

8. Li H, Dyer N, Keller J, Crawford TB. Newly recognized herpesvirus causing malignant catarrhal fever in white-tailed deer (Odocoileus Virginianus). J Clin Microbiol. 2000;38(4):1313–8.

9. Driemeier D, Brito MF, Traverso SD, Cattani C, Cruz CE. Outbreak of malignant catarrhal fever in brown brocket deer (Mazama Gouazoubira) in Brazil. Vet Rec. 2002;151(9):271–2.

10. Klieforth R, Maalouf G, Stalis I, Terio K, Janssen D, Schrenzel M. Malignant catarrhal fever-like disease in barbary red deer (Cervus Elaphus Barbarus) naturally infected with a virus resembling alcelaphine herpesvirus 2. J Clin Microbiol. 2002;40(9):3381–90.

11. Orr MB, Mackintosh CG. An outbreak of malignant catarrhal fever in pere David's deer (Elaphurus Davidianus). N Z Vet J. 1988;36(1):19–21.

12. Vikoren T, Li H, Lillehaug A, Jonassen CM, Bockerman I, Handeland K. Malignant catarrhal fever in free-ranging cervids associated with OvHV-2 and CpHV-2 DNA. J Wildl Dis. 2006;42(4):797–807.

13. Schultheiss PC, Van Campen H, Spraker TR, Bishop C, Wolfe L, Podell B. Malignant catarrhal fever associated with ovine herpesvirus-2 in free-ranging mule deer in Colorado. J Wildl Dis. 2007;43(3):533–7.

14. Imai K, Nishimori T, Horino R, Kawashima K, Murata H, Tsunemitsu H, Saito T, Katsuragi K, Yaegashi G. Experimental transmission of sheep-associated malignant catarrhal fever from sheep to Japanese deer (Cervus Nippon) and cattle. Vet Microbiol. 2001;79(1):83–90.

15. Tham KM. Molecular and clinicopathological diagnosis of malignant catarrhal fever in cattle, deer and buffalo in New Zealand. Vet Rec. 1997; 141(12):303–6.

16. Shulaw WP, Oglesbee M. An unusual clinical and pathological variant of malignant catarrhal fever in a white-tailed deer. J Wildl Dis. 1989;25(1):112–7.

17. Keel MK, Patterson JG, Noon TH, Bradley GA, Collins JK. Caprine herpesvirus-2 in association with naturally occurring malignant catarrhal fever in captive sika deer (Cervus Nippon). J Vet Diagn Investig. 2003;15(2):179–83.

18. Crawford TB, Li H, Rosenburg SR, Norhausen RW, Garner MM. Mural folliculitis and alopecia caused by infection with goat-associated malignant catarrhal fever virus in two sika deer. J Am Vet Med Assoc. 2002;221(6):843–847, 801.

19. Li H, Keller J, Knowles DP, Crawford TB. Recognition of another member of the malignant catarrhal fever virus group: an endemic gammaherpesvirus in domestic goats. J Gen Virol. 2001;82(1):227–32.

20. Li H, Wunschmann A, Keller J, Hall DG, Crawford TB. Caprine herpesvirus-2-associated malignant catarrhal fever in white-tailed deer (Odocoileus Virginianus). J Vet Diagn Investig. 2003;15(1):46–9.

21. Li H, Cunha CW, Abbitt B, deMaar TW, Lenz SD, Hayes JR, Taus NS. Goats are a potential reservoir for the herpesvirus (MCFV-WTD), causing malignant catarrhal fever in deer. J Zoo Wildl Med. 2013;44(2):484–6.

22. Foyle KL, Fuller HE, Higgins RJ, Russell GC, Willoughby K, Rosie WG, Stidworthy MF, Foster AP. Malignant catarrhal fever in sika deer (Cervus Nippon) in the UK. Vet Rec. 2009;165(15):445–7.

23. Grubman MJ, Baxt B. Foot-and-mouth disease. Clin Microbiol Rev. 2004; 17(2):465–93.

24. Gailbreath KL, Taus NS, Cunha CW, Knowles DP, Li H. Experimental infection of rabbits with ovine herpesvirus 2 from sheep nasal secretions. Vet Microbiol. 2008;132(1):65–73.

25. Li H, Taus NS, Lewis GS, Kim O, Traul DL, Crawford TB. Shedding of ovine herpesvirus 2 in sheep nasal secretions: the predominant mode for transmission. J Clin Microbiol. 2004;42(12):5558–64.

26. Kim O, Li H, Crawford TB. Demonstration of sheep-associated malignant catarrhal fever virions in sheep nasal secretions. Virus Res. 2003;98(2):117–22.

27. Li H, Karney G, O'Toole D, Crawford TB. Long distance spread of malignant catarrhal fever virus from feedlot lambs to ranch bison. Can Vet J. 2008; 49(2):183–5.

28. Romero Tejeda A, Aiello R, Salomoni A, Berton V, Vascellari M, Cattoli G. Susceptibility to and transmission of H5N1 and H7N1 highly pathogenic avian influenza viruses in bank voles (Myodes glareolus). Vet Res. 2015;46(1):51.

3

Strong inflammatory responses and apoptosis in the oviducts of egg-laying hens caused by genotype VIId Newcastle disease virus

Ruiqiao Li[†], Kangkang Guo[†], Caihong Liu, Jing Wang, Dan Tan, Xueying Han, Chao Tang, Yanming Zhang and Jingyu Wang[*]

Abstract

Background: Newcastle disease virus (NDV) can cause serious damage to the reproductive tracts of egg-laying hens and leads to egg production and quality reduction. However, the mechanism of severe pathological damage in the oviducts of egg-laying hens after NDV infection has not been fully elucidated. In this study, the correlation between the primary pathological lesions and viral load in the oviducts of egg-laying hens infected with the velogenic genotype VIId NDV strain was evaluated by pathological observation and virus detection. Subsequently, apoptosis, the expression of immune-related genes and lymphocyte infiltration into the infected oviducts were determined to explore the potential causes of the pathological changes.

Results: A higher viral load and severe tissue lesions and apoptotic bodies were observed in the oviduct of NDV-infected hens compared with the control. Immune-related genes, including TLR3/7/21, MDA5, IL-2/6/1β, IFN-β, CXCLi1/2, and CCR5, were significantly upregulated in the magnum and uterus. IL-2 presented the highest mRNA level change (137-fold) at 5 days post infection (dpi) in the magnum. Infection led to CD3$^+$CD4$^+$ and CD3$^+$CD8α$^+$ lymphocyte infiltration into the magnum of the oviduct. A higher viral load was found to be associated with pathological changes and the elevated expression of proinflammatory cytokines in the NDV-infected hens.

Conclusions: Our results indicate that the severe lesions and apoptosis in the oviducts of egg-laying hens caused by genotype VIId NDV strains are associated with the excessive release of inflammatory cytokines, chemokines and lymphocyte infiltration, which contribute to the dysfunction of the oviducts and the decrease of egg production in hens.

Keywords: Genotype VIId NDV, Oviduct, Apoptosis, Inflammatory responses, Lymphocyte infiltration, Egg-laying hens

Background

The Newcastle disease virus (NDV) is highly contagious and widespread among avian species and causes severe economic losses in domestic poultry, especially chickens [1]. It does not only cause the death of chickens but also causes serious damage to the reproductive tracts of egg-laying hens. This leads to an overall decrease in egg production and increases the percentage of deformed, sand-shelled, and soft-shelled eggs [2]. Since the 1950s, vaccination has been applied for the prevention and control of this disease in many countries, including China. However, it is still enzootic in some areas and is recognized as a major disease that affects poultry [3]. The genotype VII NDV is dominant in Asia and poses a serious threat to its poultry industry [4–6]. Previous studies have revealed that the significant oviduct inflammation and decline of egg production are caused by the variance [2, 7]. However, the inflammatory effects and apoptosis of local oviduct of the virulent NDV genotype VIId strains were still unclear, which may lead to oviduct dysfunction and drop in egg production.

* Correspondence: wjingyu2004@126.com
†Equal contributors
College of Veterinary Medicine, Northwest A&F University, Yangling, Shannxi 712100, China

Generally, host-virus interactions caused by NDV include a complex interplay of molecular pathways directed by the host to prevent viral replication [8, 9]. These involve intense inflammatory responses in the form of a cytokine storm that are initiated inside infected cells or tissues following viral replication and result in excessive cellular apoptosis and tissue damage. In vivo studies suggested that NDV initiated strong immune responses in the spleen, lymph nodes and peripheral blood of infected chickens, where the mRNA expression levels of proinflammatory and cytokines/chemokines were significantly upregulated [9–12]. Strong innate immune response and cell death were observed in chicken splenocytes infected with genotype VIId NDV [13]. The cytopathic effect (CPE) and apoptosis were observed in NDV infected cells and viral replication and caspase activation were detected [14–16]. Studies investigating the mechanism underlying severe pathological damages in the oviducts of egg-laying hens after NDV infection are limited. Furthermore, the reproductive tract is a target of both the vaccine strain and field isolates of NDV [7], so the innate immunity roles of oviduct played a critical role in viral pathogenesis, especially, the inflammatory effects and attribution of different immune cells and immune molecules.

In this study, a severe oviduct tissue damage of egg-laying hens were observed in the genotype VIId NDV strain infected hens. This damage was associated with high levels of virus replication, a strong innate immune response/inflammatory response and lymphocyte infiltration. These responses contributed to oviductal dysfunction and the decline of egg production, which played a critical role in viral pathogenesis.

Results

Clinical signs and histopathological changes

Obvious ND-related clinical signs, including ruffled feathers, depression, and drowsiness, were observed in NDV-infected birds 2 dpi. At 3 dpi, the infected hens showed typical clinical signs, including dyspnea, depression and greenish feces. The first bird died of NDV infection 3 dpi. At 4 dpi, Four infected hens died from viral challenge-associated causes. The most severe deaths occurred 5 dpi, when seven infected birds died. Furthermore, egg production of the NDV-infected hens started to plunge by approximately 40% on approximately 5 dpi, whereas the controls maintained approximately 90% egg production (Table 1). From 7 dpi, no hens died spontaneously or was moribund, and clinical signs and egg production of the remaining hens began to abate. Upon examination of the gross lesions, the experimental group hens showed proventricular hemorrhage, tracheorrhagia, brain edema and dissolved follicles, while the oviducts showed edema. Hens in the control group did not show any clinical signs or gross lesions throughout the procedure.

Histopathological examinations of infundibulum, magnum, isthmus, uterus, and vagina are shown in Fig. 1. The tissues of the laying hens of the control group appeared normal. On 1 dpi, there were scattered necrotic areas in the mucosal epithelia and lymphocyte infiltration into the infundibulum, magnum, isthmus, and uterus. Occasionally, heterophils were found in the lamina propria of the magnum. On 3 dpi, the uterus presented marked edema. Severe mucosal epithelial necrosis and numerous lymphocytes were seen in the vagina. On 5 dpi, the degree of the lesions in the isthmus was elevated, while mucosal epithelial necrosis, heterophil infiltration, and edema were also observed. Moreover, a large number of lymphocytes were also observed in the vagina. On 7 dpi, the lesions in the infundibulum became less severe. However, heterophil infiltration and mucosal epithelial necrosis persisted. On 15 dpi, some lymphocytes were found in the infundibulum. No significant changes were observed in the control group.

Table 1 Tissue processing and egg production changes in chickens infected with the velogenic genotype VIId NDV

Day	Control group			Experimental group				
	Total	Sample	Laying rate%	Total	Mortality	Autopsy	Sample	Laying rate %
0	45	0	91.1 (41/45)	55	0	0	0	90.9 (50/55)
1	45	5	87.5 (35/40)	55	0	5	5	88.0 (44/50)
3	40	5	88.6 (31/35)	50	1	5	5	66.7 (30/45)
4	35	0	91.4 (32/35)	45	4	4	0	48.8 (20/41)
5	35	5	90.0 (27/30)	41	7	7	5	38.2 (13/34)
6	30	0	90.0 (27/30)	34	3	3	0	41.9 (13/31)
7	30	5	88.0 (22/25)	31	1	5	5	46.2 (12/26)
9	25	5	90.0 (18/20)	26	0	5	5	52.4 (11/21)
11	20	5	86.7 (13/15)	21	0	5	5	56.3 (9/16)
15	15	5	90.0 (9/10)	16	0	5	5	72.7 (8/11)

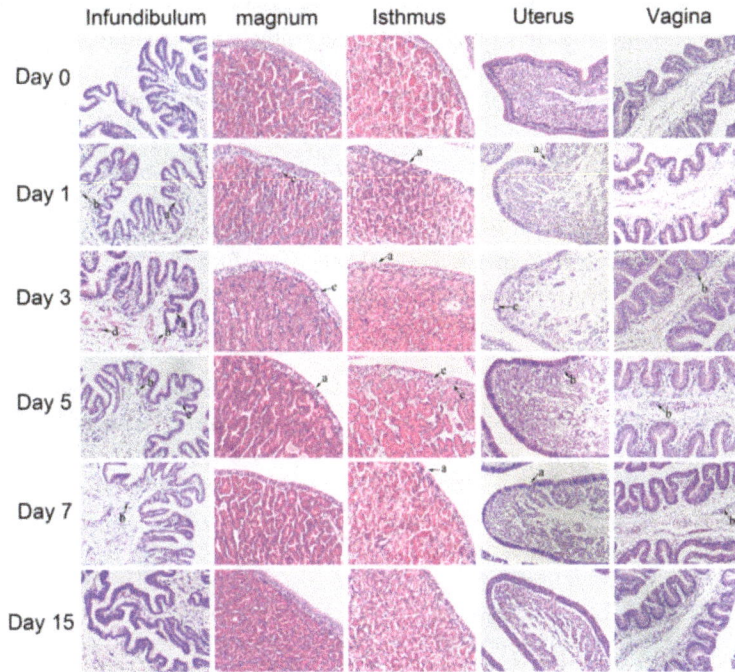

Fig. 1 Histopathology of the oviducts stained with hematoxylin and eosin (400×). (A) Cilia loss. (B) Lymphocyte infiltration. (C) Heterophil infiltration. (D) Pink-stained material in the infundibulum. (E) Mucosal epithelial necrosis

Viral loads in oviduct

To evaluate differences in viral replication in the oviduct of egg-laying hens after NDV infection, we determined the transcriptional levels of the viral matrix (M) gene in the oviducts of hens infected with the velogenic genotype VIId NDV strain by quantitative RT-PCR assay. We detected the viral load from the infundibulum, magnum, isthmus, uterus, and vagina taken at 1, 3, 5, 7 and 9 dpi. The level of viral RNA is shown in Fig. 2. The viral load was significantly increased in the infundibulum, magnum, isthmus, uterus, vagina and reached the peak at 5 dpi, gradually decreased at 7, 9 and 11 dpi. The magnum presented the highest amount of virus load (8918.6 copies/ml), followed by the uterus (2934.6 copies/ml), vagina (1890 copies/ml), isthmus (1789 copies/ml), and infundibulum (1358 copies/ml) (Fig. 2). No virus was found in healthy oviducts.

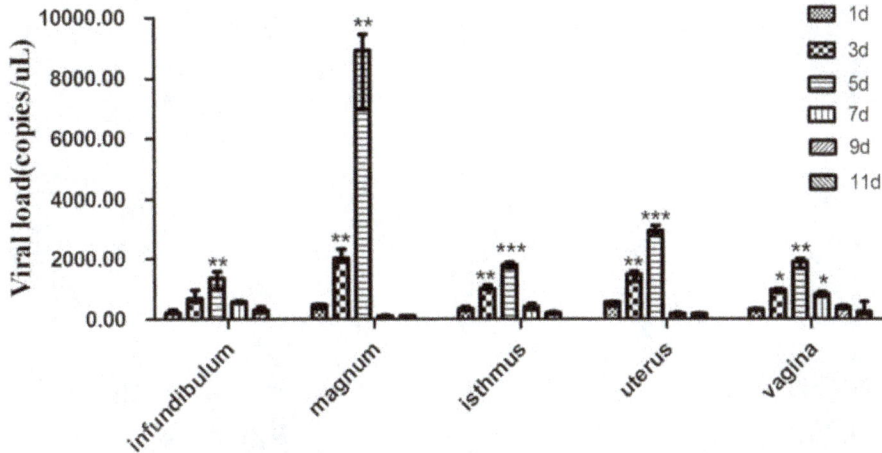

Fig. 2 NDV viral load in oviduct. Primers targeting the M gene of NDV were used to detect virus replication in the oviducts of hens infected with the velogenic genotype VIId NDV strain by quantitative RT-PCR assay with three replications. According to our standard curve, the correlation coefficient (R^2) was 0.9979 with a slope value of -3.3749 (y = -3.3749x + 37.202)

Virus distribution along the oviductal segments

Indirect immunofluorescence assays were conducted to confirm the distribution of the NDV HN protein in the oviducts at 5 dpi. Positive staining was not observed in any part of the oviducts of the control hens. However, HN proteins were observed in the glandular epithelial cells of the lamina propria and found scattered in the submucosa of the infundibulum, magnum, and isthmus. Conversely, the HN proteins were present in the submucosa and scattered in the lamina propria in the uterus. In the vagina, they were located in the mucosal epithelium (Fig. 3).

Apoptosis in the oviducts of the infected hens

In this study, caspase-3 activity and TUNEL assays were employed to detect apoptosis in the oviducts of the infected hens on 5 dpi. The caspase-3 activities in the magnum and uterus were found higher compared to the control group (2.41- and 1.98-fold increases, respectively) (Fig. 4). The results of the TUNEL assays revealed that NDV induced apoptosis in the lamia propria of the infundibulum and vagina (Fig. 5). In the magnum, isthmus, and uterus, apoptotic cells were observed in both the lamia propria and the mucosa.

Expression of TLRs, MDA5, cytokines, and chemokines

To estimate the inflammatory response in the oviducts of the infected hens, a real-time quantitative PCR assay was used to quantify the expression profiles of immune-related genes in the magnum and uterus of the oviduct of SPF egg-laying hens infected with genotype VIId NDV on 1, 3, 5, 7, 9, 11 and 15 dpi. The mRNA levels of TLR3, 7, and 21 were altered in the oviducts of the infected laying hens (Fig. 6) and were upregulated in the magnum and uterus tissues. The mRNA expression levels in the magnum segments peaked at 5 dpi (42.6-, 98.9-, and 27.1-fold, respectively), while in the uterus they peaked on 7 dpi. Moreover, TLR expression levels in the magnum were significantly greater compared to the uterus. This result suggests that the magnum segments were most sensitive to NDV. Another intracellular pattern recognition receptor (PRR), MDA5, was also upregulated, with its highest fold changes of 8.23 and 7.29 detected in the magnum and uterus, respectively (Fig. 6d).

The expression levels of IL-2/6/1β and IFN-α/β were also examined at the transcript level (Fig. 7). IL-2/6/β and IFN-β were significantly upregulated at two instances during the infection process, but the ranges of upregulation were different. IL-2 presented the highest mRNA level

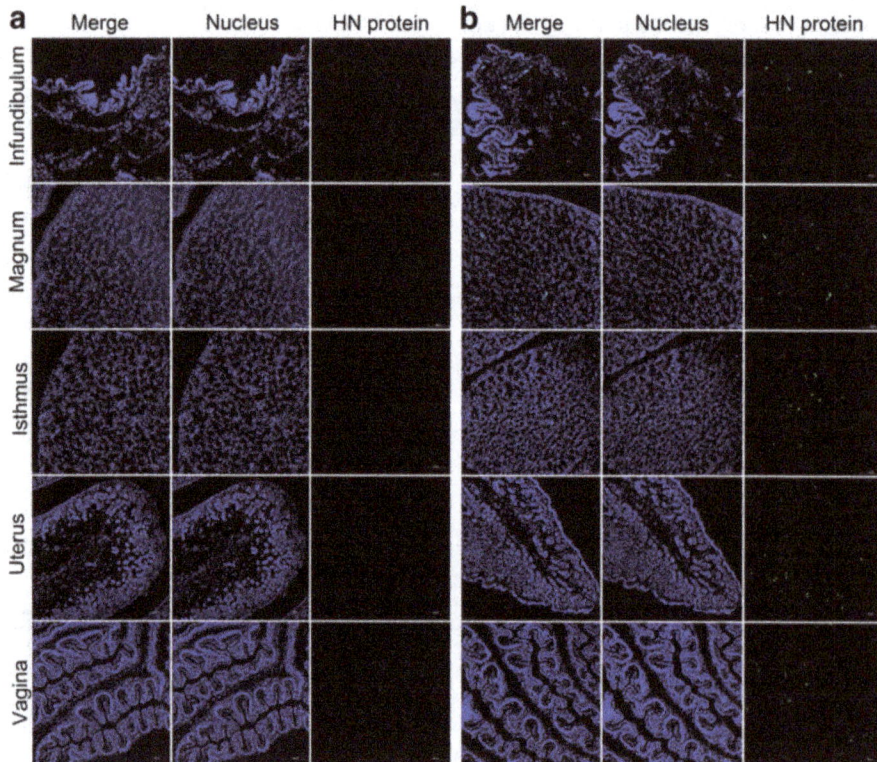

Fig. 3 Detection of the NDV HN protein in the oviducts by IFA at 5 dpi. Distribution of the NDV HN protein was detected by staining with a mouse anti-HN monoclonal antibody and fluorescein isothiocyanate (FITC)-conjugated goat anti-mouse secondary antibody (green fluorescence); the nuclei were stained with Hoechst 33342 (blue fluorescence) (400×). **a** Control hens. **b** Infected hens

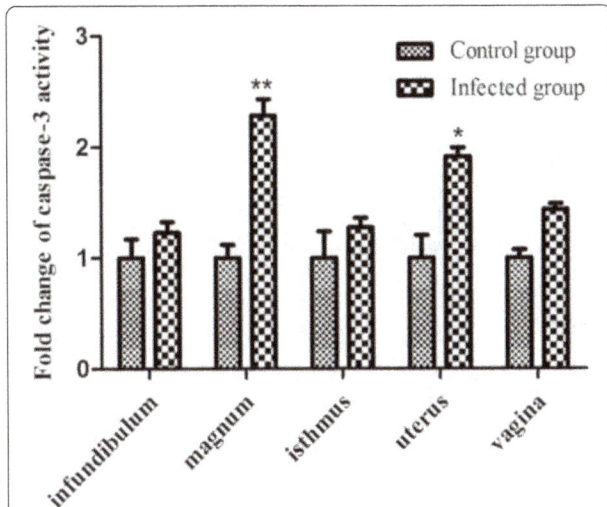

Fig. 4 Caspase-3 activity in the oviducts following NDV infection. Total protein was extracted from infected and uninfected chickens 5 dpi ($n = 3$ per group), and caspase-3 activity was detected using a specific detection kit. The data represent the mean ± SEM of three chickens. Each column represents the fold change of caspase-3 activity obtained by comparison with the data from uninfected chickens. *P <0.05, **P <0.01, ***P <0.001

change of 137-fold on 5 dpi (Fig. 7b), followed by IFN-β (95.1-fold) and IL-1β (15.9-fold). In contrast, IFN-α was downregulated by 0.19-fold 3 dpi and subsequently upregulated by 1.78-fold 5 dpi in the magnum, but not in the uterus (Fig. 7d). Furthermore, IFN-α expression was significantly upregulated and peaked 5 dpi with a 4.56-fold increase in the uterus (Fig. 7d).

We examined three chemokines (CXCLi1/2 and CCR5) (Fig. 8). The mRNA expression level of CXCLi1 was increased in the magnum throughout the infection process, reaching its peak of 20.5-fold on 5 dpi. It was also significantly upregulated (5.48-fold) in the uterus on 5 dpi (Fig. 8a). The chemokine receptor for CXCLi1 (CCR5) was also upregulated during the infection process. This upregulation was higher than CXCLi1 in the magnum (22.1-fold) and uterus (25.5-fold) (Fig. 8b). CXCLi2 was also significantly upregulated and peaked at 5 dpi with a 78.3-fold increase in the magnum (Fig. 8c).

8Dynamic changes in CD3$^+$CD4$^+$ and CD3$^+$CD8α$^+$ lymphocytes in the oviducts

To study the dynamic changes in the CD3$^+$CD4$^+$ and CD3$^+$CD8α$^+$ lymphocytes in the oviducts of hens after NDV infection, we selected the magnum, which was the segment with the highest level of viral replication and

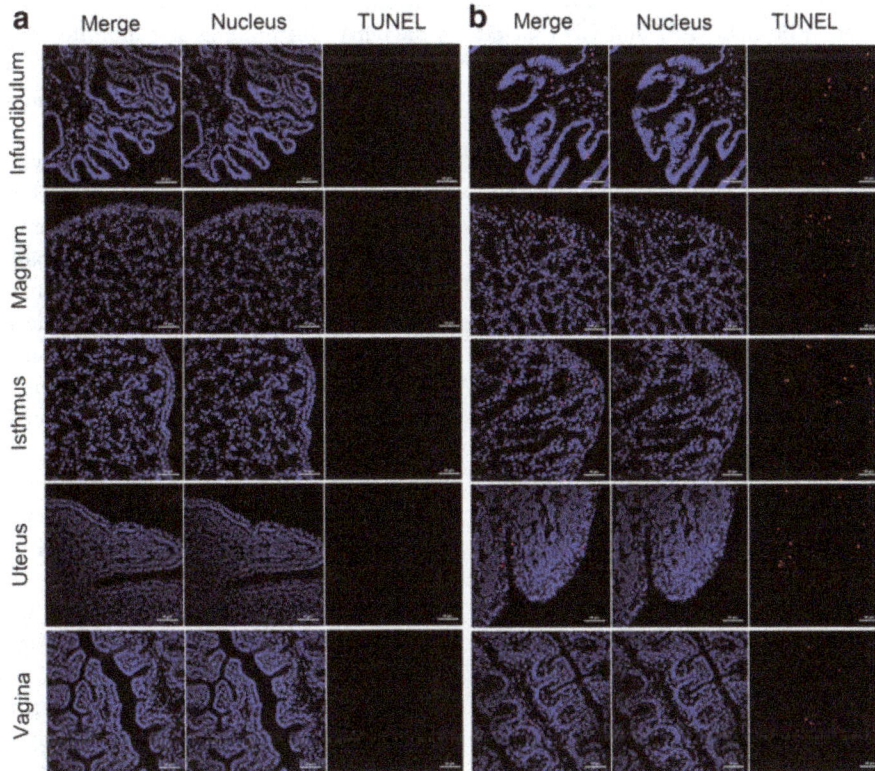

Fig. 5 TUNEL staining of the oviduct in NDV-infected hens. Oviduct tissues were collected 5 dpi. The nuclei of oviduct sections from virus-free and NDV-infected hens were stained with Hoechst 33342; a red color indicates the positive staining of apoptotic cells (400×). Oviductal segments from virus-free hens served as the negative control. **a** Control hens. **b** Infected hens

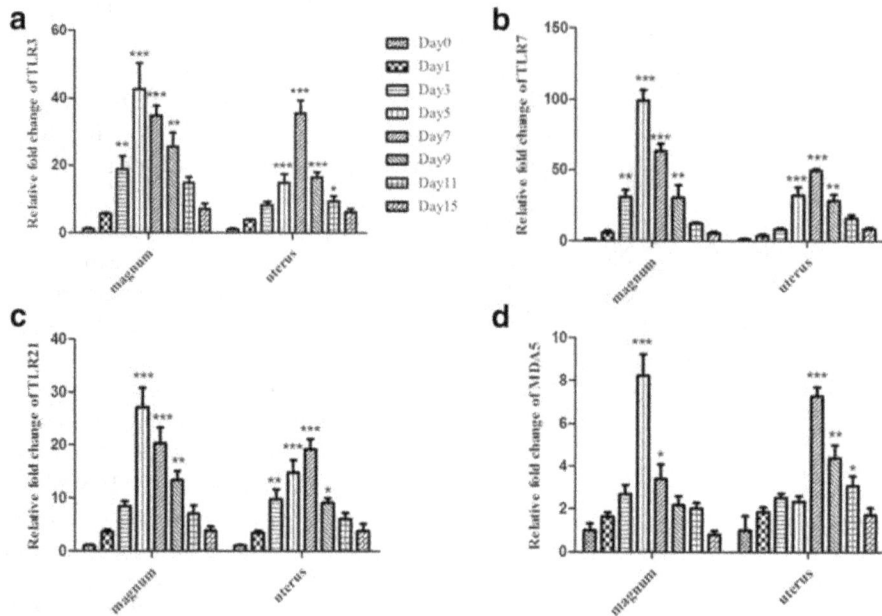

Fig. 6 mRNA expression levels of the TLRs and MDA5 in the magnum and uterus of laying hens infected with NDV. Total RNA was extracted from the infected and control groups ($n = 3$ per group). The TLRs and MDA5 were quantified using appropriate primers specific to the chicken and SYBR Green-based real-time (RT) PCR. The data represent the mean ± standard error of the mean (SEM) for three chickens. Panels **a**, **b**, **c**, and **d** depict the results for TLR3, TLR7, TLR21, and MDA5, respectively, in the magnum and uterus. Statistical analysis was performed by comparison with the healthy chickens. *$P < 0.05$, **$P < 0.01$, ***$P < 0.001$

mRNA expression of immune-related factors, for further study. The accumulation of CD3+CD4+ and CD3+CD8α+ lymphocytes in the magnum was greater in the experimental than in the control group throughout the course of infection (Figs. 9 and 10). The CD3+CD4+ and CD3+CD8α+ lymphocytes began to accumulate 1 dpi and peaked 5 and 7 dpi, respectively (Figs. 9 and 10).CD3+CD4+ and CD3+CD8α+ lymphocytes were distributed in the lamina propria of the magnum. Moreover, CD8α+ cells were recruited into the magnum more deeply than the CD4+ cells. The frequency of CD4+ cells in the lamina propria was significantly higher compared to the bottom of the mucosal epithelium. In contrast, the CD8α+ cells localized predominantly at the bottom of the mucosal epithelium. Sporadic distribution of these lymphocytes was detected in the oviducts in the control birds.

Discussion

Newcastle disease is a highly contagious viral disease that causes serious economic damage to the poultry industry [4]. The velogenic genotype VII NDV has been documented as the predominant epidemic genotype and has been responsible for frequent outbreaks on vaccinated farms in China and other Asian countries since 1990 [17, 18]. The strain used in this study was isolated from a vaccinated farm. We observed that the significant clinical signs, pathological changes, and apoptosis induced by velogenic genotype VIId NDV infection contributed to

oviduct dysfunction and the decline of egg production in the hens. Our findings also indicated that the severe tissue damage in the reproductive tract was associated with high viral replication, apoptosis, an intense inflammatory response, and lymphocyte infiltration.

In this study, examination of the viral load confirmed that NDV replication occurred in the oviducts of egg-laying hens. At 5 dpi, when the viral load was higher than 1, 3, 7,11 and 15 dpi, the histopathology were more severe, which may be due to the higher viral load in oviduct at these days. Therefore, the efficiency of viral replication in oviduct of hens appears to be associated with the elevated expression of proinflammatory cytokines in the NDV-infected hens. Furthermore, positive staining for the NDV HN protein in the oviducts revealed that the infected cells were glandular epithelial and mucosa cells. This finding was consistent with the presence of apoptotic cells in the five oviductal segments observed in the current study. We also observed the invasion of inflammatory cells and necrocytosis in the mucosal epithelia in five segments (Fig. 1). We hypothesized that the lesions aroused by NDV infection in the five segments might lead to the poor performance in egg-laying hens, thereby resulting in decreased egg production.

Apoptosis is a critical mechanism of NDV-induced tissue lesions, and caspase-3 plays a critical role in the execution of the apoptotic process [19, 20]. In our study, caspase-3 activity was significantly high in the NDV-

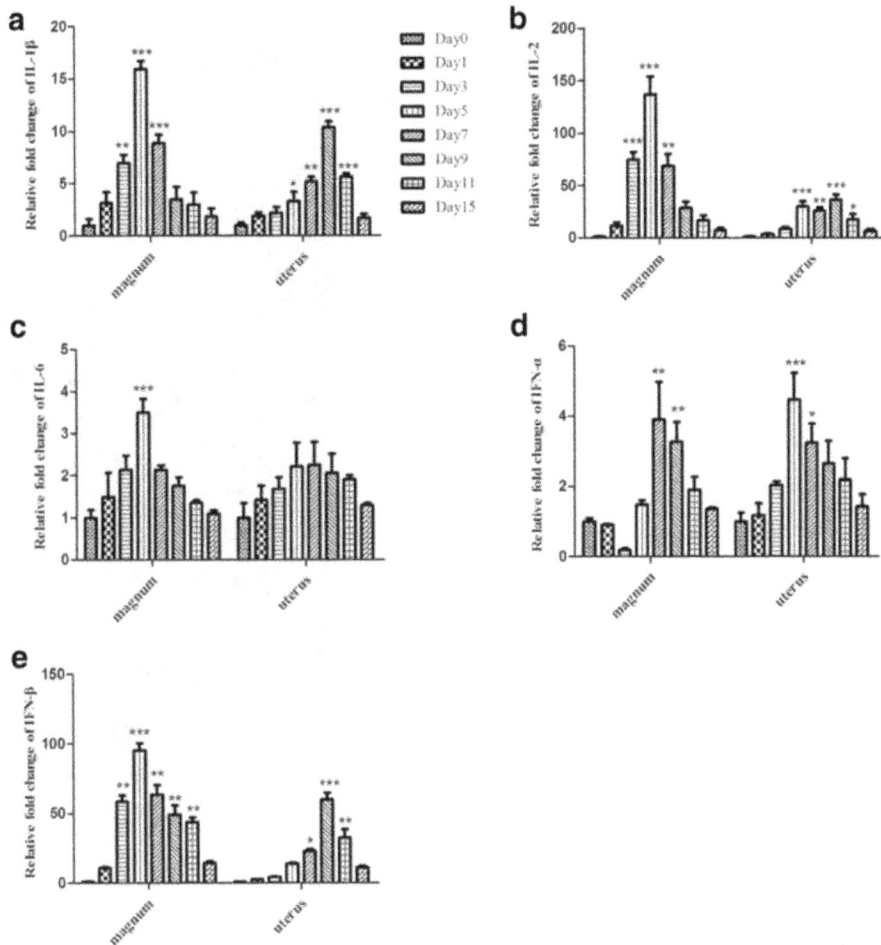

Fig. 7 Expression levels of inflammatory cytokines in the magnum and uterus of laying hens infected with NDV. IL-1β/2/6 and IFN-α/β were quantified using SYBR Green-based RT-PCR. Panels **a**, **b**, **c**, **d** and **e** depict results for IL-1β, IL-2, IL-6, IFN-α, and IFN-β, respectively, in the magnum and uterus. Statistical analysis was performed by comparison with uninfected chickens. *P <0.05, **P <0.01, ***P <0.001

infected layers, which was consistent with the results of a previous study [16]. Furthermore, our TUNEL assays also revealed apoptosis in the five oviduct segments upon NDV infection. Collectively, these results indicate that the oviductal tissues undergo cellular apoptosis both in the early and late stages of NDV infection.

We examined whether pattern recognition receptors (PRRs), inflammatory cytokines and chemokines in the oviduct exhibited altered expression in response to NDV infection. TLRs and MDA5 belong to different PRRs that recognize viral nucleic acids and are central to host antiviral defenses. TLRs are the most important family of PRRs and activate the MyD88-dependent pathway to produce cytokines, MHC molecules, and chemokines. In turn, these molecules trigger an appropriate immune response to eliminate the invading pathogens [21, 22]. In birds, the TLRs 3, 7, and 21 play major roles in the induction of antiviral IFN-β during such infections [23, 24]. In our study, the transcriptional levels of TLR3/7 were

significantly increased post-NDV infection, suggesting the anti-NDV roles of these molecules.

Cytokines and chemokines are major mediators of the host immune response during viral infection [25]. NDV infection induced strong pro-inflammatory responses that played an important role in viral pathogenesis. The IL-2/6/1β and IFN-β mRNA expression levels were increased in the oviductal tissues of NDV-infected hens; our data are consistent with previous findings [9–12]. The expression of the IL-2 gene in the peripheral blood of F48E9 NDV-infected chickens in a previous study fluctuated and did not exhibit a significant difference compared with the control group during the experimental period [26]. In contrast, the expression of IL-2 in our study increased rapidly and presented the highest mRNA level change (137-fold) at 5 dpi. IFN-α belongs to the type I interferon family and can be produced in most cells after virus infection, whereupon it localizes to the site of viral infection and inhibits viral replication. In

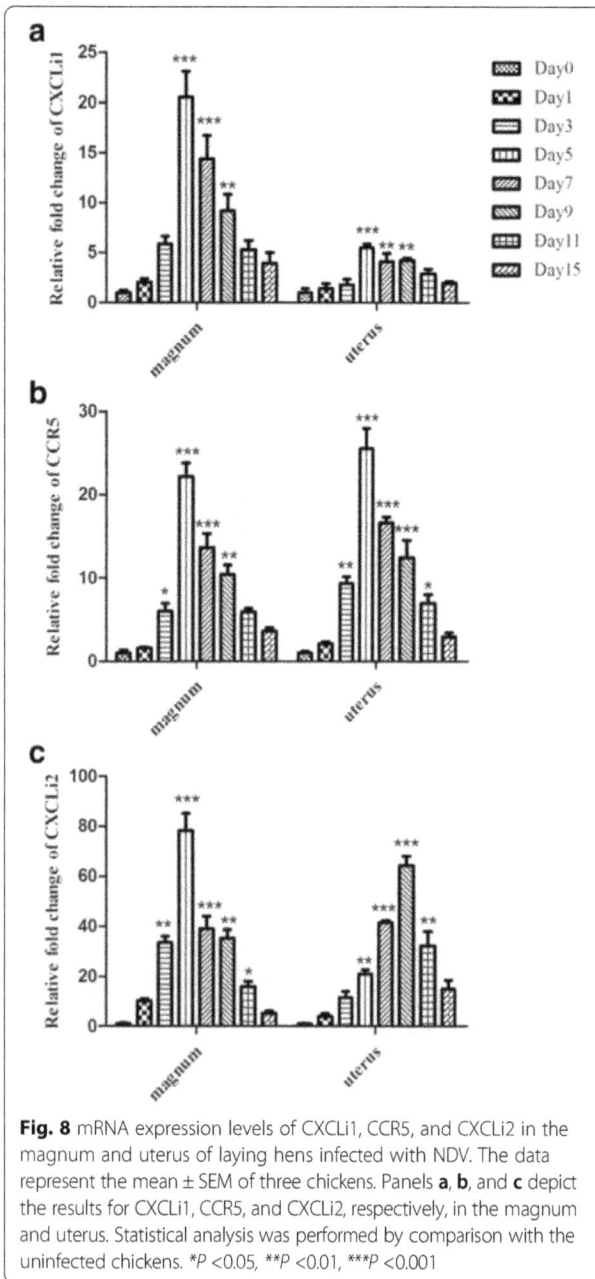

Fig. 8 mRNA expression levels of CXCLi1, CCR5, and CXCLi2 in the magnum and uterus of laying hens infected with NDV. The data represent the mean ± SEM of three chickens. Panels **a**, **b**, and **c** depict the results for CXCLi1, CCR5, and CXCLi2, respectively, in the magnum and uterus. Statistical analysis was performed by comparison with the uninfected chickens. *P <0.05, **P <0.01, ***P <0.001

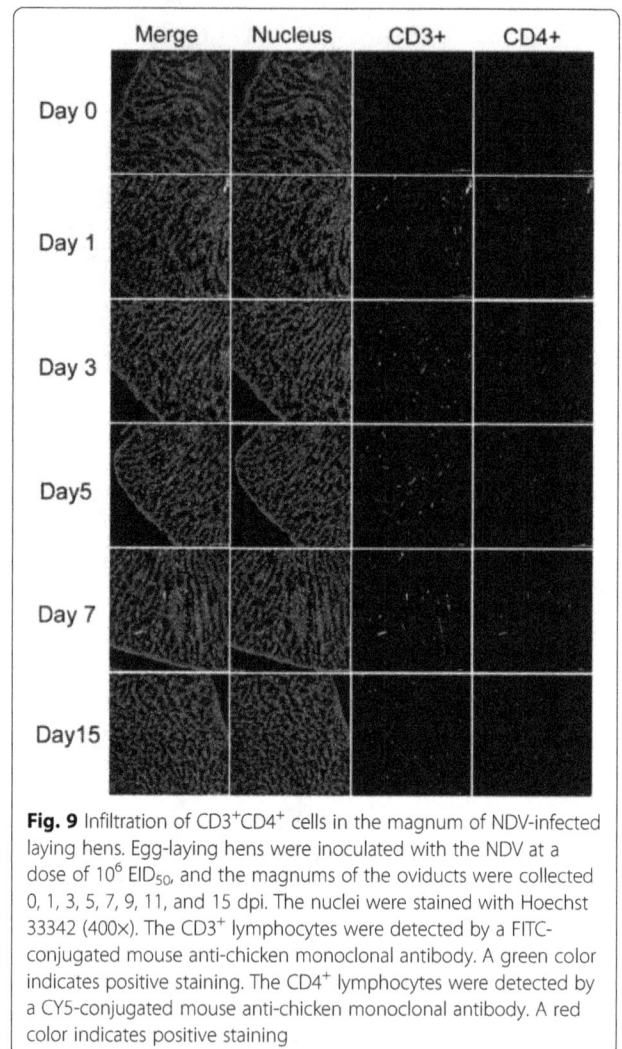

Fig. 9 Infiltration of CD3[+]CD4[+] cells in the magnum of NDV-infected laying hens. Egg-laying hens were inoculated with the NDV at a dose of 10^6 EID$_{50}$, and the magnums of the oviducts were collected 0, 1, 3, 5, 7, 9, 11, and 15 dpi. The nuclei were stained with Hoechst 33342 (400×). The CD3[+] lymphocytes were detected by a FITC-conjugated mouse anti-chicken monoclonal antibody. A green color indicates positive staining. The CD4[+] lymphocytes were detected by a CY5-conjugated mouse anti-chicken monoclonal antibody. A red color indicates positive staining

this study, IFN-α experienced a decline in the magnum of hen oviducts 3 dpi; the same phenomenon was reported in another study that demonstrated that the expression of IFN-α experienced a sharp decline in the peripheral blood of F48E9 NDV-infected chicken 7 dpi [12], which might contribute to rapid viral replication. The highest levels of expression of the examined genes corresponded with the significant clinical signs and the inflammatory responses in the oviductal tissues. In the chicken, there are two syntenic genes (CXCLi1 and CXCLi2) that mainly chemoattract heterophils and monocytes, respectively [27]. Interestingly, although both the CXCL chemokines were upregulated following

NDV infection in this study, the CXCLi2 mRNA expression levels were higher compared to CXCLi1. The upregulation of CXCLi1, CXCLi2, and CCR5 might potentially attract immune cells, thereby modulating cellular immunity.

Cell-mediated immunity is a specific adaptive immune response mediated by T lymphocytes. It is suggested to be an important factor in the development of protection in chickens vaccinated against NDV and viral clearance [28–30]. T lymphocytes are classified according to their expression of cell surface proteins. The CD3 molecule is the surface marker for mature lymphocytes; most CD4[+] lymphocytes are helper cells, while CD8[+] lymphocytes are cytotoxic cells [31]. In this study, immunofluorescence staining in the magnum demonstrated lymphocyte infiltration. CD3[+]CD4[+] and CD3[+]CD8α[+] lymphocytes were distributed in the lamina propria and mucosal epithelium of the magnum. Furthermore, the CD3[+]CD8α[+-] cells were located and infiltrated deeper into the bottom of the mucosal epithelium compared with the CD3[+]CD4[+]

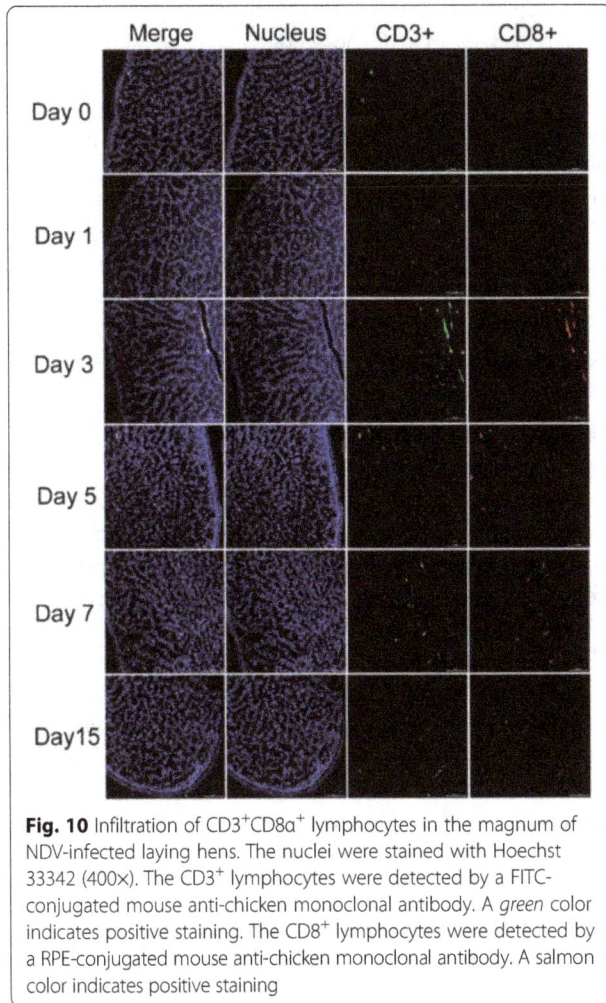

Fig. 10 Infiltration of CD3⁺CD8α⁺ lymphocytes in the magnum of NDV-infected laying hens. The nuclei were stained with Hoechst 33342 (400×). The CD3⁺ lymphocytes were detected by a FITC-conjugated mouse anti-chicken monoclonal antibody. A *green* color indicates positive staining. The CD8⁺ lymphocytes were detected by a RPE-conjugated mouse anti-chicken monoclonal antibody. A salmon color indicates positive staining

(CaBP-D28k) (data not shown), all of which may contribute to the reduced egg production. Given the findings that the magnum is the most sensitive segment to NDV, this sensitivity may contribute to the poor albumen quality of eggs laid by NDV-infected birds. Further studies are needed to determine the mechanisms underlying the apoptosis and immune responses in infected hen oviductal cell lines in vitro.

Conclusions

Infection with the velogenic genotype VIId NDV triggered apoptosis and production of cytokines and chemokines in oviduct, accompanied by inflammatory responses. It further led to infiltration of CD3 + CD4+ and CD3 + CD8α + lymphocytes in the segments of the oviduct. Our results indicate that the severe lesions in the oviducts of egg-laying hens caused by genotype VIId NDV strains is associated with excessive release of inflammatory cytokines and chemokines and lymphocytes infiltration, which contribute to the dysfunction of oviducts and the decline of egg production in hens.

Methods
Virus

The velogenic genotype VIId NDV strain (NDV/Chicken/TC/9/2011) was isolated from Shanxi province, China, in 2011 [33]. The mean death time (MDT) and intracerebral pathogenicity index (ICPI) were previously determined to be 48 h and 1.69, respectively . The virus was cultured in 10-day-old specific-pathogen-free (SPF) embryonated chicken eggs (Green Biological Engineering Co., Yangling, China).

Animals and experimental design

One hundred forty-week-old specific-pathogen-free (SPF) White Leghorn egg-laying hens (Green Biological Engineering Co., Yangling, China) were housed in isolators under negative pressure with food and water provided ad libitum. They were randomly divided into the control group and experimental group. The forty-five hens of control group were used as the NDV-free controls. The fifty-five hens in the experimental group were infected with a total of 0.5 mL (10^6 median embryo infective doses) of the velogenic genotype VIId NDV strain through combined intraocular and intranasal routes. The infected hens were monitored for clinical signs of disease and egg production. All animals were checked at least twice daily to ensure their health and welfare.

Animals displaying severe clinical distress were euthanized with intravenous sodium pentobarbital at a dose of 100 mg/kg. Five birds from each group was humanely euthanized on days 1, 3, 5, 7, 9, 11 and 15 post-infection (dpi), and various parts of the oviduct were sectioned. Birds that died as a result of infection were included

cells. Thus, infiltration of CD3⁺CD8α⁺ lymphocytes may be responsible for the aggravated tissue lesions and mediate viral clearance. A previous study suggested that T cell recruitment in association with the expression of proinflammatory cytokines and the CXCLi2 chemokine were upregulated in response to LPS in the lower part of the oviduct [32], our data are consistent with their findings. The high degree of lymphocyte infiltration in the NDV-infected chickens might contribute to the elevated oviductal damage observed during the early stage of infection.

Overall, NDV is a continuing threat to the poultry industry. Our findings revealed that the oviducts of laying hens were potential target tissues for NDV infection. Our results suggest that the strong inflammatory responses and apoptosis induced by NDV infection lead to severe local damage in the oviducts of egg-laying hens. We also measured the eggshell and albumin quality in another study, and our results suggested that NDV caused eggshell thinning and decreases in the expression of ovalbumin (OVA) and calcium-binding protein D28K

with the sampled birds on the indicated days. Birds that died on other days were observed by autopsy only. The criteria for euthanasia were somnolence, akinesia and dyspnea. Tissue processing and egg production changes were shown in Table 1. All experiments were performed in laboratories with biosafety level 2 facilities.

Tissue samples of the infundibulum, magnum, isthmus, uterus, and vagina from the control and experimental groups were collected immediately and placed on ice. Some parts of the tissues were preserved in liquid nitrogen for SYBR Green-based real-time (RT)-PCR, and others were fixed in 10% phosphate-buffered formalin for more than 24 hs [34]. The blocks were sectioned at 5-μm thicknesses for histopathological lesion observation, terminal deoxnuceotidyl transferase-mediated d-UTP biotin nick end labeling (TUNEL) staining, hemagglutinin neuraminidase (HN) protein detection, and CD3$^+$CD4$^+$ and CD3$^+$CD8α$^+$ lymphocyte detection by immunofluorescence.

Histopathological examination

The infundibulum, magnum, isthmus, uterus, and vagina were collected and fixed in 10% phosphate-buffered formalin, embedded in paraffin wax and cut into sections using routine protocols. Next, the paraffin sections were stained with hematoxylin and eosin, and the tissue changes were observed under a light microscope.

Detection of viral loads in the oviducts

The SYBR Green I (TransGen biotech, China) RT-PCR assay was used to detect the viral loads in the infundibulum, magnum, isthmus, uterus, and vagina. The M gene of NDV was used as the target gene and as a reference for the expression analyses. The primers used are as follows: Forward: 5'-ATCCGAAGAGCCCGTTAG-3' (3939-3956) and Reverse: 5'-ACATCACTGAGCCCG ACA-3' (4113-4096). Total RNA was extracted from the reproductive tract tissues using the TRIzol reagent (Takara Biotechnology, Japan). The concentration of RNA in each sample was measured spectrophotometrically at 260 nm versus 280 nm. Next, the RNA was reverse-transcribed using the EasyScript First-Strand cDNA Synthesis SuperMix (TransGen Biotech) according to the manufacturer's instructions. Construction of standard curve was done according to the procedure of [34].

Detection of the viral hemagglutinin-neuraminidase protein in the oviduct by indirect immunofluorescence

The paraffin sections of the oviducts were examined by IFA to detect the viral HN protein. Briefly, the tissue sections were deparaffinized with xylene for 10 min, followed by hydration through a graded series of alcohol to xylol (100%, 95%, 90%, 80%, and 70%) for 3 min at each step. Then, the sections were blocked with 5% BSA for 1 h at 37 °C. Next, the sections were incubated with1 mg/ml mouse anti avian monoclonal antibody specific for HN of NDV (Santa Cruz Biotechnology) in PBS overnight at 4 °C. After washing thrice with PBS, the sections were incubated with FITC-conjugated goat anti-mouse IgG (Santa Cruz Biotechnology, 1:400) in PBS for 1 h at 37 °C, and then washed thrice with PBS once again and counter stained with Hoechst 33342 (Sigma, USA) for 5 min. Finally, the treated sections were observed under a confocal laser scanning microscope (Nikon, Japan).

Determination of apoptosis in the oviduct
Detection of caspase-3 activity

A total of 0.1 g of oviduct tissue was cut into small pieces, added to 1 mL of radio immunoprecipitation assay lysis buffer, and homogenized in a glass pestle on ice. The tissue homogenates were transferred into 1.5 mL centrifugetubes and incubated on ice for 5 min. Then, they were centrifuged at 20,000 rpm for 10 min, and the supernatants were collected into pre-cooled centrifuge tubes. The caspase-3 enzyme activity was measured using the detection kit (KeyGen Biotech), and its concentration was determined by the Bradford Protein Assay Kit (Beyotime Institute of Biotechnology, Nanjing, China).

Quantification of apoptosis by the TUNEL assay

Apoptosis was examined using the one-step measurement of terminal deoxnuceotidyl transferase mediated d-UTP biotin nick end labeling (TUNEL) in the oviducts of the infected hens on 5 dpi. Tissue sections were deparaffinized and rehydrated through an alcohol gradient then rinsed in deionized water. The following steps corresponded to the standard operation staining procedures provided by manufacturer (Vazyme Biotech, Nanjing, China).

Quantification of TLRs, MDA5, and proinflammatory cytokines by RT-PCR

Total RNA extraction was performed as described in the detection of viral loads in the oviducts section. RT-PCR reactions were performed with equal amounts of cDNA samples from all egg-laying hens using a thermal cycler (Bio-Rad, America). Table 2 showed the primers used for PCR, which were designed using Primer Premier 5 software. The specificity of each primer was checked using the NCBI blast program. The expression levels of TLRs, MDA5, and proinflammatory cytokines and chemokines were normalized to chicken β-actin and amplification efficiency were consistent with β-actin. The results of real-time PCR were quantified by the

Table 2 PCR primers used for mRNA expression analysis

Genes	Primer sequences	Acession no.	Location	Product size
IL-2	F: 5'-CTGTATTTCGGTAGCAATG-3'	NM_204153.1	87-247	161
	R: 5'-ACTCCTGGGTCTCAGTTG-3'			
IL-1β	F: 5'-ACCCGCTTCATCTTCTACCG-3'	NM_204524.1	663-836	174
	R: 5'-TCAGCGCCCACTTAGCTTG-3'			
IL-6	F: 5'-GGCATTCTCATTTCCTTCT-3'	NM_204628.1	853-1051	199
	R: 5'-CTGGCTGCTGGACATTTT-3'			
IFN-α	F: 5'-GGACATGGCTCCCACACTAC-3'	XM_004937097.1	2896-3044	149
	R: 5'-GGCTGCTGAGGATTTTGAAGA-3'			
IFN-β	F: 5'-CACCACCACCTTCTCCT-3'	NM_001024836.1	66-321	256
	R: 5'-TGTGCGGTCAATCCAGT-3'			
CXCLi2	F: 5'-CATCATGAAGCATTCCATCT-3'	NM_205498.1	124-223	100
	R: 5'-CTTCCAAGGGATCTTCATTT-3'			
CXCLi1	F: 5'-AACTCCGATGCCAGTG-3'	NM_205018.1	176-380	205
	R: 5'-TTGGTGTCTGCCTTGT-3'			
CCR5	F: 5'-GTGGTCAACTGCAAAAAGCA-3'	NM_001045834.1	1126-1315	190
	R: 5'-GCCCGTTCAACTGTGTCG-3'			
MDA-5	F: 5'-GGACGACCACGATCTCTGTGT-3'	NM_001193638.1	457-535	79
	R: 5'-CACCTGTCTGGTCTGCATGTTATC-3'			
TLR3	F: 5'-GCTATTGAGCAAAGTCGAGA -3'	NM_001011691.3	2491-2693	203
	R: 5'-ACAGGGGGCACTTTACTATT-3'			
TLR7	F: 5'-TCTGGACTTCTCTAACAACA-3'	NM_001011688.2	1824-2010	187
	R: 5'-AATCTCATTCTCATTCATCATCA-3'			
TLR21	F: 5'-AGTTGTGTCCTGTGCTGAGAG-3'	NM_001030558.1	1035-1164	130
	R: 5'-AGCAGGTTGTGTTCCACTGTC-3'			
β-actin	F: 5'- AGACATCAGGGTGTGATGGTTGGT-3'	NM_205518.1	183-300	118
	R: 5'-TGGTGACAATACCGTGTTCAATGG-3'			

comparative threshold analysis after deductions of data from uninfected chickens.

Detection of T lymphocytes in the magnum by immunofluorescence

The dewaxing and rehydrating process and BSA treatment was performed as described in the HN protein detection section. After removing the blocking solution, tissue sections were incubated for 1 h at 37 °C with 1 mg/ml mouse monoclonal antibody specific for chicken CD markers, FITC conjugated CD3, CY5 conjugated CD4, or RPE labeled CD8α (Southern biotech, USA) to detect CD3CD4 and CD3CD8α glycoprotein of lymphocytes. Following steps were performed as described in the HN protein detection section.

Statistical analyses

The data are presented as the means ± SEM of three independent experiments with three replicates per experiment. Analysis of the gene expression levels was performed using one-way analysis of variance with Dunnett's post-test in GraphPad Prism version 5.0 for Windows. P values less than 0.05 were considered statistically significant.

Acknowledgements
Non applicable.

Funding
This work was financially supported by the Shaanxi Science and Technology Overall Innovation Project Plan (Grant No. 2015KTCL02-16). The funders had no role in study design, data collection, analysis and interpretation, decision to publish, or preparation of the manuscript.

Authors' contributions
RL and KG performed the majority of the experiments and were involved in manuscript preparation. CL, DT participated in collection of samples and data analysis. Jing W, CT and XH performed immunofluorescence experiment. Jingyu W conceived of the study and participated in its design and coordination. YZ helped with experiments, provided valuable

discussion and modified the final manuscript. All authors read and approved the final manuscript.

Competing interests

The authors declare that they have no competing interests.

Consent for publication

Not applicable.

References

1. Alexander DJ. Newcastle disease and other avian paramyxoviruses. Rev Sci Tech. 2000;19(2):443–62.
2. Raghul J, Raj GD, Manohar BM, Balachandran C. Protection of the reproductive tract of young chicks by Newcastle disease virus-induced haemagglutinationinhibition antibodies. Vet Res Commun. 2006;30(1): 95–102.
3. Czegledi A, Ujvari D, Somogyi E, Wehmann E, Werner O, Lomniczi B. Third genome size category of avian paramyxovirus serotype 1 (Newcastle disease virus) and evolutionary implications. Virus Res. 2006;120(1-2):36–48.
4. Qin ZM, Tan LT, Xu HY, Ma BC, Wang YL, Yuan XY, Liu WJ. Pathotypical characterization and molecular epidemiology of Newcastle disease virus isolates from different hosts in China from 1996 to 2005. J Clin Microbiol. 2008;46(2):601–11.
5. Wu S, Wang W, Yao C, Wang X, Hu S, Cao J, Wu Y, Liu W, Liu X. Genetic diversity of Newcastle disease viruses isolated from domestic poultry species in Eastern China during 2005-2008. Arch Virol. 2011;156(2):253–61.
6. Ebrahimi MM, Shahsavandi S, Moazenijula G, Shamsara M. Phylogeny and evolution of Newcastle disease virus genotypes isolated in Asia during 2008-2011. Virus Genes. 2012;45(1):63–8.
7. Bwala DG, Clift S, Duncan NM, Bisschop SP, Oludayo FF. Determination of the distribution of lentogenic vaccine and virulent Newcastle disease virus antigen in the oviduct of SPF and commercial hen using immunohistochemistry. Res Vet Sci. 2012;93(1):520–8.
8. Kumar R, Kirubaharan JJ, Chandran ND, Gnanapriya N. Transcriptional response of chicken embryo cells to Newcastle disease virus (D58 strain) infection. Indian J Virol. 2013;24(2):278–83.
9. Rasoli M, Yeap SK, Tan SW, Moeini H, Ideris A, Bejo MH, Alitheen NB, Kaiser P, Omar AR. Alteration in lymphocyte responses, cytokine and chemokine profiles in chickens infected with genotype VII and VIII velogenic Newcastle disease virus. Comp Immunol Microbiol Infect Dis. 2014;37(1):11–21.
10. Ecco R, Brown C, Susta L, Cagle C, Cornax I, Pantin-Jackwood M, Miller PJ, Afonso CL. In vivo transcriptional cytokine responses and association with clinical and pathological outcomes in chickens infected with different Newcastle disease virus isolates using formalin-fixed paraffin-embedded samples. Vet Immunol Immunopathol. 2011;141(3-4):221–9.
11. Rue CA, Susta L, Cornax I, Brown CC, Kapczynski DR, Suarez DL, King DJ, Miller PJ, Afonso CL. Virulent Newcastle disease virus elicits a strong innate immune response in chickens. J Gen Virol. 2011;92(Pt 4):931–9.
12. Hu Z, Hu J, Hu S, Song Q, Ding P, Zhu J, Liu X, Wang X, Liu X. High levels of virus replication and an intense inflammatory response contribute to the severe pathology in lymphoid tissues caused by Newcastle disease virus genotype VIId. Arch Virol. 2015;160(3):639–48.
13. Hu Z, Hu J, Hu S, Liu X, Wang X, Zhu J, Liu X. Strong innate immune response and cell death in chicken splenocytes infected with genotype VIId Newcastle disease virus. Virol J. 2012;9:208.
14. Schwartzman RA, Cidlowski JA. Apoptosis: the biochemistry and molecular biology of programmed cell death. Endocr Rev. 1993;14(2):133–51.
15. Ravindra PV, Tiwari AK, Ratta B, Chaturvedi U, Palia SK, Subudhi PK, Kumar R, Sharma B, Rai A, Chauhan RS. Induction of apoptosis in Vero cells by Newcastle disease virus requires viral replication, de-novo protein synthesis and caspase activation. Virus Res. 2008;133(2):285–90.
16. Ravindra PV, Tiwari AK, Ratta B, Chaturvedi U, Palia SK, Chauhan RS. Newcastle disease virus-induced cytopathic effect in infected cells is caused by apoptosis. Virus Res. 2009;141(1):13–20.
17. Yu L, Wang Z, Jiang Y, Chang L, Kwang J. Characterization of newly emerging Newcastle disease virus isolates from the People's Republic of China and Taiwan. J Clin Microbiol. 2001;39(10):3512–9.
18. Tan SW, Ideris A, Omar AR, Yusoff K, Hair-Bejo M. Sequence and phylogenetic analysis of Newcastle disease virus genotypes isolated in Malaysia between 2004 and 2005. Arch Virol. 2010;155(1):63–70.
19. Elankumaran S, Rockemann D, Samal SK. Newcastle disease virus exerts oncolysis by both intrinsic and extrinsic caspase-dependent pathways of cell death. J Virol. 2006;80(15):7522–34.
20. Bian J, Wang K, Kong X, Liu H, Chen F, Hu M, Zhang X, Jiao X, Ge B, Wu Y, et al. Caspase- and p38-MAPK-dependent induction of apoptosis in A549 lung cancer cells by Newcastle disease virus. Arch Virol. 2011;156(8):1335–44.
21. Brownlie R, Allan B. Avian toll-like receptors. Cell Tissue Res. 2011;343(1): 121–30.
22. Cheng J, Sun Y, Zhang X, Zhang F, Zhang S, Yu S, Qiu X, Tan L, Song C, Gao S, et al. Toll-like receptor 3 inhibits Newcastle disease virus replication through activation of pro-inflammatory cytokines and the type-1 interferon pathway. Arch Virol. 2014;159(11):2937–48.
23. Guo X, Wang L, Cui D, Ruan W, Liu F, Li H. Differential expression of the Toll-like receptor pathway and related genes of chicken bursa after experimental infection with infectious bursa disease virus. Arch Virol. 2012;157(11):2189–99.
24. Raj GD, Rajanathan TM, Kumanan K, Elankumaran S. Changes in the cytokine and toll-like receptor gene expression following infection of indigenous and commercial chickens with infectious bursal disease virus. Indian J Virol. 2011;22(2):146–51.
25. Swaggerty CL, Pevzner IY, Kaiser P, Kogut MH. Profiling pro-inflammatory cytokine and chemokine mRNA expression levels as a novel method for selection of increased innate immune responsiveness. Vet Immunol Immunopathol. 2008;126(1-2):35–42.
26. Martinez-Sobrido L, Zuniga EI, Rosario D, Garcia-Sastre A, de la Torre JC. Inhibition of the type I interferon response by the nucleoprotein of the prototypic arenavirus lymphocytic choriomeningitis virus. J Virol. 2006; 80(18):9192–9.
27. Poh TY, Pease J, Young JR, Bumstead N, Kaiser P. Re-evaluation of chicken CXCR1 determines the true gene structure: CXCLi1 (K60) and CXCLi2 (CAF/interleukin-8) are ligands for this receptor. J Biol Chem. 2008;283(24):16408–15.
28. Merz DC, Scheid A, Choppin PW. Immunological studies of the functions of paramyxovirus glycoproteins. Virology. 1981;109(1):94–105.
29. Cannon MJ, Russell PH. Secondary in vitro stimulation of specific cytotoxic cells to Newcastle disease virus in chickens. Avian Pathol. 1986;15(4):731–40.
30. Sharma JM, Zhang Y, Jensen D, Rautenschlein S, Yeh HY. Field trial in commercial broilers with a multivalent in ovo vaccine comprising a mixture of live viral vaccines against Marek's disease, infectious bursal disease, Newcastle disease, and fowl pox. Avian Dis. 2002;46(3):613–22.
31. Kapczynski DR, Afonso CL, Miller PJ. Immune responses of poultry to Newcastle disease virus. Dev Comp Immunol. 2013;41(3):447–53.
32. Nii T, Sonoda Y, Isobe N, Yoshimura Y. Effects of lipopolysaccharide on the expression of proinflammatory cytokines and chemokines and the subsequent recruitment of immunocompetent cells in the oviduct of laying and molting hens. Poult Sci. 2011;90(10):2332–41.
33. Wang JY, Liu WH, Ren JJ, Tang P, Wu N, Liu HJ. Complete genome sequence of a newly emerging newcastle disease virus. Genome Announc. 2013;1(3): e00261–13.
34. Wang JY, Chen ZL, Li CS, Cao XL, Wang R, Tang C, Huang JJ, Chang CD, Liu HJ. The distribution of sialic acid receptors of avian influenza virus in the reproductive tract of laying hens. Mol Cell Probes. 2015;29(2):129–34.

4

Development of duplex PCR for differential detection of goatpox and sheeppox viruses

Zhixun Zhao, Guohua Wu, Xinmin Yan, Xueliang Zhu, Jian Li, Haixia Zhu, Zhidong Zhang[*] and Qiang Zhang[*]

Abstract

Background: Clinically, sheeppox and goatpox have the same symptoms and cannot be distinguished serologically. A cheaper and easy method for differential diagnosis is important in control of this disease in endemic region.

Methods: A duplex PCR assay was developed for the specific differential detection of Goatpox virus (GTPV) and Sheeppox virus (SPPV), using two sets of primers based on viral E10R gene and RPO132 gene.

Results: Nucleic acid electrophoresis results showed that SPPV-positive samples appear two bands, and GTPV-positive samples only one stripe. There were no cross-reactions with nucleic acids extracted from other pathogens including foot-and-mouth disease virus, Orf virus. The duplex PCR assay developed can specially detect SPPV or GTPV present in samples ($n = 135$) collected from suspected cases of Capripox.

Conclusions: The duplex PCR assay developed is a specific and sensitive method for the differential diagnosis of GTPV and SPPV infection, with the potential to be standardized as a detection method for Capripox in endemic areas.

Keywords: Goatpox virus, Sheeppox virus, Duplex PCR assay, Differential diagnosis

Background

Goatpox virus (GTPV) and sheeppox virus (SPPV) are members of the genus Capripoxvirus (CaPV) of the family Poxviridae, which also contains the lumpy skin disease virus (LSDV). GTPV and SPPV can cause systemic poxviral diseases of domesticated small ruminants [1–3], which are classified as notifiable diseases by World Organization for Animal Health (OIE). Morbidity and mortality varied with different animal species [4]. Mild infections are usual among native breeds in endemic areas and the morbidity ranges from 1% to 75% or higher [5]. In contrast, more severe clinical symptoms are observed in animals with concurrent infections, stressed or young animals, or animals from regions where the poxviral diseases have not happened for long time. The mortality was close to 100% in these highly susceptible animals [6]. The genome of CaPV is about 150 kbp of double-stranded DNA, which shares more than 147 putative genes, including conserved poxvirus structural and replicative genes, and genes likely involved in host range and virulence [7]. Sheeppox and goatpox is an economically important disease in goat and sheep-producing areas of the world [8]. Sheeppox and goatpox have the same symptoms and are clinically distinguishable. For accurately and promptly controlling any outbreak of sheeppox and goatpox, the foremost requirement is a specific tool for differential detection of the causative agents. Digestion of CaPV p32 gene with Hinf I and sequence alignment of GpCR gene were developed to discriminate GTPV and SPPV [9, 10]. In this study, a duplex polymerase chain reaction (PCR) assay based on the E10R and RPO132 gene was developed to distinguish GTPV and SPPV. The performance of the assay was evaluated with clinical samples. In comparison with the PCR-RFLP, the newly established duplex PCR assay is a time-efficient and simple alternative for differential diagnosis of GTP and SPP that could be used as a diagnostic tool in clinical samples.

Methods

Gene sequences and primers design

PCR primers were designed using Primer Premier 5.0 software based on E10R gene and RPO132 gene of

[*] Correspondence: zhangzhidong@caas.cn, zhangqiang@caas.cn
Key Laboratory of Animal virology of the Ministry of Agriculture, State Key Laboratory of Veterinary Etiological Biology, Lanzhou Veterinary Research Institute, CAAS, Lanzhou, People's Republic of China

GTPV and SPPV E10R gene. Selection of these targets for primer designs were based upon previous bioinformatics analyses of CaPV genomes and corresponding homologs from other near-neighbor viruses listed in the NCBI (National Center for Biotechnology Information) database (data not shown). Nucleotide sequences of the primers are shown in Table 1 and Fig. 1. A nonspecific sequence highlighted in bold was added at 5′ end of each primer (E10R-f primer: "CCGCTCGAGGCCACC"; E10R-r primer: "CGCGGATCCCGC" RPO132-f: "CGCGG ATCCCGC" and RPO132-r: "CCGCTCGAGGC CACC").

Viruses

SPPV Gulang/2009 strain was isolated and adapted in Vero cell culture. GTPV AV40 vaccine strains adapted in Vero cell culture was obtained from the Institute of Veterinary Drug Control, Beijing, China. The nucleic acid extracted from foot-and-mouth disease virus (FMDV) isolates was provided by the National Foot-and-Mouth Disease Reference Laboratory, LVRI, and the nucleic acids extracted from *Mycoplasma ovipneumoniae* (*M. ovipneumoniae*), *Chlamydia psittaci, L. interrogans, Toxoplasma gondii* and *Babesia* were provided by the Key Laboratory of Veterinary Parasitology of Gansu Province, LVRI. The clinical samples collected from suspected cases of goatpox and sheeppox were stored in PBS, pH 7.4 and a 10% (*w/v*) suspension was made for extraction of total gDNA using commercial DNA extraction kit (TaKaRa, Dalian, China) according to the manufacture's instruction.

Reaction mixtures and optimal duplex PCR conditions

Duplex PCR reactions were carried out in a volume of 25 μL containing 10 μL premix Taq (TaKaRa, Dalian, China), 0.2 μM of each of the E10R-f and E10R-r primers, 1.0 μM of each of the RPO132-f and RPO132-r primers and 2 μL (100 ng) of extracted SPPV or GTPV gDNA as template. The duplex PCR assay was performed at the following conditions: initial denaturizing at 94 °C for 4 min, followed by 34 cycles of 94 °C for 30 s, 50 °C for 30 s, 72 °C for 45 s and final extension of 72 °C for 5 min.

Analysis of duplex PCR products

Products were separated electrophoretically in 1% agarose gel (Gelrose TM, Life Technologies, USA) containing 0.5 g/ml ethidium bromide, followed by visualization under ultraviolet (UV) light. 6 μL of duplex PCR products was used for visualization and the electrophoresis was done for 20 min at a constant 120 V. After visualization, the picture was documented using a gel documentation system (Peiqing Image Biosystem, Shanghai).

Duplex PCR sensitivity

The sensitivity of the duplex PCR assay was tested using 10-fold serially diluted GTPV (1.011×10^{10}–1.011×10^{0} copies per μL) or SPPV DNA (1.043×10^{10}–1.043×10^{0} copies per μL) as template.

Duplex PCR specificity

The specificity of the duplex PCR assay was evaluated by using purified the nucleic acids extracted from sheep, goats and some wild ruminants, SPPV, GTPV, Orf, FMDV, *M.ovippneumoniae, Chlamydia psittaci, L.interrogans, Toxoplasma gondii, Babesia* and a control without template was also included as a negative control in each test. The duplex PCR products were sequenced to further confirm the specificity of the assay.

Evaluation of the duplex PCR assay in clinical samples

The performance of the duplex PCR assay was evaluated with 135 clinical samples preserved in our laboratory. These samples were pre-screened by PCR-RFLP [9, 10] and LAMP [11] as previously reported. Among them 48 samples were collected from goats infected with GTPV and 87 from sheep infected with SPPV. Animal experimentations were performed inside the biosafety facilities of the Lanzhou Veterinary Research Institute, Chinese Academy of Agricultural Sciences (LVRI, CAAS), in compliance with the regulations of the Animal Ethics Procedures and Guidelines of the People's Republic of China (AEPGPRC).

Results

Primers and gene sequences

Several GTPV and SPPV genomic sequences were downloaded from GenBank and aligned with each other using the MegAlign. The most conserved segments within the E10R and RPO132 genes of GTPV and SPPV were selected as the targets. All primers were designed by the software Primer Premier 5.0. Two pairs of primers were used for the duplex PCR assay, i.e., E10R primers (E10R-

Table 1 Primer designed to detect goatpox and sheeppox virus by duplex PCR

Primer name	Length	Sequence(5′-3′)	Notes
E10R-f	41	CCGCTCGAGGCCACCATGAATCCTAAACACTGGGGAAGAGC	The universal primers for GTPV and SPPV RPO40 gene, the predicted length of PCR is 285 bp.
E10R-r	36	CGCGGATCCCGAAGCGGTAATACCTTATTTAAATTG	
RPO132-f	39	CCGCTCGAGGCCACCATGAATAGGTTCAAGGAAAAGCAT	The special PCR primers for SPPV RPO132 gene, the predicted length of Lamp is 746 bp.
RPO132-r	30	CGCGGATCCCGCATTATTTTTTCATACGAT	

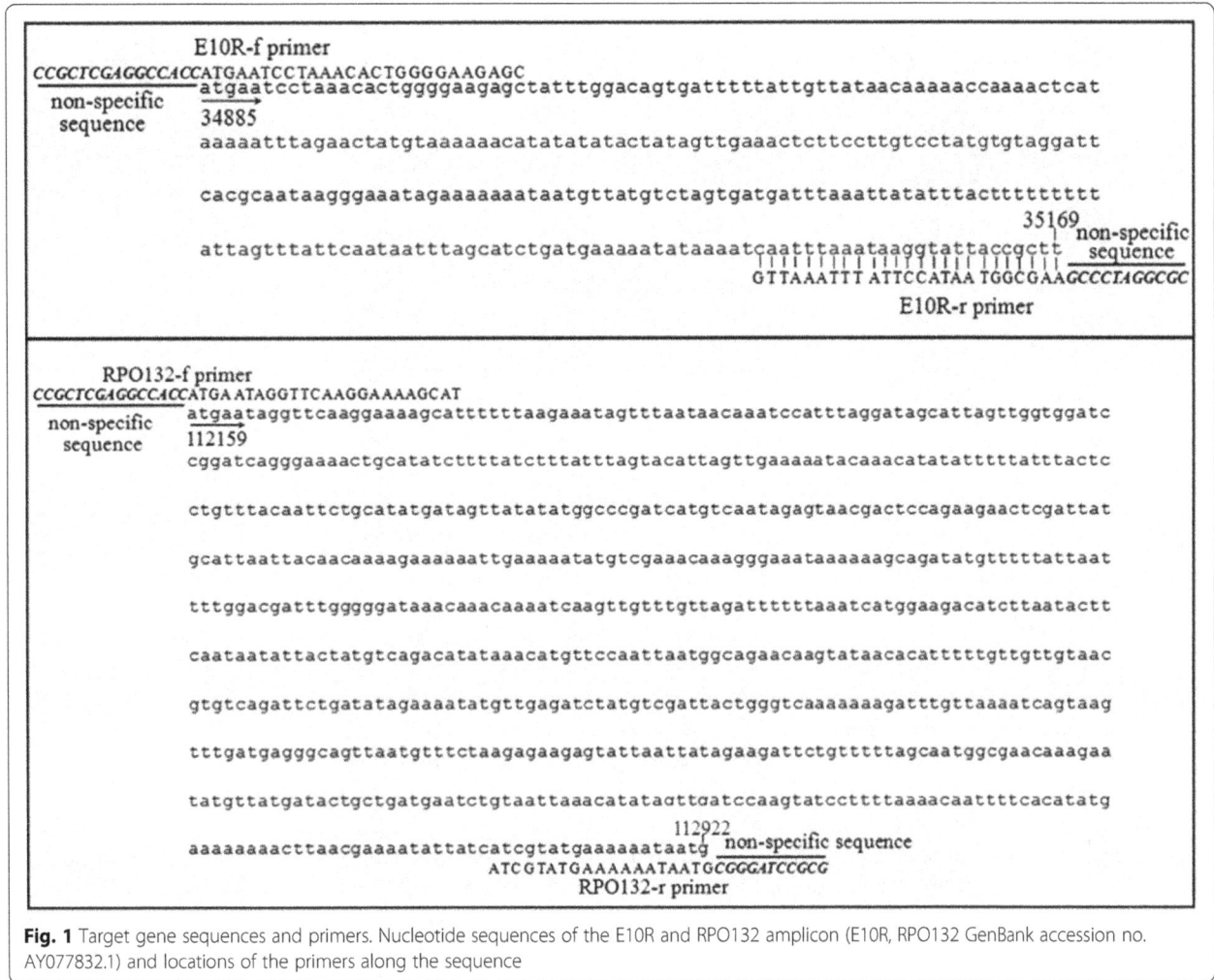

Fig. 1 Target gene sequences and primers. Nucleotide sequences of the E10R and RPO132 amplicon (E10R, RPO132 GenBank accession no. AY077832.1) and locations of the primers along the sequence

f and E10R-r) and RPO132 primers (RPO132-f and RPO132-r). The sequences of all primers were shown in Table 1.

Optimization of duplex PCR reaction conditions

The SPPV specific primers (RPO132) amplified approx. 746 bp fragments and the GTPV specific primers (E10R)

amplified approx. 285 bp fragments as expected, when visualized in 1% agarose gel using ethidium bromide staining (Fig. 2).

To determine the optimal temperature for amplification in the duplex PCR assay, a range of temperatures from 46 °C to 56 °C were tested. The results showed that the targeted GTPV and SPPV genes can be amplified at

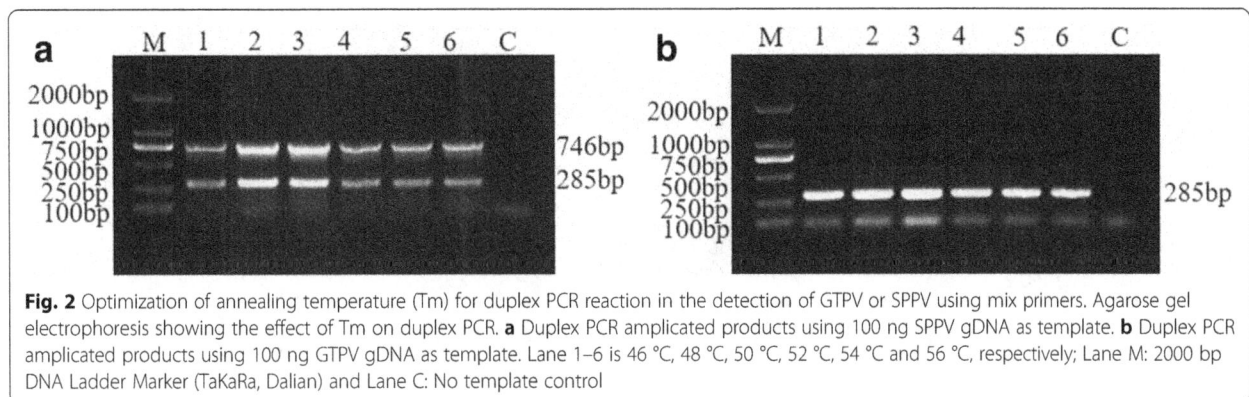

Fig. 2 Optimization of annealing temperature (Tm) for duplex PCR reaction in the detection of GTPV or SPPV using mix primers. Agarose gel electrophoresis showing the effect of Tm on duplex PCR. **a** Duplex PCR amplicated products using 100 ng SPPV gDNA as template. **b** Duplex PCR amplicated products using 100 ng GTPV gDNA as template. Lane 1–6 is 46 °C, 48 °C, 50 °C, 52 °C, 54 °C and 56 °C, respectively; Lane M: 2000 bp DNA Ladder Marker (TaKaRa, Dalian) and Lane C: No template control

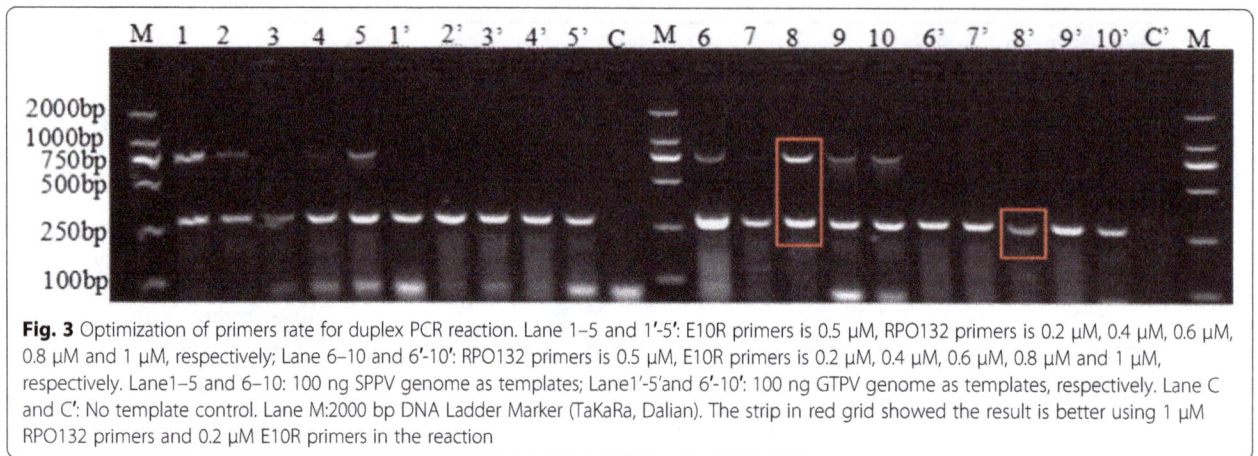

Fig. 3 Optimization of primers rate for duplex PCR reaction. Lane 1–5 and 1'–5': E10R primers is 0.5 μM, RPO132 primers is 0.2 μM, 0.4 μM, 0.6 μM, 0.8 μM and 1 μM, respectively; Lane 6–10 and 6'–10': RPO132 primers is 0.5 μM, E10R primers is 0.2 μM, 0.4 μM, 0.6 μM, 0.8 μM and 1 μM, respectively. Lane1–5 and 6–10: 100 ng SPPV genome as templates; Lane1'–5'and 6'–10': 100 ng GTPV genome as templates, respectively. Lane C and C': No template control. Lane M:2000 bp DNA Ladder Marker (TaKaRa, Dalian). The strip in red grid showed the result is better using 1 μM RPO132 primers and 0.2 μM E10R primers in the reaction

all of above annealing temperature using E10R and RPO132 mix primers, respectively. As shown in Fig. 2 the strongest amplified products were observed at 50 °C. In the optimization of ratio of E10R to RPO132 primers, it was found that

the targeted genes can be amplified at any proportion of E10R and RPO132 primers, respectively. However, the best amplification occurred when each of RPO132 primers is 1 μM and each of E10R primers is 0.2 μM, respectively (Fig. 3).

In determination of the best reaction cycles, the results showed that the targeted genes can be produced after 25 cycles to 40 cycles using the optimized concentration of the primers at 50 °C, and the amplified products at 35 cycles was clearer (Fig. 4) than other cycle conditions.

Duplex PCR sensitivity for detection of GTPV and SPPV

To determine the sensitivity of the duplex PCR assay developed, a serial dilutions of the purified genome DNA (gDNA) of SPPV and GTPV were used. The concentration of the purified viral gDNA was measured by NanoDrop 2000 (Thermo Scientific). SPPV Gulang/2009 gDNA was ten-fold diluted ranging from 1.011×10^{10}–1.011×10^{0}

copies and GTPV AV40 gDNA was diluted ranging from 1.043×10^{10}–1.043×10^{0} copies respectively. After amplification under the optimized conditions as described above 6 μL PCR amplified products e was tested by nucleic acid electrophoresis, and then was observed by UV gel imaging system. The results showed that the duplex PCR assay was able to specifically amplify the SPPV gDNA from 1.011×10^{10}–1.011×10^{4} copies, and the control group had no stripe (Fig. 5a). 1.043×10^{5} copies of GTPV gDNA template was detected (Fig. 5b, Table 2).

Duplex PCR assay specificity

In investigation of the specificity of the duplex PCR assay, the reactions were performed using a panel of genomes extracted from GTPV, SPPV, Orf virus, FMDV, *M. ovippneumoniae*, *Chlamydia psittaci*, *L.interrogans*, *Toxoplasma gondii* were tested, respectively. The duplex PCR assay was shown to be specific for the detection of GTPV and SPPV, respectively and there was no cross-reaction with genome of Orf, FMDV, *M. ovippneumoniae*, *Chlamydia psittaci*, *L.interrogans*, *Toxoplasma gondii*, *Babesia*, and without template control (Fig. 6). The duplex PCR products were then sequenced and found to match with the corresponding nucleotide sequences of E10R or RPO132

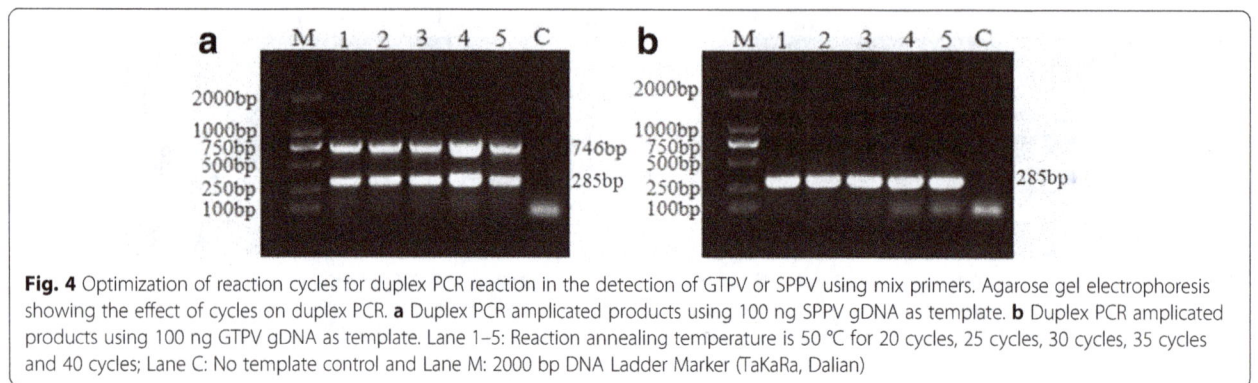

Fig. 4 Optimization of reaction cycles for duplex PCR reaction in the detection of GTPV or SPPV using mix primers. Agarose gel electrophoresis showing the effect of cycles on duplex PCR. **a** Duplex PCR amplicated products using 100 ng SPPV gDNA as template. **b** Duplex PCR amplicated products using 100 ng GTPV gDNA as template. Lane 1–5: Reaction annealing temperature is 50 °C for 20 cycles, 25 cycles, 30 cycles, 35 cycles and 40 cycles; Lane C: No template control and Lane M: 2000 bp DNA Ladder Marker (TaKaRa, Dalian)

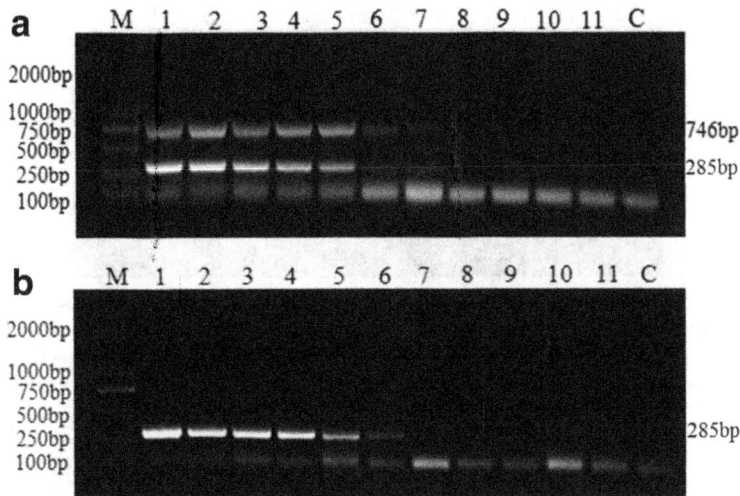

Fig. 5 Duplex PCR sensitivity. Amplification using different concentration gradient of the gene as template, nucleic acid electrophoresis test results. **a** Lane 1–10: SPPV gDNA concentration gradient for 1.011×10^{10}–1.011×10^{0} copies as template, (**b**) Lane 1–10: GTPV gDNA concentration gradient for 1.043×10^{10}–1.043×10^{0} copies as template, respectively. Lane M: 2000 bp DNA Ladder Marker (TaKaRa, Dalian) and Lane C: No template control

gene (data not shown). Phylogenetic tree analysis showed that the SPPV Gulang/2009 and other isolated SPPVs were distinctly different from GTPVs and LSDVs.

Performance of the duplex PCR assay on clinical samples

All the clinical samples (n = 135) were detected by the duplex PCR assay, and the results were then compared to the PCR-RFLP assay [9, 10] and LAMP method [11]. The results showed that 48 samples were determined to be positive for GTPV and 87 samples were positive for SPPV by the duplex PCR assay, respectively, which was 100% consistent with the LAMP method and PCR-RFLP assay (Table 3).

Discussion

GTPV and SPPV genomes are approximately 150kbp double-stranded DNA, which share at least 147 putative genes, including conserved poxvirus replication and structural genes and genes likely involved in virulence and host range [7]. Restriction endonuclease analysis and cross-hybridization studies of SPPV and GTPV indicate that

these viruses, although closely related (estimated 96 to 97% nucleotide identity), can be distinguished from one another and may undergo recombination in nature. Several PCR tests have been developed for the detection of CaPV [12–19], but only PCR-RFLP assay was developed to distinguish them in the beginning. However, PCR-RFLP assay is time consuming, and expensive and requires a high degree of laboratory experience in molecular biology. Then a LAMP method was developed for the specific differential detection of GTPV and SPPV, using three sets of LAMP primers designed on the basis of the ITRs [11]. LAMP for distinguishing GTPV and SPPV was performed at 62 °C in 45–60 min. However, the operation of the experiment requires extreme caution because of its high sensitivity leading to contamination. There is a real need for a more convenient alternative to PCR that is inexpensive, and easy to operate and maintain.

E10R is encoded by ORF40 gene and RNA polymerase subunit RPO132 is encoded by ORF116 gene. The E10R sequence of GTPV and SPPV showed high homologies

Table 2 Comparison of LAMP method [19] sensitivity with the duplex PCR

Methods	Templates	Sensitivity (copies/reaction)									
		10^9	10^8	10^7	10^6	10^5	10^4	10^3	10^2	10^1	10^0
Duplex PCR	GTPV	+	+	+	+	+	−	−	−	−	−
Duplex PCR	SPPV	+	+	+	+	+	+	−	−	−	−
GTPV LAMP	GTPV	I	+	+	+	−	−	−	−	−	−
SPPV LAMP	SPPV	+	+	+	+	+	+	−	−	−	−
GSPV LAMP	GTPV/SPPV	+	+	+	+	+	+	+	−	−	−

Note: Detection of about 100 ng GTPV AV40 or SPPV Gulang/2009 genome DNA using different methods. "+" stand for positive result and "-" stand for negative result

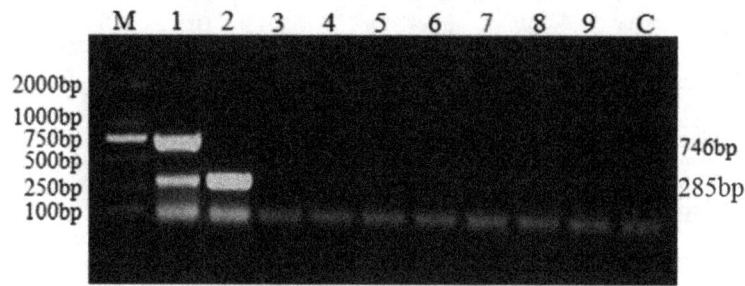

Fig. 6 Specificity of duplex PCR for detection different pathogen nucleic acid. Aboat 100 ng DNA or cDNA template of ten different sheep or goat pathogens were used in LAMP reaction. Agarose gel electrophoresis (1%) of PCR products stained with Ethidium bromide and visualized under UV transilluminator. Lane 1: SPPV; Lane 2: GTPV; Lane 3: *Orf* virus; Lane 4: FMDV O/China 99; Lane 5: *M. ovippneumoniae*; Lane 6: *Chlamydia psittaci*; Lane 7: *L.interrogans*; Lane 8: *Toxoplasma gondii*; Lane 9: *Babesia*; C: No template control and Lane M: 2000 bp DNA Ladder Marker (TaKaRa, Dalian)

to VACV E10R sequence, and RPO132 sequences of GTPV and SPPV showed high homologies to VACV A32R sequence. Initially, we found E10R primer designed based on SPPV E10R gene can amplify target gene from SPPV genome, also can amplify target gene from GTPV genome. RPO132 primer designed based on SPPV RPO132 gene only can amplify target gene from SPPV genome, but cannot amplify target gene from GTPV genome. But both annealing temperature of E10R and RPO132 primers will not be able to adjust to a consistent system. We initially planned to express E10R and RPO132, and inserted them into expression vector by enzyme restriction sites we added in the primers. So, non-specific sequences including Kozak sequence, restriction enzyme sites and protective bases or only restriction enzyme sites and protective bases sequences were added located in front of the initiation codon in upstream primer or added located behind termination codon in downstream primer, respectively. Accidentally, we found the above primers added non-specificity sequence can amplify the specificity target and the annealing temperature can be consistent at the same time in one system. E10R and RPO132 primers adding the nonspecific sequence can differential detect SPPV and GTPV obviously (Fig. 7). That is to say, added nonspecific sequence E10R and RPO132 primers mixture can specific amplification two target genes sizes obvious differences from SPPV genome, respectively, and can only amplify one gene from GTPV genome in this study (Fig. 7). Related experiments

showed that the sequences of the duplex PCR primers can specificity amplify two target genes from SPPV nucleic acid, but only one target gene can be amplified from GTPV nucleic acid under the same conditions. Further experiments proved that this duplex PCR analysis can fully distinguish between GTPV and SPPV.

One hundred thirty-five epidemic materials preserved in our laboratory were tested using duplex PCR diagnosis method, the results were consistent with the laboratory tested results (Table 3), which suggests the duplex PCR is able to detect the clinal samples.

Conclusion

The present study showed that the duplex PCR method of differential detection of GTPV and SPPV is highly specific, and sensitive. Thus, it might be the optimal detection system for field detection and differential diagnosis of GTP and SPP. It is a promising assay for extensive application for the diagnosis of GTPV and SPPV infection in the laboratory and field, especially in countries that lack the resources needed for molecular diagnostic techniques.

Table 3 Results of different methods detection with clinic samples

Methods	Sample size	Positive result	Negative result	Positive detection rates (%)	Sample category
PCR-RFLP [9, 10]	135	135	0	100	48 GTPV samples and 87 SPPV samples
GSPV lamp [19]	135	135	0	100	
GTPV lamp [19]	135	48	87	100	
SPPV lamp [19]	135	86	49	98.8	
Duplex PCR	135	135	0	100	

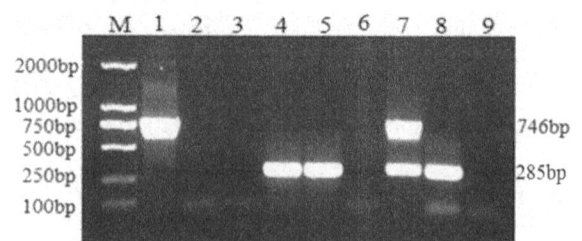

Fig. 7 Single PCR and duplex PCR for detection different GTPV and SPPV nucleic acid using E10R primers and RPO132 primers or mix primers. Lane 1–3: single PCR detect 100 ng SPPV gDNA, 100 ng GTPV gDNA and no template using RPO132 primers; Lane 4–6: single PCR detect 100 ng SPPV gDNA, 100 ng GTPV gDNA and no template using E10R primers; Lane 7–9: duplex PCR detect 100 ng SPPV gDNA, 100 ng GTPV gDNA and no template using mixture of RPO132 primers and E10R primers. Lane M: 2000 bp DNA Ladder Marker (TaKaRa, Dalian)

Abbreviations

CaPV: Capripoxvirus; gDNA: Genome DNA; GTPV: Goatpox virus; LSDV: lumpy skin disease virus; ORF: Open reading frame; PCR-RFLP; RFLP-PCR: Polymerase chain reaction-restriction fragment length polymorphism; SPPV: Sheeppox virus; UV: Ultraviolet light

Acknowledgments

The work was supported by the National Key Research and Development Program of China (2016YFD0500907), the Central Public-interest Scientific Institution basal Research Fund (1610312016015) the National Natural Science Foundation of China (31201892, 31502096) and International Science and Technology Cooperation Program of Gansu Province (2016GS08219), and the National Natural Science Foundation of Gansu Province (1506RJYA150) and the Public-Sector Research Special Foundation of AQSIQ (no.201310093).

Authors' contributions

ZXZ assisted in experimental design of the study, and wrote this manuscript and prepared figures for publication. Most of the experiments were conducted by ZXZ who developed and optimized duplex PCR assays; GW helped to construct partial plasmids and analyzed data; XY finished the RFLP-PCR experiments. XZ and QZ provided reference viruses and field isolates of GTPV and SPPV. QZ and ZZ analyzed data and modified the manuscript. The final manuscript was read and approved by all the authors.

Ethics approval

This study was approved by the Animal Ethics Committee of Lanzhou Veterinary Research Institute, Chinese Academy of Agricultural Sciences (approval number LVRIAEC 2012–018). Goats and sheep, from which tissues samples were collected, were handled with good animal practices required by the Animal Ethics Procedures and Guidelines of the People's Republic of China (AEPGPRC). The collection of tissues samples was performed as part of routine process of disease monitoring and surveillance for these goats and sheep. The owners of goats and sheep had given permission for the collection of tissues samples.

Consent for publication

Not applicable.

Competing interests

The authors declare that they have no competing interests.

References

1. Coetzer JAW, Guthrie AJ. African horse sickness. In: Coetzer JAW, Tustin RC, editors. Infectious Diseases of Livestock, vol. 3 vols. Cape Town: Oxford University Press; 2004. p. 1231–46.
2. Esposito JJ, Fenner F. Poxviruses in fields. In: Howley PM, Knipe DM, editors. Fields Virology. Philadelphia, PA, USA: Lippincott Williams & Wilkins Publishers; 2001. p. 2885–921.
3. Fields BN, Knipe DM, Howley PM, Chanock RM, Melnick JL, Monathy TP, Roizman B, Straus SE (ed.). Fields virology, 4th ed. Lippincott, Williams and Wilkins, Philadelphia, Pa., 45 Munz, E.& K. Dumbell. Sheeppox and goatpox. 1994; p.613–615.
4. Yeruham I, Yadin H, Van Ham M. Economic and epidemiological aspects of an outbreak of sheeppox in a dairy sheep flock. Vet Rec. 2007;160:236–7.
5. Babiuk S, Bowden TR, Boyle DB, Wallace DB, Kitching RP. Capripoxviruses: an emerging worldwide threat to sheep, goats and cattle. Transbound Emerg DIs. 2008;55:263–72.
6. Sadri R. Prevalence and economic significance of goat pox virus disease in semi-arid provinces of Iran. Iran J Vet Med. 2012;6:187–90.
7. Tulman ER, Afonso CL, Lu Z, Zsak L, Sur JH, Sandybaev NT, Kerembekova UZ, Zaitsev VL, Kutish GF, Rock DL. The genomes of sheeppox and goatpox viruses. J Virol. 2002;76:6054–61.

8. Senthilkumar V, Thirunavukkarasu M. Economic losses due to sheep pox in sheep farms in Tamil Nadu. Tamil Nadu Journal of Veterinary and Animal Sciences. 2010;6:88–94.
9. Yan XM, Zhang Q, Wu GH, Li J, Zhu HX. Discrimination of Goatpox virus and Sheeppox virus by digestion of p32 gene with Hif I andsequence alinment of GpCR gene. Animal husbandry and feed Sci. 2010;2:32–34,38.
10. Yan XM, Chu YF, Wu GH, Zhao ZX, Li J, Zhu HX, Zhang Q. An outbreak of sheep pox associated with goat poxvirus in Gansu province of China. Vet Microbio. 2011;156:425–8.
11. Zhao Z, Fan B, Wu G, Yan Y, Li Y, Zhou X, Yue H, Dai D, Zhu H, Tian B, Li J, Zhang Z. Development of loop-mediated isothermal amplification assay for specific and rapid detection of differential goat Pox virus and Sheep Pox virus. BMC Microbiol. 2014;14:10.
12. Gershon PD, Black DN. The nucleotide sequence around the Capripoxvirus thymidine kinase gene reveals a gene shared specifically with leporipoxvirus. J Gen Virol. 1989;70:525–33.
13. Gershon PD, Kitching RP, Hammond JM, Black DN. Poxvirus genetic recombination during natural virus transmission. J Gen Virol. 1989;70:485–9.
14. Kitching RP, Bhat PP, Black DN. The characterization of African strains of Capripoxvirus. Epidemiol Infect. 1989;102:335–43.
15. Mirzaiel K, Barani MS, Bokaie S. A review of sheep pox and goat pox: perspective of their control and eradication in Iran. J Adv Vet Anim Res. 2015;2(4):373–81.
16. Rao TVS, Bandyopadhyay SK. A comprehensive review of goat pox and sheep pox and their diagnos. Anim Health Res Rev. 2000;1:127–36.
17. Zheng M, Liu Q, Jin N. A duplex PCR assay for simultaneous detection and differentiation of Capripoxvirus and Orf virus. Mol Cell Probes. 2007;21:276–81.
18. Tian H, Wu J, Zhang K. Development of a SYBR green real-time PCR method for rapid detection of sheep pox virus. Virol J. 2012;9:1–4.
19. Venkatesan G, Balamurugan V, Bhanuprakash V. Multiplex PCR for simultaneous detection and differentiation of sheeppox, goatpox and orf viruses from clinical samples of sheep and goats. J Virol Methods. 2014;195:1–8.

5

Whole genome sequence and a phylogenetic analysis of the G8P[14] group A rotavirus strain from roe deer

Urska Jamnikar-Ciglenecki[1*] (iD), Urska Kuhar[2], Andrej Steyer[3] and Andrej Kirbis[1]

Abstract

Background: Group A rotaviruses (RVA) are associated with acute gastroenteritis in children and in young domestic and wild animals. A RVA strain was detected from a roe deer for the first time during a survey of game animals in Slovenia in 2014. A further RVA strain (SLO/D110–15) was detected from a roe deer during 2015. The aim of this study was to provide a full genetic profile of the detected RVA strain from roe deer and to obtain additional information about zoonotic transmitted strains and potential reassortments between human rotavirus strains and zoonotic transmitted rotavirus strains. The next generation sequencing (NGS) analysis on Ion Torrent was performed and the whole genome sequence has been determined together with a phylogenetic analysis.

Results: The whole genome sequence of SLO/D110–15 was obtained by NGS analyses on an IonTorrent platform. According to the genetic profile, the strain SLO/D110–15 clusters with the DS-1-like group and expresses the G8-P[14]-I2-R2-C2-M2-A3-N2-T6-E2-H3 genome constellation. Phylogenetic analysis shows that this roe deer G8P[14] strain is most closely related to RVA strains found in sheep, cattle and humans. A human RVA strain with the same genotype profile was detected in 2009 in Slovenia.

Conclusions: The G8P[14] genotype has been found, for the first time, in deer, a newly described host from the order *Artiodactyla* for this RVA genotype. The finding of a rotavirus with the same genome segment constellation in humans indicates the possible zoonotic potential of this virus strain.

Keywords: Group A rotavirus, Wildlife, Deer, Zoonotic transmission, Phylogenetic analysis, Next generation sequencing, NGS

Background

Group A rotaviruses (RVA) are members of the genus *Rotavirus* belonging to the highly diverse *Reoviridae* family, whose members are capable of infecting various host species (mammals, reptiles, fish, birds, fungi, plants and insects) [1]. Their genome consists of eleven double-stranded RNA segments that encode six structural proteins (VP1 to VP4, VP6 and VP7) and six non-structural proteins (NSP1–6) [2]. From a medical and veterinary perspective, RVA is the most important member

of the genus and is associated with acute gastroenteritis in children and in young domestic and wild animals [2–4].

The RVA are classified according to the two outer capsid proteins VP7 and VP4, and at least 35 G-types and 50 P-types [5], respectively, have been characterized in humans and animals with more than 60 G-P combinations. In order to provide a better insight into RVA diversity and evolution, a new whole genome genotyping system was established and proposed by the Rotavirus Classification Working Group (RCWG) [6]. Under this classification system, the notation Gx–P[x]–Ix-Rx-Cx-Mx–Ax–Nx–Tx–Ex–Hx ("x" denotes the genotype number) has been used to represent the complete genotype constellation (VP7–VP4–VP6–VP1–VP2–VP3–NSP1–

* Correspondence: urska.jamnikar@vf.uni-lj.si
[1]Institute of Food safety, Feed and Environment, Veterinary Faculty, University of Ljubljana, Gerbičeva 60, 1000 Ljubljana, Slovenia
Full list of author information is available at the end of the article

NSP2–NSP3–NSP4–NSP5 genes) of a RVA strain [7]. According to the genome segments' genotype constellation, there are three main rotavirus genogroups, Wa-like (G1P[8]), DS-1-like (G2P[4]) and AU-1-like (G3P[9]), identified in humans and animals, respectively [8]. It was shown previously, that human Wa-like strains are related to porcine Wa-like strains and that human and bovine DS-1-like strains are also closely related. In addition, reports on bovine-like G6, G8 and G10 strains from the DS-1-like genogroup were frequently reported in human infections indicating the possible zoonotic transmission from animals to humans [9]. Whole genome sequencing is particularly important for the study of zoonotic transmitted strains and potential reassortments between human and animal RVA strains [10].

Rotaviruses have been reported in many ungulates, including deer [11], but only a few genomes of these RVA strains have been studied. Until recently, there have been no reports describing the genomes of RVA strains in deer. In our previous report [12], the first genome of the roe deer RVA strain was described as having the G6-P[15]-I2-R2-C2-M2-A3-N2-T6-E2-H3 genotype constellation. Later that year a red deer RVA strain from the USA with G8-P[1]-I2-R2-C2-M2-A3-N2-T6-E2-H3 was reported [13]. These two strains share the same bovine DS1-like genetic backbone but have different G/P combinations. Here we report a second roe deer RVA strain with the genotype constellation G8-P[14]-I2-R2-C2-M2-A3-N2-T6-E2-H3.

Methods

Sample collection and molecular detection of RVA

Initial findings from a survey conducted in 2014 and 2015 to screen certain game animals as a potential source of rotaviruses have previously been reported [12]. Screening of a further 15 samples from roe deer using specific RT-PCR and real-time RT-PCR [12] identified a further RVA-positive sample (SLO/D110–15). The sample was collected in October 2015, in Lahovče (hunting family Krvavec) from a one-year-old roe deer of

Table 1 Genome genotype constellation of the 11 segments of RVA/Roe deer-wt/SLO/D110–15/G8P[14], the closest nucleotide identities from GenBank and identities shared with Slovenian RVA/Hu-wt/SVN/SI-2987/09/G8P[14] strain

Gene	Strain SLO/D110–15		Strains in the GenBank with the closest nucleotide identity			
	Genotype	Accession no.	Strain	Accession no.	Nt identity (%)	Host
VP7	G8	KY426808	OVR762	EF554153	96.9	Ovine
			SI-2987/09	KY972333	85.7[a]	Human
VP4	P[14]	KY426812	Tottori-SG	AB853893	94.7	Bovine
			SI-2987/09	KY972331	83.7[a]	Human
VP6	I2	KY426813	Tottori-SG	AB853894	94.1	Bovine
			SI-2987/09	KY972332	87.0[a]	Human
VP1	R2	KY426809	B10925	EF554115	94.8	Human
			SI-2987/09	KY972328	85.5[a]	Human
VP2	C2	KY426810	182-02	KU508381	92.3	Human
			SI-2987/09	KY972329	95.4[a]	Human
VP3	M2	KY426811	SI-R56	JX094030	98.6	Human
			SI-2987/09	KY972330	91.1[a]	Human
NSP1	A3	KY426803	NCDV	GU808570	97.2	Bovine
			SI-2987/09	KY972323	99.7[a]	Human
NSP2	N2	KY426804	1604	JN831216	96.6	Bovine
			SI-2987/09	KY972324	99.3[a]	Human
NSP3	T6	KY426805	UCD	GQ428138	97.8	Giraffe
			SI-2987/09	KY972325	93.1[a]	Human
NSP4	E2	KY426806	BEF06018	KU128901	97.8	Human
			SI-2987/09	KY972326	97.0[a]	Human
NSP5	H3	KY426813	BEF06018	KU128902	99.7	Human
			SI-2987/09	KY972327	95.1[a]	Human

[a]the identity scores were calculated with the partial nucleotide sequences of the rotavirus strain SI-2987/09

appropriate weight for its age and exhibiting no specific clinical signs.

RNA extraction, NGS and analysis of sequence reads

To determine the whole genome of the RVA strain SLO/D110–15, the sample was prepared for the NGS. Total RNA was extracted with Trizol reagent (Invitrogen, Carlsbad, USA) in combination with the RNeasy Mini Kit (Qiagen, Hilden, Germany), using the protocol with on-column DNase I digestion according to the manufacturer's instructions. The RNA was used as the template for cDNA synthesis with the cDNA Synthesis System (Roche, Manheim, Germany) according to the Genome Sequencer Rapid RNA library protocol (Roche). The cDNA was fragmented with a Covaris M220 focused ultrasonicator, targeting peak fragments with lengths of 200–300 bp. The fragmented cDNA was used for library preparation using GeneRead DNA Library L Core Kit (Qiagen,

Hilden, Germany). Purification and size selection of the library were performed with Ampure XP magnetic beads (Beckman Coulter, Brea, CA, USA). The library was quantified with the GeneRead Library Quant Kit (Qiagen, Hilden, Germany) and a Qubit 3.0 fluorometer (Thermo Fisher Scientific, Carlsbad, CA, USA). Emulsion PCR and enrichment were carried out using the Ion PGM™ Template OT2 200 Kit (Thermo Fisher Scientific, Carlsbad, CA, USA). The library was sequenced on the Ion PGM platform using the Ion PGM HiQ Sequencing Kit and Ion 314 Chip v2 (Thermo Fisher Scientific, Carlsbad, CA, USA). Sequenced reads were quality checked and trimmed using Ion Torrent Suite version 5.0.4 and assembled into contigs by de novo assembly, using the Genome Sequencer software version 2.9 (Roche, Basel, Switzerland). The contigs were compared to the GenBank non-redundant nucleotide database (BLASTn) to determine the contigs that represent the rotavirus

Table 2 Comparison of the genotype constellation of Slovenian RVA roe deer SLO/D110–15 with other RVA complete genome sequences from GenBank

Strain	Origin	VP7 G	VP4 [P]	VP6 I	VP1 R	VP2 C	VP3 M	NSP1 A	NSP2 N	NSP3 T	NSP4 E	NSP5 H
RVA/Human-wt/HUN/BP1879/2003/G6P[14]	Hu	6	14	2	2	2	2	11	2	6	2	3
RVA/Human-wt/JAP/KF17/2010/G6P[9]	Hu	6	9	2	2	2	2	3	2	3	3	3
RVA/Human-wt/ITA/PR1300/2004/G8P[14]	Hu	8	14	2	2	2	2	3	2	6	2	3
RVA/Human-wt/HUN/182-02/2002/G8P[14]	Hu	8	14	2	2	2	2	11	2	6	2	3
RVA/Human-wt/HUN/Hun5/1997/G6P[14]	Hu	6	14	2	2	2	2	11	2	6	2	3
RVA/Human-wt/ITA/PR457/2009/G10P[14]	Hu	10	14	2	2	2	2	11	2	6	2	3
RVA/Human-wt/ITA/PR1973/2009/G8P[14]	Hu	8	14	2	2	2	2	3	2	6	2	3
RVA/Human-tc/EGY/AS970/2012/G8P[14]	Hu	8	14	2	2	2	2	11	2	6	2	3
RVA/Human-WT/KEN/AK26/1982/G2P[4]	Hu	2	4	2	2	2	2	2	1	2	2	2
RVA/Human-wt/MWI/1473/2001/G8P[4]	Hu	8	4	2	2	2	2	2	2	2	2	2
RVA/Sheep-tc/ESP/OVR762/2002/G8P[14]	Ov	8	14	2	2	2	2	11	2	6	2	3
RVA/Human-wt/BEL/B10925/1997/G6P[14]	Hu	6	14	2	2	2	2	3	2	6	2	3
RVA/Human-wt/ITA/111-05-27/2005/G6P[14]	Hu	6	14	2	2	2	2	3	2	6	2	3
RVA/Human-tc/GBR/A64/1987/G10P[14]	Hu	10	14	2	2	2	2	3	2	6	2	3
RVA/Human-wt/HUN/BP1062/2004/G8P[14]	Hu	8	14	2	2	2	2	11	2	6	2	3
RVA/Human-wt/BEL/B1711/2002/G6P[6]	Hu	6	6	2	2	2	2	2	2	2	2	2
RVA/Human-tc/USA/DS-1/1976/G2P[4]	Hu	2	4	2	2	2	2	2	2	2	2	2
RVA/Human-wt/THA/SKT-27/2012/G6P[14]	Hu	6	14	2	2	2	2	3	2	6	2	3
RVA/Human-wt/BEL/B4106/2000/G3P[14]	Hu	3	14	2	2	2	3	9	2	6	5	3
RVA/Human-wt/BEL/BE5028/2012/G3P[14]	Hu	3	14	2	2	2	3	9	2	6	5	3
RVA/Rabbit-tc/ITA/30-96/1996/G3P[14]	Rab	3	14	2	2	2	3	9	2	6	5	3
RVA/Human-tc/ITA/PA169/1988/G6P[14]	Hu	6	14	2	2	2	2	3	2	6	2	3
RVA/Bovine-wt/JPN/Dai-10/2008/G24P[33]	Bo	24	33	2	2	2	2	13	2	9	2	3
RVA/Human-tc/USA/Se584/1998/G6P[9]	Hu	6	9	2	2	2	2	3	2	1	2	3
RVA/Human-wt/AUS/V585/2011/G10P[14]	Hu	10	14	2	2	2	2	11	2	6	2	3
RVA/Human-wt/AUS/MG6/1993/G6P[14]	Hu	6	14	2	2	2	2	11	2	6	2	3
RVA/Cow-wt/JPN/Tottori-SG/2013/G15P[14]	Bo	15	14	2	2	2	2	3	2	6	2	3
RVA/Human-wt/GTM/2009726790/2009/G8P[14]	Hu	8	14	2	2	2	2	13	2	6	2	3
RVA/Guanaco-wt/ARG/Chubut/1999/G8P[14]	La	8	14	2	5	2	2	3	2	6	12	3
RVA/Cow-wt/ARG/B383/1998/G15P[11]	Bo	15	11	2	5	2	2	13	2	6	12	3
RVA/Human-wt/AUS/RCH272/2012/G3P[14]	Hu	3	14	2	3	3	3	9	2	6	2	3
RVA/Rabbit-tc/CHN/N5/1992/G3P[14]	Rab	3	14	17	3	3	3	9	1	1	3	2
RVA/Vicuna-wt/ARG/C75/2010/G8P[14]	Vi	8	14	2	2	2	2	-	2	6	3	-
RVA/Human-wt/US/2012841174/2012/G8P[14]	Hu	8	14	1	1	1	1	8	1	1	1	1
RVA/Human-wt/BRB/2012821133/2012/G4P[14]	Hu	4	14	1	1	1	1	8	1	1	1	1
RVA/Human-wt/USA/Wa/1974/G1P[8]	Hu	1	8	1	1	1	1	1	1	1	1	1
RVA/roe_deer-wt/SLO/D38-14/2014/G6P[15]	Roe	6	15	2	2	2	2	3	2	6	2	3
RVA/Roe deer-wt/SLO/D110-15/2015/G8P[14]	**Roe**	**8**	**14**	**2**	**2**	**2**	**2**	**3**	**2**	**6**	**2**	**3**

genome segments. To eliminate assembly errors, all sequenced reads were mapped against the concatenated segments of the assembled RVA strain SLO/D110–15 genome with the Genome Sequencer software version 2.9 (Roche). The Geneious software suite v 9.0.5 (Biomatters LtD, Auckland, New Zeland) was used for visualization and final data analysis. To obtain the genomic constellation of the RVA strain, the RotaC v 2.0 online automated genotyping tool [14] was used to assign the genotype of each genome segment.

Phylogenetic analysis of the genome segments

Selected sequences of RVA deposited in GenBank, with complete genome and genotype relevant to our strain, were used in the phylogenetic analyses. In addition, some of the most nearly identical sequences were added

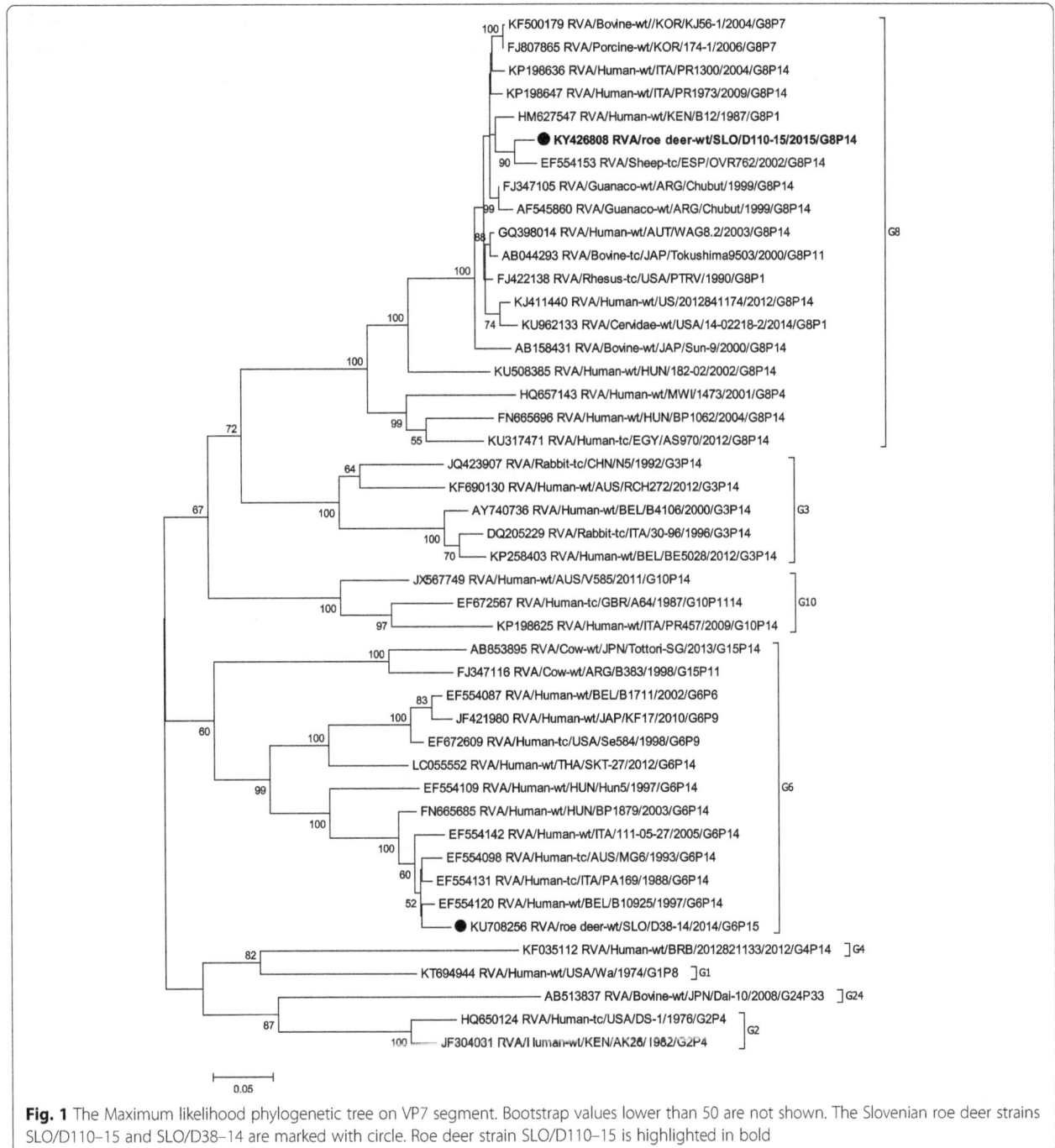

Fig. 1 The Maximum likelihood phylogenetic tree on VP7 segment. Bootstrap values lower than 50 are not shown. The Slovenian roe deer strains SLO/D110–15 and SLO/D38–14 are marked with circle. Roe deer strain SLO/D110–15 is highlighted in bold

according to the BLAST search on each segment. The Slovenian RVA strain SI-2987/09, with genome profile G8-P[14]-I2-R2-C2-M2-A3-N2-T6-E2-H3 was also included in the genome nucleotide sequence identity analysis (Table 1). The strain was detected in a child with gastroenteritis during the RVA survey in 2009 described in the study by Steyer et al. [15]. As the strain SI-2987/09 was available only with partial nucleotide sequences of the genome segments, it was not included in the phylogenetic analysis.

Nucleotides were aligned using ClustalW implemented in MEGA 7.0.21 [16]. Nucleotide identities were calculated according to the p-distances implemented in MEGA 7.0.21 [16]. Phylogenetic trees were constructed using the Maximum-likelihood method based on the Kimura-2 parameter model. Branch statistics were calculated by bootstrap analysis of 1000 replicates.

Results

NGS analyses and construction of the complete genome sequence

The complete genome sequence of all 11 segments of the RVA/roe deer-wt/SLO/D110–15/2015/G8P[14] strain was obtained using the Ion Torrent PGM platform. The nucleotide sequences of all eleven genome segments were deposited in GenBank and are available under the following accession numbers: KY426809 (VP1), KY426810 (VP2), KY426811 (VP3), KY426812 (VP4), KY426813

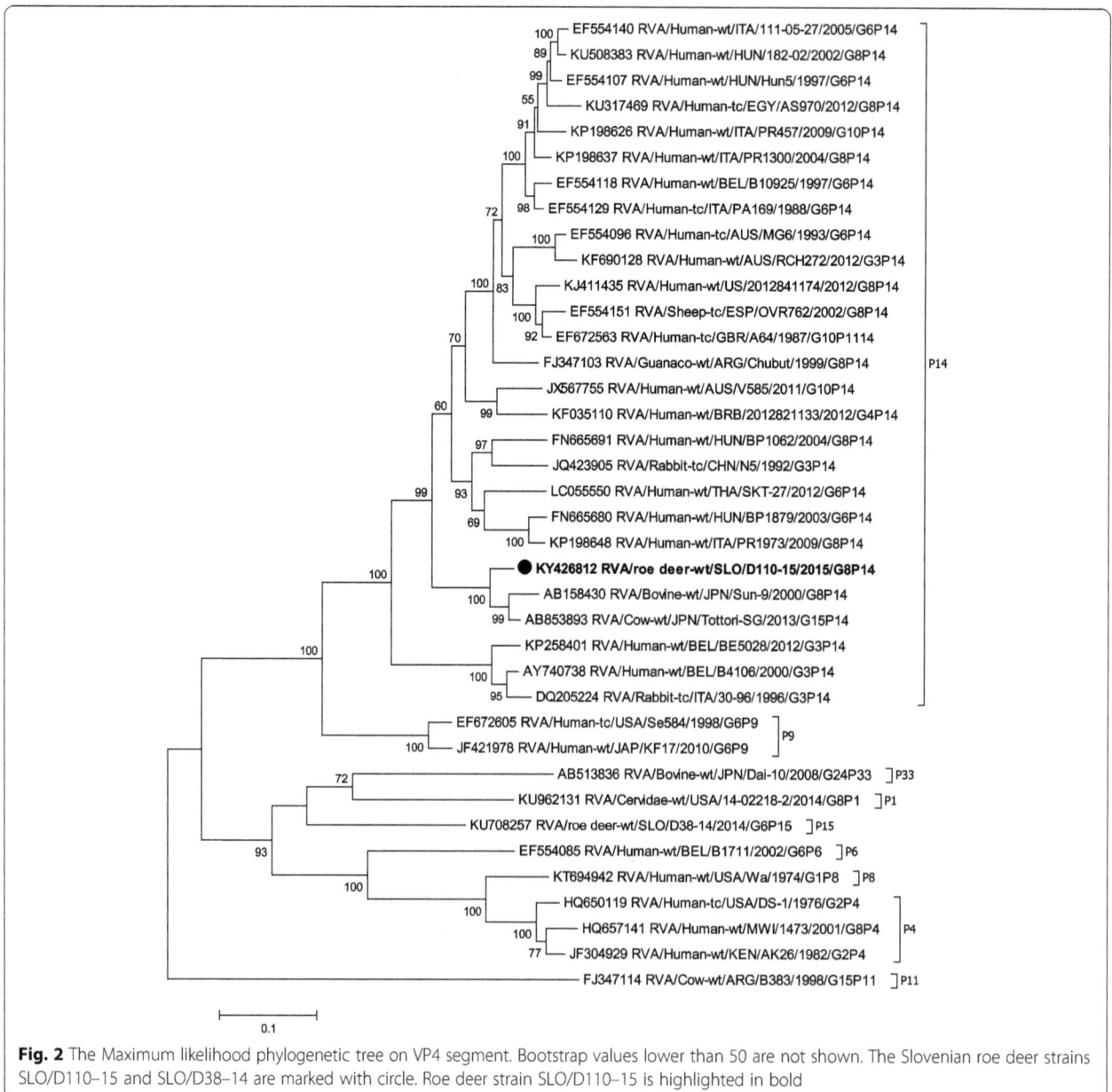

Fig. 2 The Maximum likelihood phylogenetic tree on VP4 segment. Bootstrap values lower than 50 are not shown. The Slovenian roe deer strains SLO/D110–15 and SLO/D38–14 are marked with circle. Roe deer strain SLO/D110–15 is highlighted in bold

(VP6), KY426808 (VP7), KY426803 (NSP1), KY426804 (NSP2), KY426805 (NSP3), KY426806 (NSP4) and KY426807 (NSP5 and NSP6).

In total, 98,441 reads were generated and 25 large contigs (lengths >500 nt) obtained by de novo assembly. The BLAST search revealed that 11 of the contigs belong to the 11 RVA genome segments and constituted the complete genome of the RVA strain SLO/D110–15. Mapping against the concatenated segments of the RVA strain SLO/D110–15 genome resulted in 66,153 mapped reads (81.4% of all reads) with an average depth of 569.6 and average map length of 158 nt. Using the RotaC classification tool, the whole genome constellation of the RVA strain SLO/D110–15 was determined to be G8-P[14]-I2-R2-C2-M2-A3-N2-T6-E2-H3. The complete genotype constellation of the SLO/D110–15 strain was compared with those of RVA G8P[14] strains and other representative strains from humans and animals (Table 2). The genotype constellation of the SLO/D110–15 strain was identical to those of the two Italian human strains ITA/PR1300 and ITA/PR1973, and of the Slovenian human strain

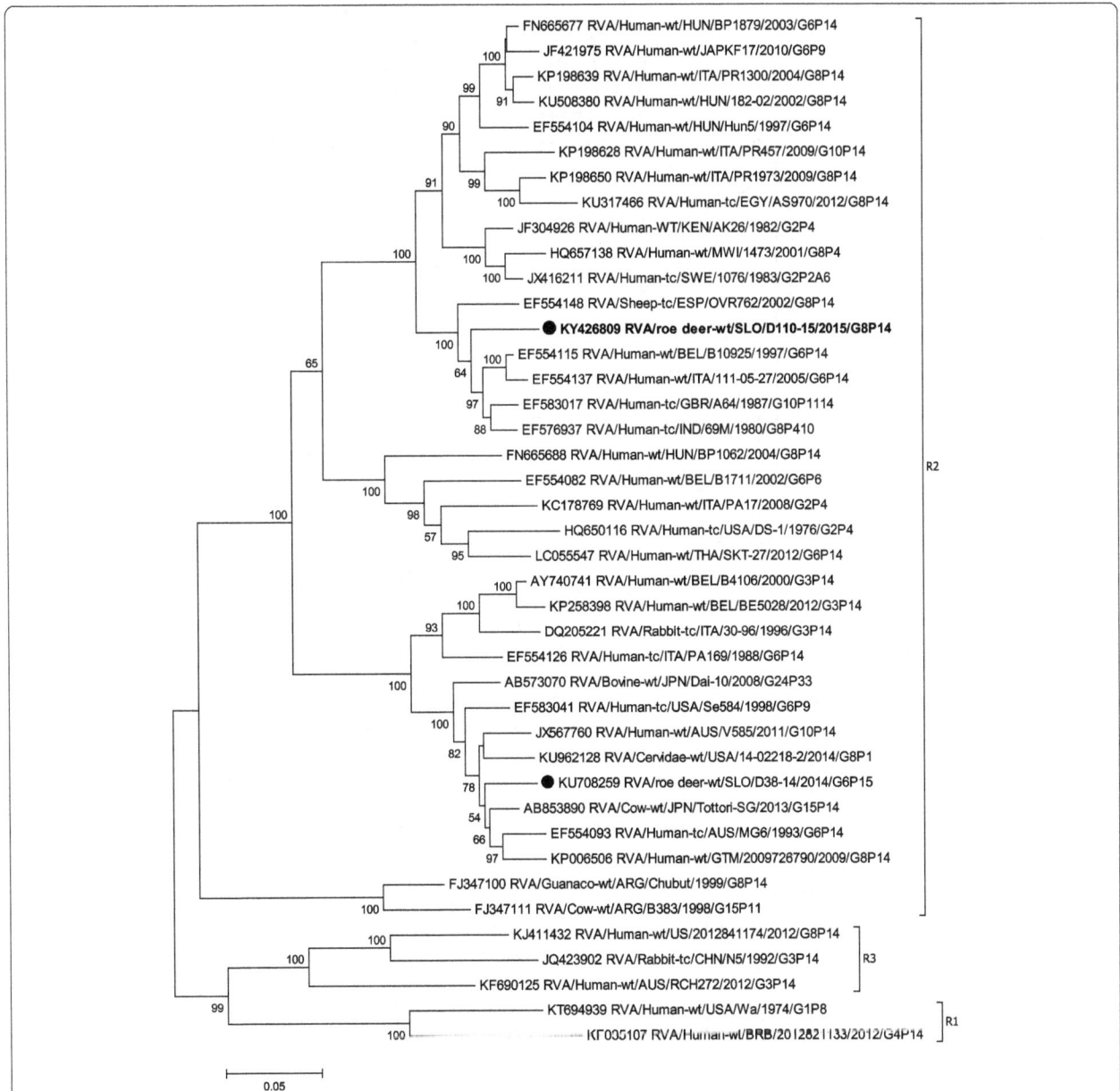

Fig. 3 The Maximum likelihood phylogenetic tree on VP1 segment. Bootstrap values lower than 50 are not shown. The Slovenian roe deer strains SLO/D110–15 and SLO/D38–14 are marked with circle. Roe deer strain SLO/D110–15 is highlighted in bold

SI-2987/09, namely G8-P[14]-I2-R2-C2-M2-A3-N2-T6-E2-H3.

Nucleotide sequence identity and phylogenetic analysis of the genome segments

p-Distances were calculated for each of the 11 segments of SLO/D110–15 and for selected strains from Gen-Bank. The highest level of nucleotide identity, 99.7%, was observed on the NSP5 segment with strain BEL/BEF06018 and on the NSP1 segment with strain SI-2987/09. The lowest nucleotide identity was 92.3% and was observed on the VP2 segment with strain HUN/182–02 (Table 1). The roe deer SLO/D110–15 and human SI-2987/09 strains shared fewer than 90% nucleotide identities for partial sequences of structural genes VP1, VP6, VP7 and fragment VP8* of the VP4 gene,

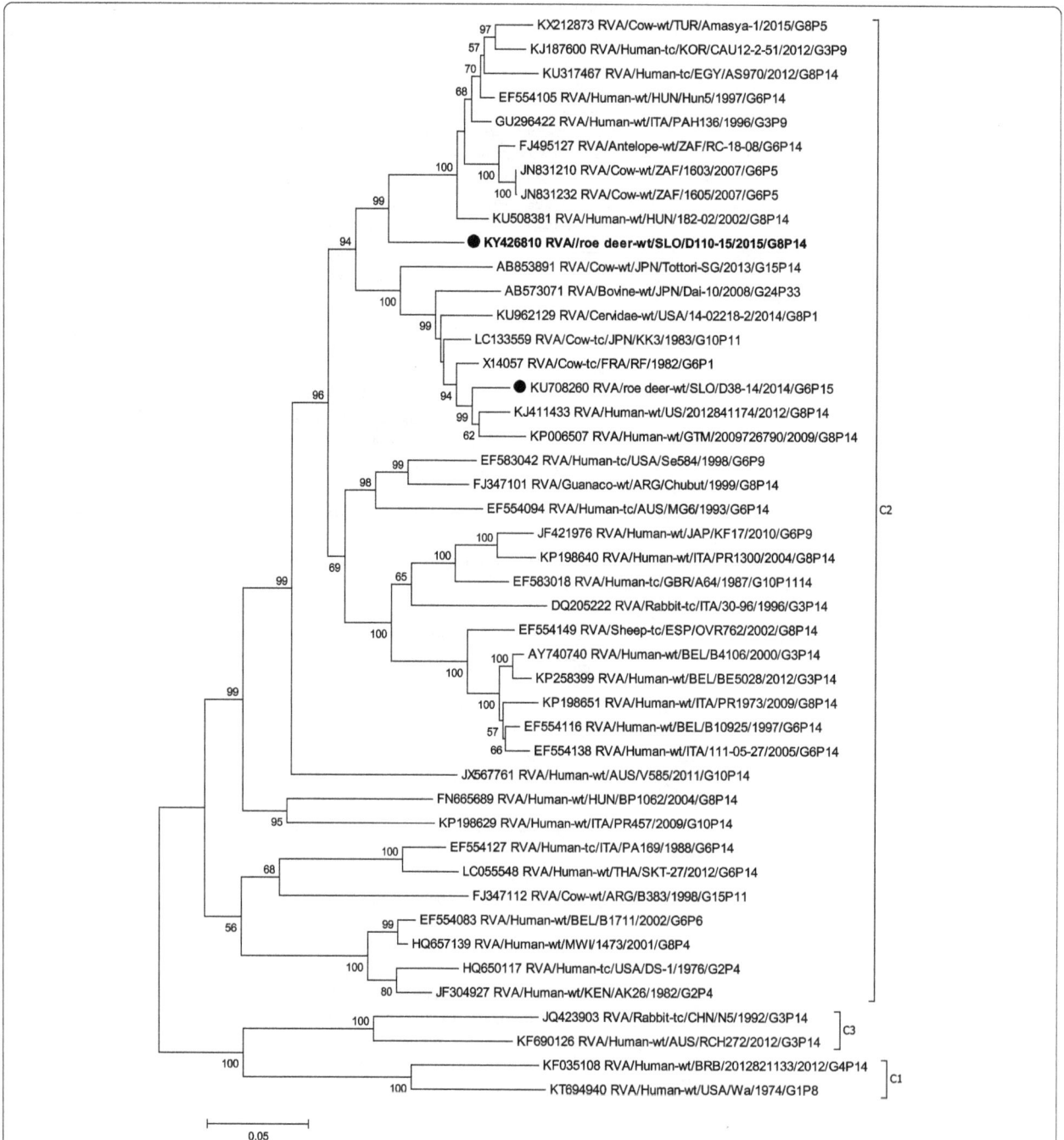

Fig. 4 The Maximum likelihood phylogenetic tree on VP2 segment. Bootstrap values lower than 50 are not shown. The Slovenian roe deer strains SLO/D110–15 and SLO/D38–14 are marked with circle. Roe deer strain SLO/D110–15 is highlighted in bold

although sharing the same genotype. In contrast, the partial sequences of the non-structural protein genes shared higher degrees of identity, ranging from 93.1% to 99.7%.

In the VP7 phylogenetic tree, the strain SLO/D110–15 was most closely related to the Spanish ovine strain ESP/OVR762, with a nucleotide identity of 96.6%

(Table 1, Fig. 1) and, when compared to other G8 strains, the lowest nucleotide identity of 81.8%. Comparison of the VP7 segment with the other known roe deer sample, RVA SLO/D38–14, led to an identity of 85.8%.

Phylogenetic analysis of the VP4 gene showed that the strain SLO/D110–15 formed a cluster with the Japanese

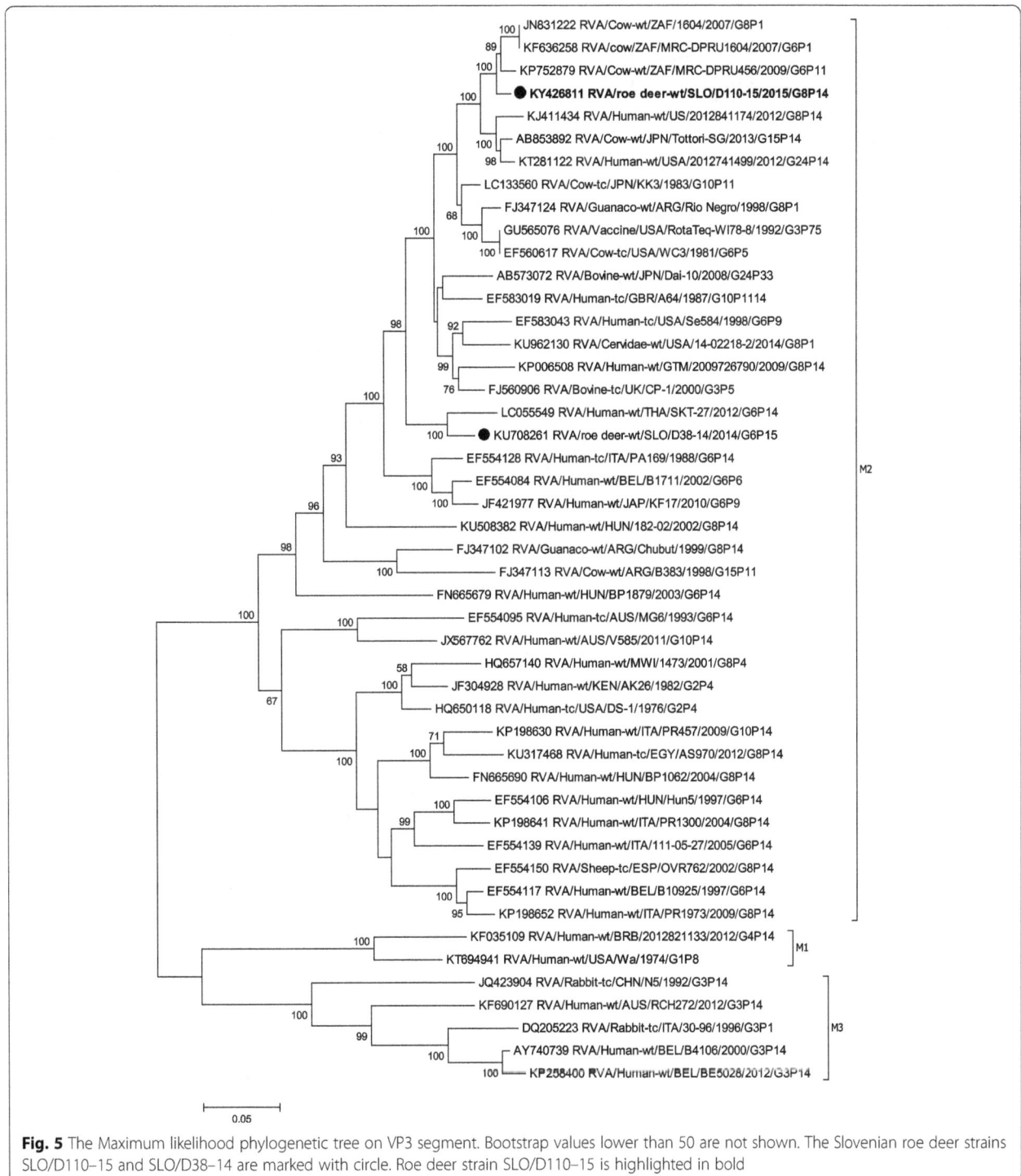

Fig. 5 The Maximum likelihood phylogenetic tree on VP3 segment. Bootstrap values lower than 50 are not shown. The Slovenian roe deer strains SLO/D110–15 and SLO/D38–14 are marked with circle. Roe deer strain SLO/D110–15 is highlighted in bold

P[14] bovine strains JPN/Tottori-SG and JPN/Sun9, with nucleotide identities of 94.8% and 93.2%, respectively (Table 1, Fig. 2). The lowest degree of nucleotide identity, when compared with other P[14] strains, was 80.9%. On the VP4 segment the nucleotide identity between roe deer RVA SLO/D110–15 and SLO/D38–14 was, as expected, low (67.3%), as they represent different P genotypes.

Phylogenetic analysis of VP1-VP3, VP6 and NSP1-NSP5 gene segments revealed that the SLO/D110–15 strain clusters with bovine and bovine-like RVA strains, sharing the same non-G/P genotype constellation (genetic backbone) I2-R2-C2-M2-A3-N2-T6-E2-H3 (Table 2) and with the highest degree of nucleotide identity in the range of 92.2% to 99.7% (Table 1, Figs. 3, 4, 5, 6, 7, 8, 9, 10 and 11).

When comparing segments from the backbone of roe deer SLO/D110–15 with those of roe deer SLO/D38–14,

the highest degree of nucleotide identity ranged only from 85.8% to 96.8%, even though they share the same genetic backbone.

Discussion

There are only two reports describing the complete genome constellation of RVA in deer. The first detection and complete genome characterization of a roe deer rotavirus was in 2015 [12]. Here, we describe the second detection of RVA in roe deer and the first RVA strain with the G8-P[14]-I2-R2-C2-M2-A3-N2-T6-E2-H3 genotype constellation. Important insights into the complete genetic makeup of a deer rotavirus strain are provided in this report and consequently new knowledge about the host range for this RVA genotype, together with their strain diversity.

The sequencing result of the SLO/D110–15 roe deer sample investigated in this study revealed a large

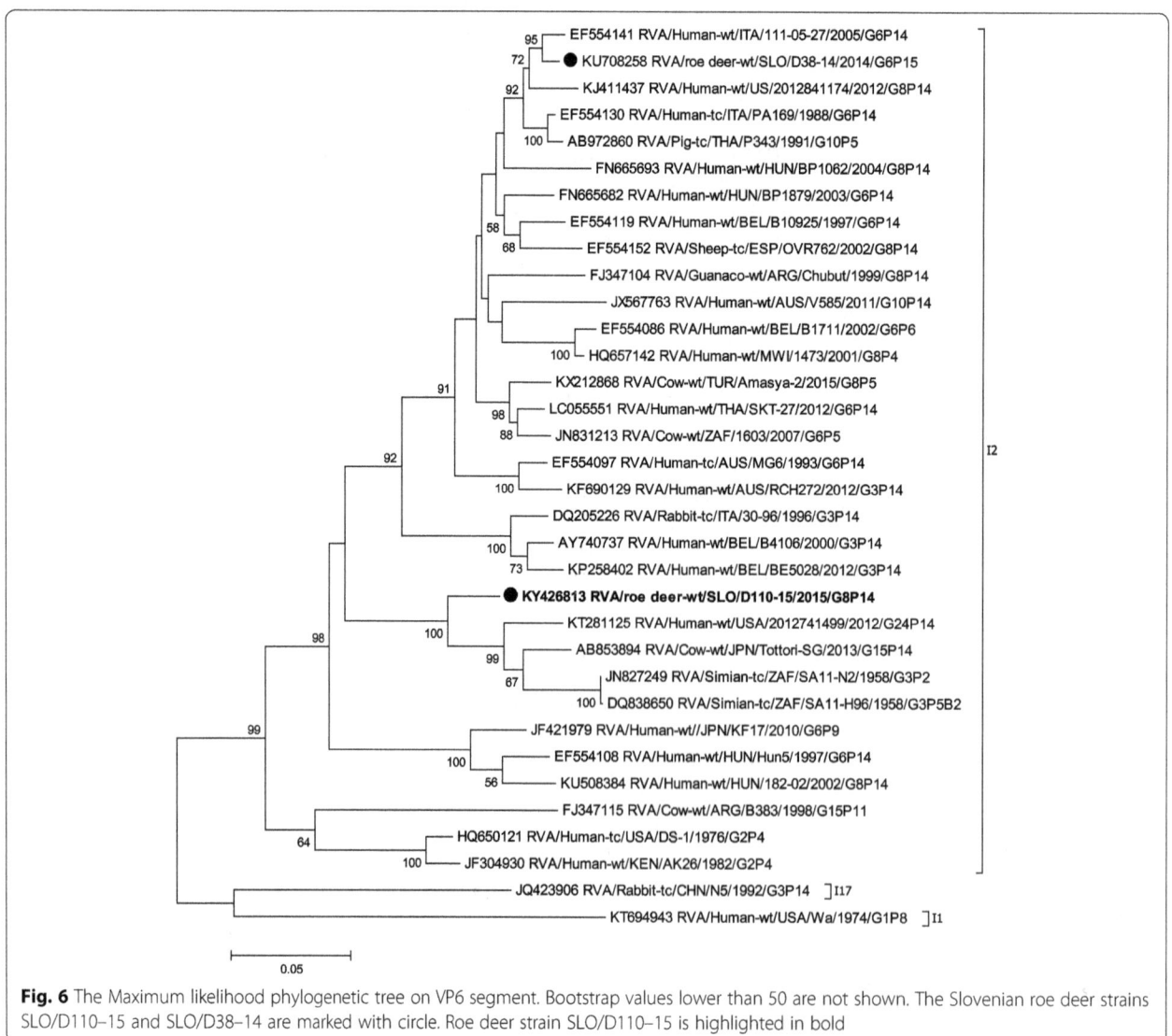

Fig. 6 The Maximum likelihood phylogenetic tree on VP6 segment. Bootstrap values lower than 50 are not shown. The Slovenian roe deer strains SLO/D110–15 and SLO/D38–14 are marked with circle. Roe deer strain SLO/D110–15 is highlighted in bold

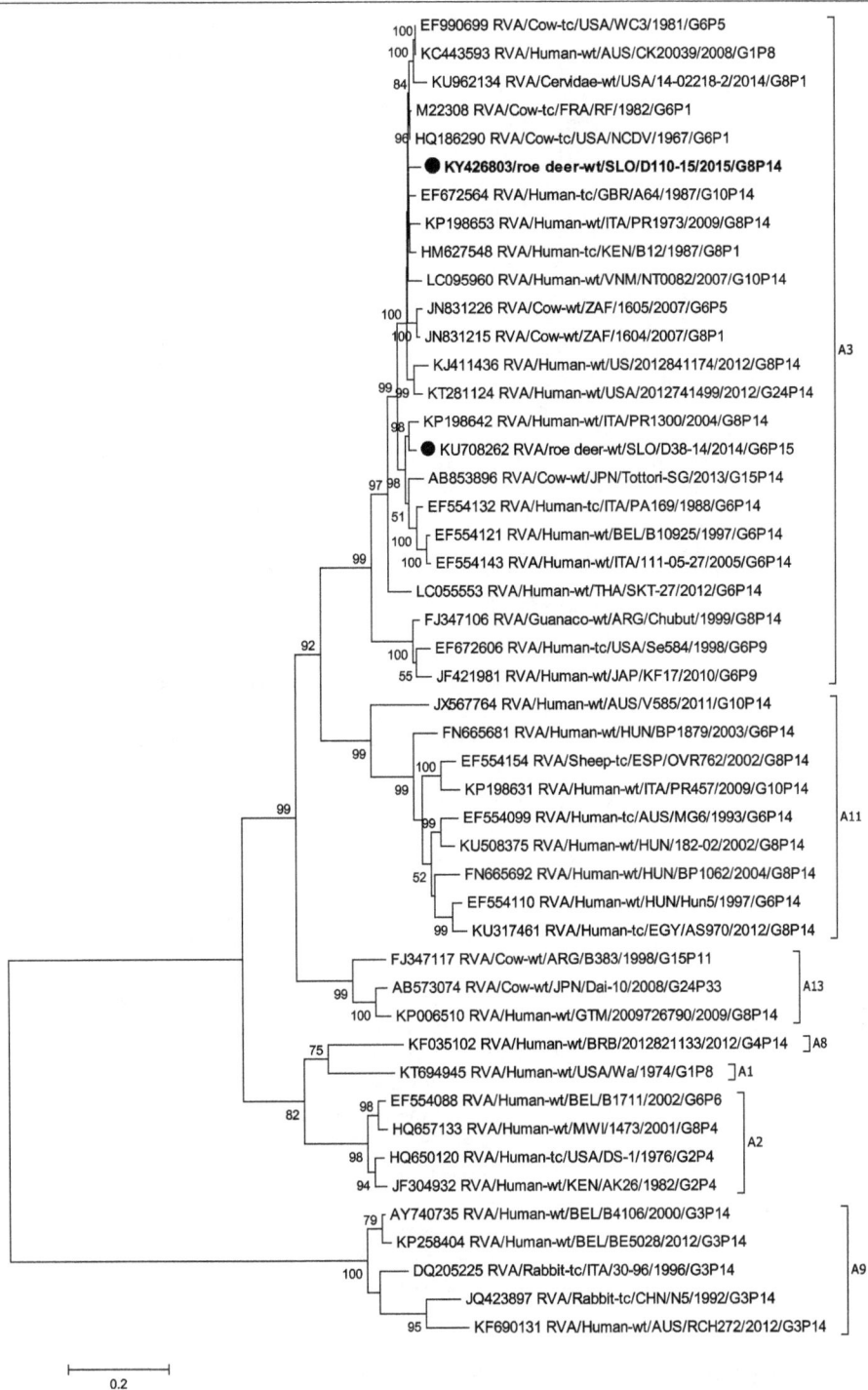

Fig. 7 The Maximum likelihood phylogenetic tree on NSP1 segment. Bootstrap values lower than 50 are not shown. The Slovenian roe deer strains SLO/D110–15 and SLO/D38–14 are marked with circle. Roe deer strain SLO/D110–15 is highlighted in bold

number of RVA sequencing reads (81.4% of all sequencing reads), using an unbiased protocol for sample and library preparation, and with no observed clinical signs in the respective animal. The NGS analyses show that, unlike the first positive roe deer sample collected in 2014 in Slovenia, SLO/D38–14, that belongs to the G6P[15] genotype, the roe deer sample SLO/D110–15 belongs to the G8P[14] genotype. They share from 67.3% to 96.8% genome segment nucleotide identity.

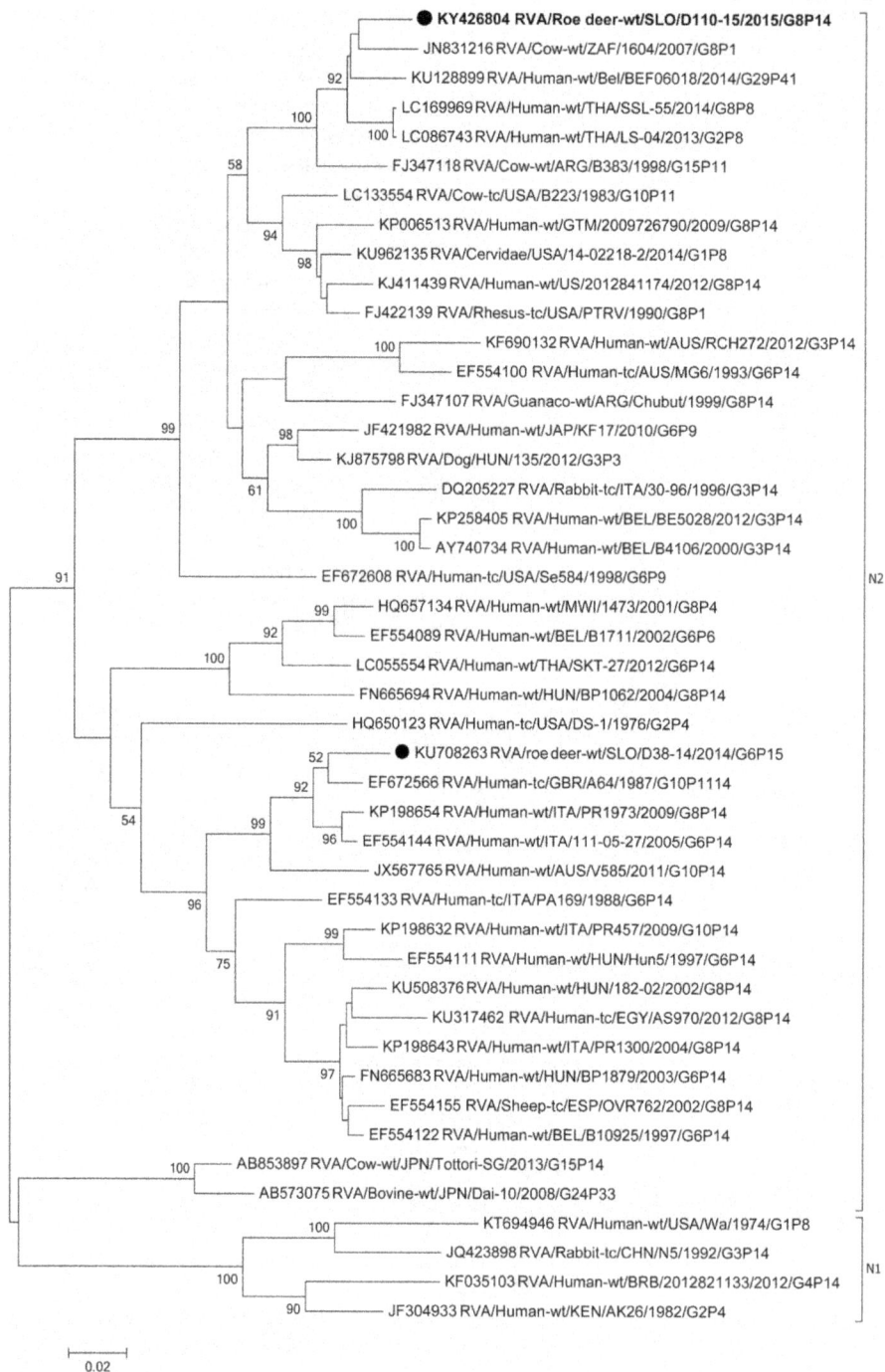

Fig. 8 The Maximum likelihood phylogenetic tree on NSP2 segment. Bootstrap values lower than 50 are not shown. The Slovenian roe deer strains SLO/D110–15 and SLO/D38–14 are marked with circle. Roe deer strain SLO/D110–15 is highlighted in bold

The strain SLO/D110–15 has the typical bovine DS1-like genetic backbone found in cattle and other animals from the order *Artiodactyla* together with a G/P combination also found in zoonotic human RVA strains. Detailed phylogenetic analysis of the 11 genome segments revealed the closest relatedness of the

SLO/D110–15 strain to RVA strains having the bovine-like genotype constellation from humans and animals.

RVA strains of the G8P[14] genotype, combined with the bovine DS1-like genetic backbone, are detected sporadically in cattle, sheep, guanaco, vicuna and in

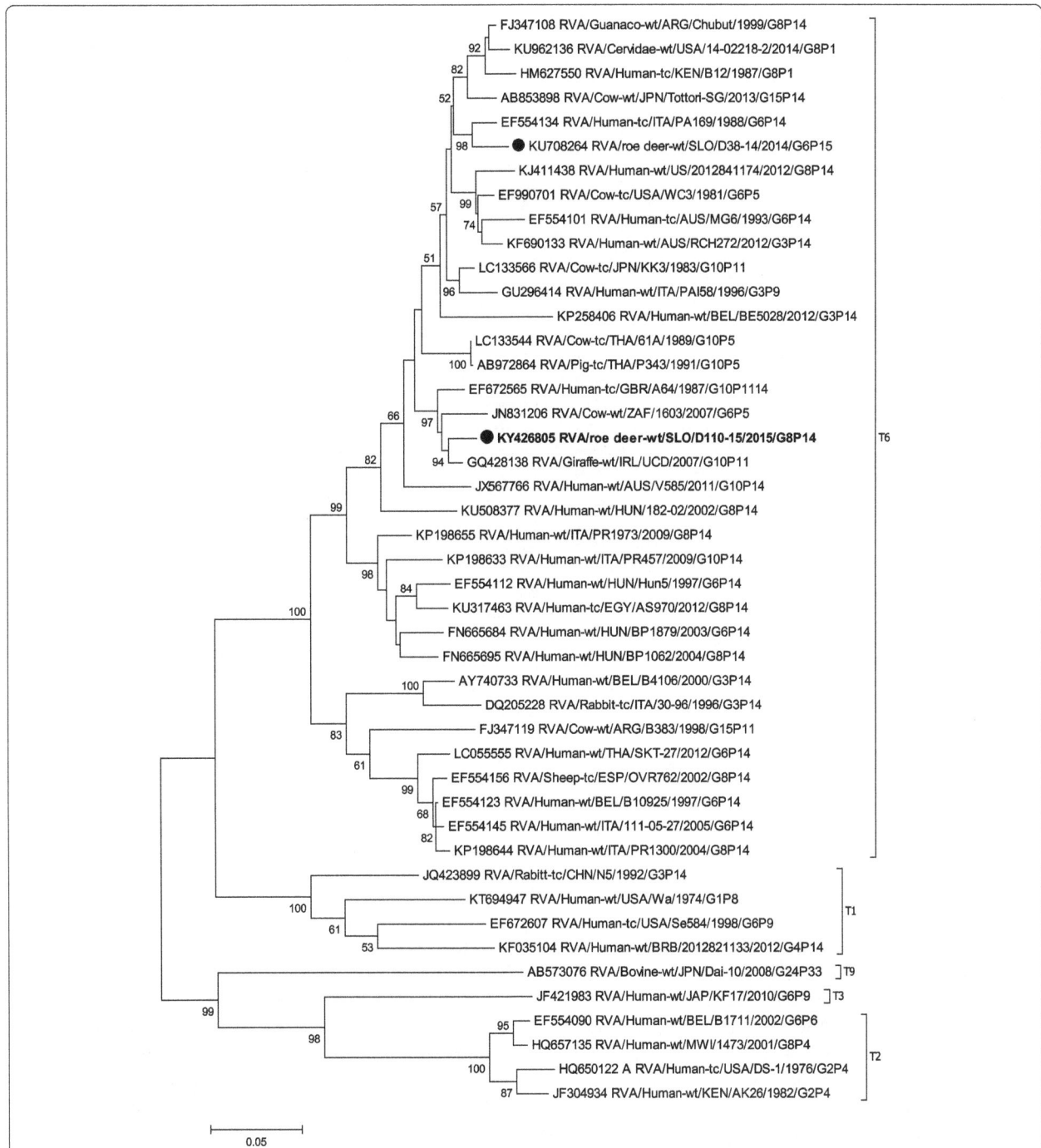

Fig. 9 The Maximum likelihood phylogenetic tree on NSP3 segment. Bootstrap values lower than 50 are not shown. The Slovenian roe deer strains SLO/D110–15 and SLO/D38–14 are marked with circle. Roe deer strain SLO/D110–15 is highlighted in bold

humans [17–23]. It was suggested that the P[14] geno-types are less virulent for the ruminant host species, and thus more probably shed by animals with subclinical infections [20]. This was supported by our finding that the SLO/D110–15 was detected in an animal without evident clinical signs. However, the real pathogenic

potential of these rotaviruses still has to be explored. Virus isolation in MA104 cell line was attempted for the strain SLO/D110–15 but was not successful, even after three passages.

The original source of RVA strains with G8P[14] geno-type most probably includes multiple human to animal

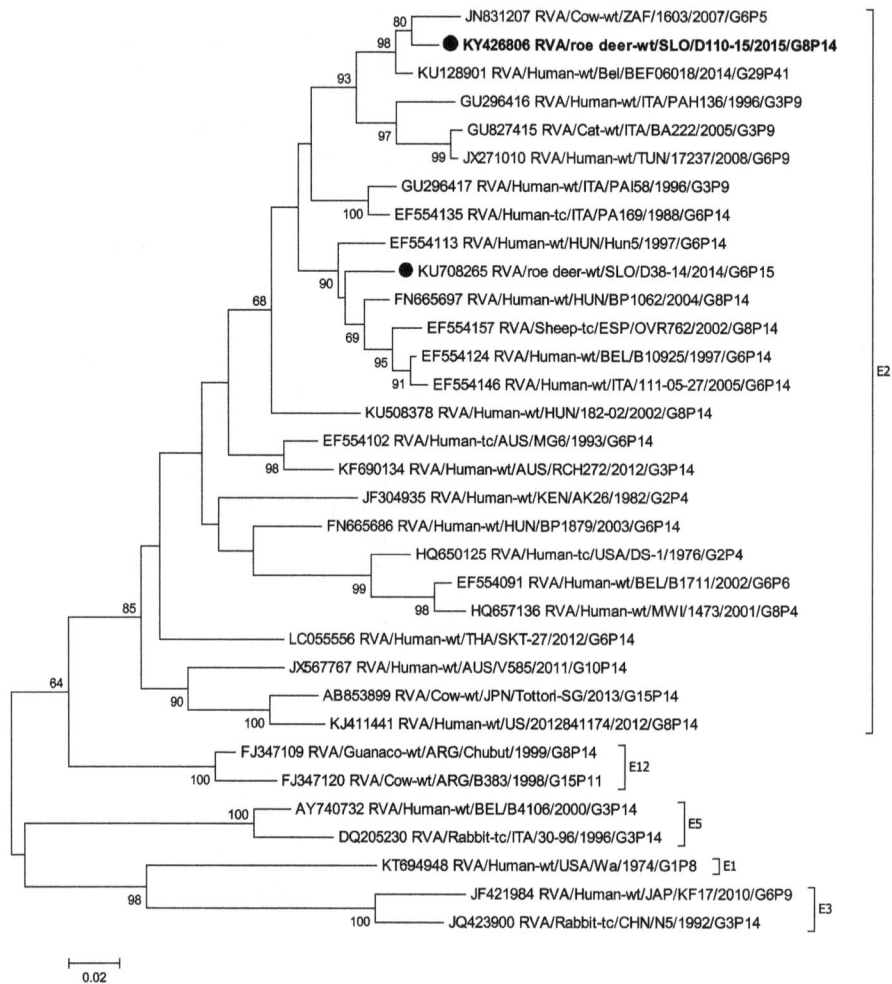

Fig. 10 The Maximum likelihood phylogenetic tree on NSP4 segment. Bootstrap values lower than 50 are not shown. The Slovenian roe deer strains SLO/D110–15 and SLO/D38–14 are marked with circle. Roe deer strain SLO/D110–15 is highlighted in bold

and animal to human transmission events [24]. It has been proposed that human P[14] strains are derived from interspecies transmission of RVA from humans and ungulates [17]. Other studies suggest zoonotic transmissions of the G8P[14] RVA strains with the bovine DS1-like genetic backbone [20]. Two human strains from Italy, PR1300 and PR1973 have a full genome constellation identical to that of our roe deer strain (G8-P[14]-I2-R2-C2-M2-A3-N2-T6-E2-H3) [19]. For these two human strains it was suggested that they are zoonotic and transmitted to humans from an animal belonging to the order *Artiodactyla* [19]. In addition, one G8P[14] strain with a genotype constellation identical to that of the roe deer strain, was observed in a child with gastroenteritis during the RVA survey in Slovenia in 2009 [15]. Although the genome segment sequences of this RVA human strain are not complete, it was shown that human and roe deer strains from Slovenia are not closely related as the VP8* fragment and partial VP7 nucleotide

sequence identities were both less than 90%. Relatively low identities between these two strains were shown also for other genes coding for structural proteins. For genes coding non-structural proteins the identities were much higher, indicating the reassortant nature of G8P[14], probably as a result of circulating in different hosts. Solving the riddle of the evolutionary path of G8P[14] strains requires the analysis of many more strains. Surveillance of the RVA in animals and humans should be continued to gain a clearer molecular and epidemiological history of zoonotic RVA strains.

Conclusions

The G8P[14] genotype has been found, for the first time, in deer, a newly described host from the order *Artiodactyla* for this RVA genotype. The finding of a RVA strain with the same genome segment constellation in humans indicates the possible zoonotic potential of this virus strain.

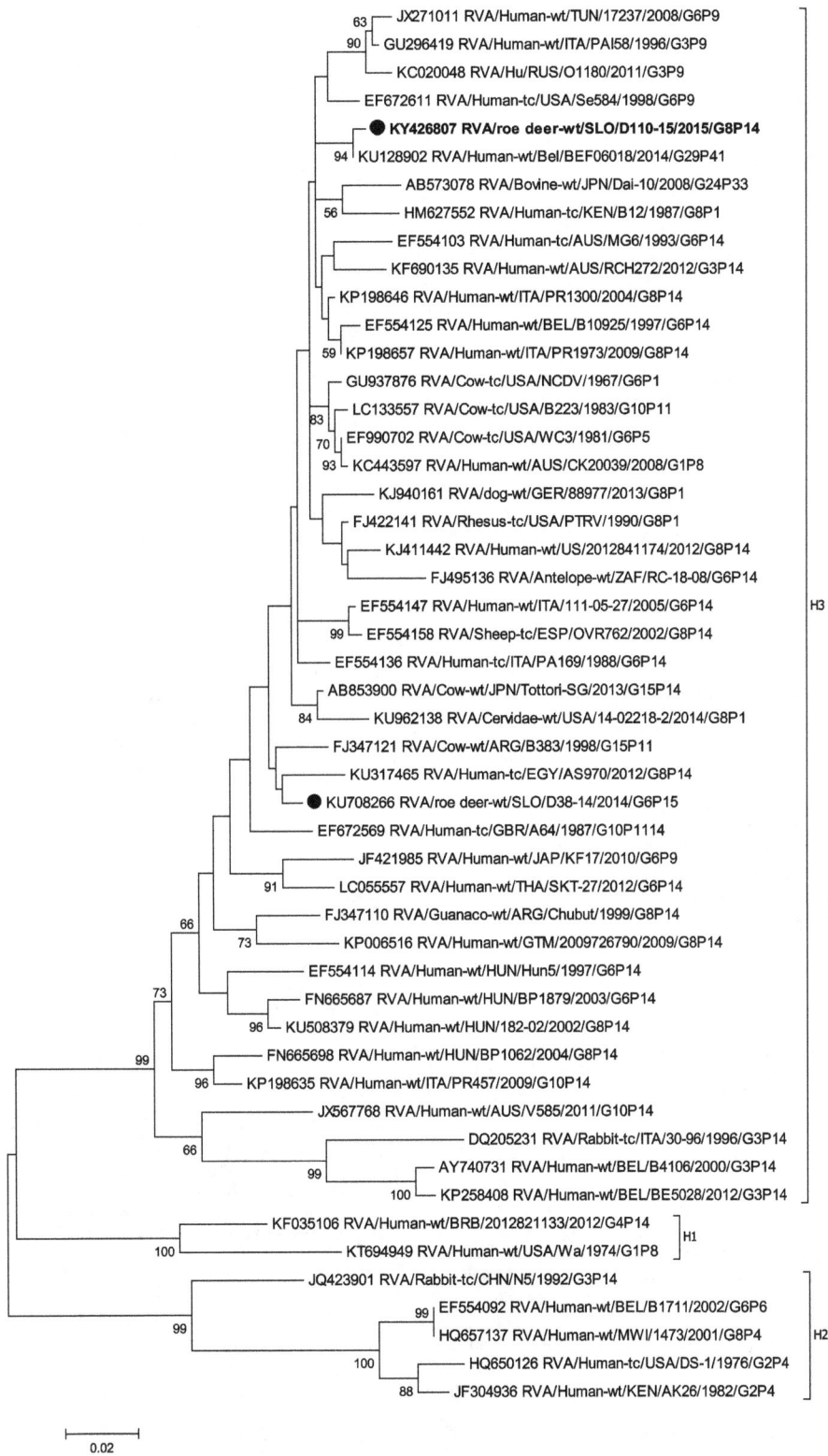

Fig. 11 The Maximum likelihood phylogenetic tree on NSP5/NSP6 segment. Bootstrap values lower than 50 are not shown. The Slovenian roe deer strains SLO/D110–15 and SLO/D38–14 are marked with circle. Roe deer strain SLO/D110–15 is highlighted in bold

Abbreviations
NGS: next generation sequencing technology; RCWG: Rotavirus Classification Working Group; RVA: Group A rotaviruses

Acknowledgements
Not applicable.

Competing interests
We declare no competing interests, financial or otherwise.

Funding
This work was supported by the Slovenian Ministry for Higher Education, Science and Technology (Research program P4–0092).

Authors' contributions
UJC processed faecal sample and isolated RNA. UK performed the NGS and analysed the results. UJC, UK, AS and AK performed the phylogenetic analysis on all 11 segments of the RVA. All authors contributed in writing the manuscript. All authors read and approved the final manuscript.

Consent for publication
The manuscript contains no individual person's data and does not require consent for publication.

Author details
[1]Institute of Food safety, Feed and Environment, Veterinary Faculty, University of Ljubljana, Gerbičeva 60, 1000 Ljubljana, Slovenia. [2]Institute of Microbiology and Parasitology, Veterinary Faculty, University of Ljubljana, Gerbičeva 60, 1000 Ljubljana, Slovenia. [3]Institute of Microbiology and Immunology, Faculty of Medicine, University of Ljubljana, Zaloška 4, 1000 Ljubljana, Slovenia.

References

1. Schiff LA, Nibert ML, Tyler KL: Orthoreoviruses and their replication. In: *Fields virology*. 5th edn. Edited by Knipe DM, Howley PM, Griffin DE, Lamb RA, Martin MA, Roizman B, Straus SE. Philadelphia: Lippincott, Williams & Wilkins; 2007: 1854–1915.

2. Estes MK, Greenberg HB: Rotaviruses. In: *Fields virology*. 6th edn. Edited by Fields BN, Knipe DM, Howley PM. Philadelphia: Lippincott, Williams & Wilkins; 2013: 1347–1401.

3. Martella V, Bányai K, Matthijnssens J, Buonavoglia C, Ciarlet M. Zoonotic aspects of rotaviruses. Vet Microbiol. 2010;140(3–4):246–55.

4. Matthijnssens J, Taraporewala ZF, Yang H, Rao S, Yuan L, Cao D, Hoshino Y, Mertens PP, Carner GR, McNeal M, et al. Simian rotaviruses possess divergent gene constellations that originated from interspecies transmission and reassortment. J Virol. 2010;84(4):2013–26.

5. Li K, Lin XD, Huang KY, Zhang B, Shi M, Guo WP, Wang MR, Wang W, Xing JG, Li MH, et al. Identification of novel and diverse rotaviruses in rodents and insectivores, and evidence of cross-species transmission into humans. Virology. 2016;494:168–77.

6. Matthijnssens J, Ciarlet M, Rahman M, Attoui H, Bányai K, Estes MK, Gentsch JR, Iturriza-Gómara M, Kirkwood CD, Martella V, et al. Recommendations for the classification of group a rotaviruses using all 11 genomic RNA segments. Arch Virol. 2008;153(8):1621–9.

7. Matthijnssens J, Ciarlet M, McDonald SM, Attoui H, Bányai K, Brister JR, Buesa J, Esona MD, Estes MK, Gentsch JR, et al. Uniformity of rotavirus strain nomenclature proposed by the rotavirus classification working group (RCWG). Arch Virol. 2011;156(8):1397–413.

8. Matthijnssens J, Ciarlet M, Heiman E, Arijs I, Delbeke T, McDonald SM, Palombo EA, Iturriza-Gómara M, Maes P, Patton JT, et al. Full genome-based classification of rotaviruses reveals a common origin between human Wa-like and porcine rotavirus strains and human DS-1-like and bovine rotavirus strains. J Virol. 2008;82(7):3204–19.

9. Steyer A, Poljsak-Prijatelj M, Barlic-Maganja D, Marin J. Human, porcine and bovine rotaviruses in Slovenia: evidence of interspecies transmission and genome reassortment. J Gen Virol. 2008;89(Pt 7):1690–8.

10. Ghosh S, Kobayashi N. Whole-genomic analysis of rotavirus strains: current status and future prospects. Future Microbiol. 2011;6(9):1049–65.

11. Ghosh S, Kobayashi N. Exotic rotaviruses in animals and rotaviruses in exotic animals. Virusdisease. 2014;25(2):158–72.

12. Jamnikar-Ciglenecki U, Kuhar U, Sturm S, Kirbis A, Racki N, Steyer A. The first detection and whole genome characterization of the G6P[15] group a rotavirus strain from roe deer. Vet Microbiol. 2016;191:52–9.

13. Anbalagan S, Peterson J. Detection and whole-genome characterization of a G8P[1] group a rotavirus strain from deer. Genome Announc. 2016;4(6)

14. Maes P, Matthijnssens J, Rahman M, Van Ranst M. RotaC: a web-based tool for the complete genome classification of group a rotaviruses. BMC Microbiol. 2009;9:238.

15. Steyer A, Sagadin M, Kolenc M, Poljsak-Prijatelj M. Molecular characterization of rotavirus strains from pre- and post-vaccination periods in a country with low vaccination coverage: the case of Slovenia. Infect Genet Evol. 2014;28:413–25.

16. Kumar S, Stecher G, Tamura K. MEGA7: molecular evolutionary genetics analysis version 7.0 for bigger datasets. Mol Biol Evol. 2016;33(7):1870–4.

17. Matthijnssens J, Potgieter CA, Ciarlet M, Parreño V, Martella V, Bányai K, Garaicoechea L, Palombo EA, Novo L, Zeller M, et al. Are human P[14] rotavirus strains the result of interspecies transmissions from sheep or other ungulates that belong to the mammalian order Artiodactyla? J Virol. 2009; 83(7):2917–29.

18. Bányai K, Papp H, Dandár E, Molnár P, Mihály I, Van Ranst M, Martella V, Matthijnssens J. Whole genome sequencing and phylogenetic analysis of a zoonotic human G8P[14] rotavirus strain. Infect Genet Evol. 2010;10(7):1140–4.

19. Medici MC, Tummolo F, Bonica MB, Heylen E, Zeller M, Calderaro A, Matthijnssens J. Genetic diversity in three bovine-like human G8P[14] and G10P[14] rotaviruses suggests independent interspecies transmission events. J Gen Virol. 2015;96(Pt 5):1161–8.

20. Marton S, Dóró R, Fehér E, Forró B, Ihász K, Varga-Kugler R, Farkas SL, Bányai K. Whole genome sequencing of a rare rotavirus from archived stool sample demonstrates independent zoonotic origin of human G8P[14] strains in Hungary. Virus Res. 2017;227:96–103.

21. Badaracco A, Matthijnssens J, Romero S, Heylen E, Zeller M, Garaicoechea L, Van Ranst M, Parreño V. Discovery and molecular characterization of a group a rotavirus strain detected in an Argentinean vicuña (Vicugna Vicugna). Vet Microbiol. 2013;161(3–4):247–54.

22. Gautam R, Mijatovic-Rustempasic S, Roy S, Esona MD, Lopez B, Mencos Y, Rey-Benito G, Bowen MD. Full genomic characterization and phylogenetic analysis of a zoonotic human G8P[14] rotavirus strain detected in a sample from Guatemala. Infect Genet Evol. 2015;33:206–11.

23. Ciarlet M, Estes M, Britton G: Rotaviruses: basic biology, epidemiology and methodologies. In: *Encyclopedia of environmental microbiology*. Edn. New York: John Wiley & Sons; 2002.

24. Swiatek DL, Palombo EA, Lee A, Coventry MJ, Britz ML, Kirkwood CD. Characterisation of G8 human rotaviruses in Australian children with gastroenteritis. Virus Res. 2010;148(1–2):1–7.

6

High detection rate of dog circovirus in diarrheal dogs

Han-Siang Hsu[1†], Ting-Han Lin[1†], Hung-Yi Wu[2], Lee-Shuan Lin[1,3], Cheng-Shu Chung[1,3], Ming-Tang Chiou[1,4*] and Chao-Nan Lin[1,4*] [ORCID]

Abstract

Background: Diarrhea is one of the most common clinical symptoms reported in companion animal clinics. Dog circovirus (DogCV) is a new mammalian circovirus that is considered to be a cause of alimentary syndromes such as diarrhea, vomiting and hemorrhagic enteritis. DogCV has previously only been identified in the United States, Italy, Germany (GeneBank accession number: KF887949) and China (GeneBank accession number: KT946839). Therefore, the aims of this study were to determine the prevalence of DogCV in Taiwan and to explore the correlation between diarrhea and DogCV infection. Clinical specimens were collected between 2012 and 2014 from 207 dogs suffering from diarrhea and 160 healthy dogs.

Results: In this study, we developed a sensitive and specific SYBR Green-based real-time PCR assays to detected DogCV in naturally infected animals. Of the analyzed fecal samples from diarrheal dogs and health dogs, 58 (28.0 %) and 19 (11.9 %), respectively, were DogCV positive. The difference in DogCV prevalence was highly significant ($P = 0.0002755$) in diarrheal dogs.

Conclusions: This is the first study to reveal that DogCV is currently circulating in domestic dogs in Taiwan and to demonstrate its high detection rate in dogs with diarrhea.

Keywords: Dog circovirus, Diarrhea, DogCV, Real-time PCR

Background

Gastrointestinal disorders are one of the most common diseases reported in companion animal clinics. They can be caused by a number of viral, bacterial and parasitic pathogens. The most common viral gastrointestinal-pathogens are canine parvovirus [1, 2] and coronavirus. However, other agents, such as dog circovirus (DogCV), have recently been considered to be related to enteric disorders in dogs [3, 4]. DogCV was first identified in dogs with vasculitis and/or hemorrhagic gastroenteritis in the United States in 2012 [4]. DogCV is a non-enveloped, circular, single-stranded DNA virus containing a circular genome approximately 2 kb in length. It belongs to the genus *Circovirus*, together with porcine circovirus type 1 (PCV1), porcine circovirus type 2 (PCV2), canary circovirus, beak and feather disease virus and other viruses of domestic and wild birds [3].

PCV2 causes clinical conditions including systemic, lung, enteric, reproductive and skin diseases [5]. In recent years, a possible association between DogCV and canine enteritis has been suggested [3, 4]. DogCV has also been reported to cause necrotizing lymphadenitis [4] and vasculitis, which are also caused by porcine circovirus type 2 infections in pigs [5]. Previous studies have shown that DogCV is associated with hemorrhagic enteritis in dogs [3, 4], however, limited information is available to determine the direct correlation between the severity of diarrhea and DogCV infections.

DogCV has only been detected in the US [4, 6], Italy [3], Germany (GeneBank accession number: KF887949) and China (GeneBank accession number: KT946839). In the present study, we determined the previously unidentified DogCV and its prevalence in Taiwanese household dogs and clarified the correlation between diarrhea and DogCV infection.

* Correspondence: mtchiou@mail.npust.edu.tw; cnlin6@mail.npust.edu.tw
†Equal contributors
1Department of Veterinary Medicine, College of Veterinary Medicine, National Pingtung University of Science and Technology, Neipu, Pingtung, Taiwan
Full list of author information is available at the end of the article

Methods

Ethics and consents

The study did not involve any animal experiment. The Institutional Animal Care and Use Committee (IACUC) of National Pingtung University of Science and Technology did not deem it necessary for this research group to obtain formal approval to conduct this study. The dog owners gave his/her written consent for sample collection and data publication. Besides, according to Dr. Baneux' recommendations "Privately owned animals that are recruited for clinical studies (not Public Health Service funded) do not need to be subjected to IACUC oversight as long as their involvement includes only procedures that are consistent with the standard of care provided to patients with the same diagnosis that are not included in the clinical study" [7].

Specimen collection and DNA extraction

Clinical specimens, including rectal swabs or feces, were collected from dogs presenting at animal hospitals in Taiwan between June 2012 and June 2014. A total of 367 household dogs were enrolled in this study. These included 160 healthy dogs with no symptoms of diarrhea (for routine health check) and 207 dogs with diarrhea. The age, breed, sex of each dog and the date of sampling were recorded. The range of age span for dogs with and without diarrhea was from 2 months - 13 years old and 2.5 months - 18 years old, respectively. Fecal scores were determined and recorded according to the Waltham Fecal Scoring System. For detection of DogCV, fecal samples from these dogs with and without diarrhea were collected by inserting the swab into the rectum (~2 cm), rotating the swab and then placing the swab in the nuclease-free water. To ensure the diarrhea only results from DogCV infection, fecal samples taken from diarrheal dogs were also tested for CPV-2, CCV, giardia (CPV/CCV/Giardia Ag Test Kit, BIONOTE, Republic of Korea) and CDV (CDV Ag Test Kit, BIONOTE, Republic of Korea). All clinical specimens were sent to the animal disease diagnostic center at the National Pingtung University of Science and Technology. The weight of the stool sample was diluted 1:10 (w/v) with PBS, and vortexed for 5 min at maximum speed to generate a stool suspension. Viral DNA was extracted from clinical samples using the Genomic DNA Mini Kit (Geneaid Biotech, Ltd., Taipei, Taiwan) according to the manufacturer's protocol. GAPDH was used for DNA quality control of all samples. All of specimens were screened using real-time PCR for CPV-2, as described by Lin et al [8] and for DogCV, as described below.

Primer design for DogCV

Based on the sequence homology of PCV1, PCV2 and DogCV, a conserved region of the replicase gene of PCV1, PCV2 and DogCV was identified in nucleotide sequences available from GenBank (PCV1: AY660574; PCV2: AY424401; DogCV: JQ821392, NC020904, KC241983, KC241984, KF887949, [9] KJ530972 and KT946839). The sequences were aligned with the Clustal W method using the MegAlign program (DNASTAR, Madison, WI). Figure 1 shows the nucleotide sequence alignment of the partial replicase genes of PCV1, PCV2 and DogCVs.

Construction of plasmid DNA standard curves

The oligonucleotide primers used for the amplification of PCV1, PCV2 and DogCV were Rep CircoV-F, 5'-TGG TGG GAY GGH TAY SAT GG-3' and Rep CircoV-R, 5'-TAH CGR TCA CAB ART CTC A-3'; the sequences of these primers correspond to base pair positions 607-626 and 701-683, respectively, of DogCV strain NY214 (GenBank accession number JQ821932) [6]. Using these primers, 95 bp fragments of the replicase gene were amplified from PCV1, PCV2 and DogCV. The PCR products were cloned using the TA cloning kit (Yeastern Biotech Co., Ltd. Taipei, Taiwan) and sequenced. The plasmids containing the circovirus sequences were purified using a plasmid miniprep purification kit (GMbiolab Co, Ltd. Taichung, Taiwan) and quantified by the measurement of optical density (OD) at 260 nm with a spectrophotometer (Hitachi U2900, Dallas, TX, USA). A standard curve was generated using 10-fold dilutions (10^2-10^8 copies/μl) of standard plasmid DNA to determine the detection limits of the SYBR Green-based real-time PCR assay. Intra- (within-run) and inter- (between-

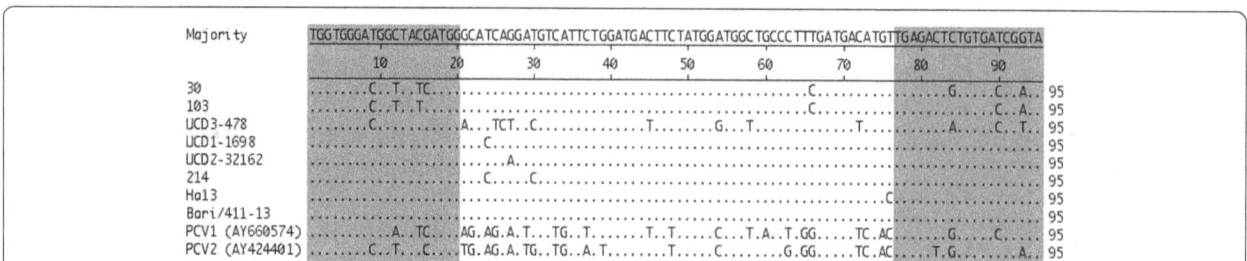

Fig. 1 Nucleotide sequence alignment of the 95-bp fragments of the replicase gene from Dog circovirus, porcine circovirus type 1 and porcine circovirus type 2. Only nucleotides differing from the overall majority sequence are shown in uppercase. The primer set is shown in light gray

run) assay reproducibility was evaluated using the standard plasmid DNA in triplicate on three different days.

Real-time PCR for the detection of DogCV

SYBR Green-based real-time PCR assays were performed using a LightCycler Nano (Roche Diagnostics, Mannheim, Germany). Each 10-μl reaction mixture contained 0.2 μM of forward and reverse primers, 5 μl of Master buffer (Roche Diagnostics, Mannheim, Germany) and 3 μl of DNA extract. The thermocycling conditions consisted of 10 min at 95 °C, 45 cycles of 10 s each at 95 °C, 10 s at 51 °C and 10 s at 72 °C. Each run included serial 10-fold dilutions of the standard plasmid DNA as a positive control and for the construction of a standard curve. A negative control without the DNA template was included to detect possible cross-contamination.

Replicase gene sequencing and sequence analysis

The DNA fragments were purified and sequenced, as described by Lin et al [10]. The partial replicase DNA sequence of our samples were compared to those of reference DogCV (JQ821392, NC020904, KC241983, KC241984, KF887949 and KJ530972), PCV1 (AY660574) and PCV2 (AY424401). Multiple alignments of the nucleic acid sequences were performed with the Clustal W method, using the MegAlign program (DNASTAR, Madison, WI).

Statistical analysis

The chi-square test with Yate's correction was used to evaluate the correlation between the samples that were positive for DogCV and different clinical observation groups. The paired t-test was used to determine whether there is a statistically significant difference in age between diarrheal and healthy groups. A Wilcoxon Rank-Sum test was used to evaluate the correlation between the severity of diarrhea and samples positive for DogCV. Differences were considered to be statistically significant and highly significant if the associated P-value was <0.05 and <0.01, respectively.

Results

Real-time PCR amplification efficiency and limits of detection

The optimized PCR assay targeting a conserved region of the replicase gene permitted the detection of PCV1, PCV2 and DogCV (data not shown). The standard curve generated using standard plasmid DNA had a linear range of seven orders of magnitude (10^2 to 10^8 copies/μl). The correlation coefficient (R^2) between the number of quantification cycles (Cq) and the logarithm of the plasmid copy number was 0.999 (slope = -4.18), based on triplicate runs (data not shown). The average Tm value of DogCV was approximately 80 °C (data not shown). To assess the detection limits of the SYBR Green-based real-time PCR assays, standard plasmid DNA was serially diluted 10-fold ranging from 10^2 to 10^8 copies/μl, and was used as a template. At 10^3 copies per reaction, 100 % of the replicates were found to be positive using real-time PCR, whereas only 90 % of the replicates were positive at 10^2 copies (Table 1). The coefficients of variation of the within-run and between-run mean Cq values for standard plasmid DNA ranged from 0.41 to 1.33 % and from 2.45 to 6.84 %, respectively (data not shown). To test the specificity of the SYBR Green-based real-time PCR assay, we analyzed canine distemper virus (CDV), canine coronavirus (CCV) and canine parvovirus type 2 (CPV-2). No specific amplification was detected for any of these samples (data not shown).

Overall detection of gastrointestinal pathogens using commercial kits and real-time PCR

Between June 2012 and June 2014, a total of 207 specimens were obtained from 207 enrolled dogs. Of these 207 cases, 74 (35.7 %) were positive for at least one gastrointestinal pathogen, as determined by a commercial kit and real-time PCR. Among those dogs with positive results, most were positive for DogCV (58/74, 78.4 %), followed by CPV-2 (22/74, 29.7 %), giardia (8/74, 10.8 %), CCV (4/74, 5.4 %) and CDV (3/74, 4.1 %). The correlations between the severity of diarrhea and positive tests for the different canine gastrointestinal pathogens are summarized in Table 2.

Correlation of diarrhea with positive test for DogCV

To understand the prevalence of DogCV in clinically healthy dogs, DogCV was also screened for in fecal samples from 160 healthy dogs using real-time PCR. Nineteen (11.9 %) samples were positive for DogCV. Using a chi-square test, the correlation of the diarrhea with a positive test for DogCV was calculated. Surprisingly, among the 367 dogs analyzed, a positive DogCV test had a highly significant correlation with the presence of diarrhea in dogs ($P = 0.0002755$)(Table 3). The odds for diarrheal and healthy groups were 0.39 and 0.13, respectively (Table 3),

Table 1 Efficiency of the Dog circovirus (DogCV) SYBR Green-based real-time PCR assay

Estimated number of DogCV plasmid DNA copies	Number of positive/ number of tested	Mean Cq ± SD
10^8	10/10	14.64 ± 0.62
10^7	10/10	18.75 ± 0.83
10^6	10/10	23.06 ± 1.24
10^5	10/10	27.61 ± 1.53
10^4	10/10	32.11 ± 1.71
10^3	10/10	36.58 ± 2.28
10^2	9/10	40.19 ± 1.29

Table 2 Correlation between the severity of diarrhea and the different canine gastrointestinal pathogens

Pathogen	The Waltham fecal scoring system				Total
	5	4.5	4	3.5	
DogCV only	15	11	16	1	43
DogCV + other pathogens	14	1	0	0	15
Other pathogens (without DogCV)	12	3	1	0	16
All negative	30	50	51	2	133
Total	71	65	68	3	207

indicating that dogs with diarrhea were 3 times more likely to be DogCV-positive than those which are clinically healthy. In addition, we evaluated the correlation between the severity of diarrhea and a positive test for DogCV. According to the Waltham Fecal Scoring System and the results of the DogCV tests, the difference in the prevalence of DogCV was not significant between different severities of diarrhea. ($P = 0.2286$, Wilcoxon Rank-Sum test).

Correlation of age with positive test for DogCV
Of the 58 DogCV-positive dogs in diarrheal group, 25 (43.1 %) were less than 1 year old, 25 (43.1 %) were between 1-7 years old, and 8 (13.8 %) were more than 7 years old. The range of age spans for DogCV-positive dogs was from 2 months - 13 years old (median is 1 year old). Of 19 DogCV-positive dogs in the health group, 3 (15.8 %) were less than 1 year old, 8 (42.1 %) were between 1-7 years old, and 8 (42.1 %) were more than 7 years old. The range of age spans for DogCV-positive dog was from 3 months - 18 years old (median is 5 year old). By statistic analysis, it was found that there was a statistically significant difference in age between diarrheal and healthy groups with respect to the DogCV-positive dogs ($P = 0.005$). That is, the DogCV was more frequently identified from younger dogs in diarrheal group.

Genetic comparison of the DogCVs
Comparison of the partial replicase nucleotide sequences revealed 97.9 % and 82.1-93.7 % homology within the analyzed local DogCV isolates and between the local DogCV and reference strains, respectively (Table 4). The same nucleotide sequences for local DogCV strains also

Table 3 Correlation of diarrhea with dog circovirus (DogCV) infection

Clinical status	Results of DogCV detection		Odds	Total
	Negative	Positive		
Diarrheal	149 (51.4 %)	58 (75.3 %)	0.39	207
Healthy	141 (48.6 %)	19 (24.7 %)	0.13	160
Total	290	77		367

exhibited 91.6-93.7 % homology with those for Italian DogCV strain Bari/411-13 (Table 4).

Discussion
In the present study we determine the prevalence of DogCV in Taiwan. Besides, our results suggest that the positive test results for DogCV appear to have a highly significant correlation with the presence of diarrhea ($P = 0.0002755$) and that dogs with diarrhea were 3 times more likely to be DogCV-positive than those which are clinically healthy. In contrast, a study conducted in the US using real-time PCR reported that the difference in the prevalence of DogCV between healthy and diarrheal dogs was not significant [4]. These differing results might be due to the following: i) geographic distribution, (i.e., USA vs. Taiwan); or ii) primer design (i.e., the specific primer for DogCV used in the previous study (Additional file 1: Figure S1) vs. the degenerate primer for mammal circovirus in our study). In the present study, we established a SYBR-Green-based PCR assay to detect DogCV. This novel, real-time PCR was specific, sensitive and reliable for the amplification of DogCV. Both intra- and inter-assay CVs were satisfactorily low.

Forty-three out of the 58 (74.1 %) DogCV-positive dogs were found to be positive for DogCV alone. The fecal scoring of those dogs was observed to range from "very moist stool" to "watery diarrhea". This agrees with US and Italian studies, which showed a similar proportion of diarrheal dogs to be infected by DogCV alone [3, 4]. Diarrheal symptoms caused by circovirus have been observed in PCV2-associated enteritis [5]. DogCV DNA has been detected in several lymphoid tissues, including Peyer's patches and mesenteric lymph nodes [4]. A moderate depletion of lymphocytes was found in DogCV-infected dogs, which might suggest immunosuppression in DogCV-affected dogs [4]. However, the pathogenesis of DogCV infection remains to be elucidated.

A previous study reported that co-infection with other canine pathogens was found in DogCV-positive dogs with diarrhea [4]. Co-infection with canine gastrointestinal pathogens was also observed in the present study. Surprisingly, twelve of the 15 DogCV-positive diarrheal dogs were also found to be CPV2-positive. Co-infection with DogCV and CPV-2 was not observed in the US or Italian studies [3, 4]. We further investigated the antigenic variants of CPV-2, and only CPV-2a or 2b were found in dogs co-infected with DogCV. This might be because no CPV-2c was found in Taiwan during the study period [10]. In PCV2 infection, the expression of clinical disease in infected-pigs typically involves secondary microbial infections [9, 11]. Therefore, the role of co-infection in the pathogenesis of DogCV will require further investigation.

Table 4 Sequence homology between Taiwan and reference DogCVs

Strains									Strains (Country)
103	Bari/411-13	UCD1-1698	UCD2-32162	Ha13	NY214	UCD3-478	JZ98/2014		
97.9	91.6	90.5	90.5	90.5	89.5	84.2	82.1		30 (Taiwan)
	93.7	92.6	92.6	92.6	91.6	85.3	83.2		103 (Taiwan)
		98.9	98.9	98.9	97.9	86.3	87.4		Bari/411-13 (Italy)
			97.9	97.9	98.9	85.3	88.4		UCD1-1698 (USA)
				97.9	96.8	86.3	86.3		UCD2-32162 (USA)
					96.8	85.3	88.4		Ha13 (Germany)
						86.3	87.4		NY214 (USA)
							82.1		UCD3-478 (USA)

Our results showed that the partial replicase nucleotide sequences for local DogCV strains exhibited 91.6-93.7 % homology with those for Italian DogCV strain Bari/411-13. However, only a few complete genomes of DogCV are available in GeneBank. The patterns of the genetic evolution of DogCV need to be investigated through continuous surveillance and by analyzing the sequence of the complete genome.

Conclusions

This is the first study to reveal that DogCV is currently circulating in domestic dogs in Taiwan and to demonstrate its high detection rate in dogs with diarrhea.

Abbreviations

CCV, canine coronavirus; CDV, canine distemper virus; CPV-2, canine parvovirus type 2; Cq, quantification cycles; DogCV, Dog circovirus (DogCV); OD: optical density; PCR, polymerase chain reaction; PCV1, porcine circovirus type 1; PCV2, porcine circovirus type 2; Tm, melting temperature.

Acknowledgements
The authors would like to thank Prof. Ling-Ling Chueh for comments that greatly improved the manuscript.

Funding
No funding was obtained for this study.

Authors' contributions
HSH designed and analyzed the experiments and wrote the manuscript. THL designed and analyzed the experiments and wrote the manuscript. HYW performed and analyzed the experiments. LSL performed and analyzed the experiments. CSC performed and analyzed the experiments. MTC provided materials and reagents, analyzed and interpreted the data and drafted the manuscript. CNL provided materials and reagents, analyzed and interpreted the data and drafted the manuscript. All of the authors read and approved the final manuscript.

Competing interests
The authors declare that they have no competing interests.

Consent for publication
Not applicable.

Author details
[1]Department of Veterinary Medicine, College of Veterinary Medicine, National Pingtung University of Science and Technology, Neipu, Pingtung, Taiwan. [2]Graduate Institute of Veterinary Pathobiology, College of Veterinary Medicine, National Chung-Hsing University, South Dist, Taichung, Taiwan. [3]Veterinary Hospital, College of Veterinary Medicine, National Pingtung University of Science and Technology, Neipu, Pingtung, Taiwan. [4]Animal Disease Diagnostic Center, College of Veterinary Medicine, National Pingtung University of Science and Technology, Neipu, Pingtung, Taiwan.

References
1. Decaro N, Buonavoglia C. Canine parvovirus- a review of epidemiological and diagnostic aspects, with emphasis on type 2c. Vet Microbiol. 2012;155(1):1–12.
2. Decaro N, Buonavoglia C. An update on canine coronaviruses: viral evolution and pathobiology. Vet Microbiol. 2008;132(3-4):221–34.
3. Decaro N, Martella V, Desario C, Lanave G, Circella E, Cavalli A, Elia G, Camero M, Buonavoglia C. Genomic characterization of a circovirus associated with fatal hemorrhagic enteritis in dog, Italy. PLoS One. 2014;9(8):e105909.
4. Li L, McGraw S, Zhu K, Leutenegger CM, Marks SL, Kubiski S, Gaffney P, Dela Cruz FN, Jr., Wang C, Delwart E, et al. Circovirus in tissues of dogs with vasculitis and hemorrhage. Emerg Infect Dis. 2013;19(4):534–41.
5. Segales J. Porcine circovirus type 2 (PCV2) infections: clinical signs, pathology and laboratory diagnosis. Virus Res. 2012;164(1-2):10–9.
6. Kapoor A, Dubovi EJ, Henriquez-Rivera JA, Lipkin WI. Complete genome sequence of the first canine circovirus. J Virol. 2012;86(12):7018.
7. Baneux PJ, Martin ME, Allen MJ, Hallman TM. Issues related to institutional animal care and use committees and clinical trials using privately owned animals. ILAR J. 2014;55(1):200–9.
8. Lin CN, Chien CH, Chiou MT, Wang JW, Lin YL, Xu YM. Development of SYBR green-based real-time PCR for the detection of canine, feline and porcine parvoviruses. Taiwan Vet J. 2014;40(1):1–9.
9. Darwich L, Segales J, Mateu E. Pathogenesis of postweaning multisystemic wasting syndrome caused by Porcine circovirus 2: An immune riddle. Arch Virol. 2004;149(5):857–74.
10. Lin CN, Chien CH, Chiou MT, Chueh LL, Hung MY, Hsu HS. Genetic characterization of type 2a canine parvoviruses from Taiwan reveals the emergence of an Ile324 mutation in VP2. Virol J. 2014;11(1):39.
11. Segales J, Mateu E. Immunosuppression as a feature of postweaning multisystemic wasting syndrome. Vet J. 2006;171(3):396–7.

Prevalence of hepatitis E virus infection in wild boars from Spain: a possible seasonal pattern?

Antonio Rivero-Juarez[1]* , María A. Risalde[1], Mario Frias[1], Ignacio García-Bocanegra[2], Pedro Lopez-Lopez[1], David Cano-Terriza[2], Angela Camacho[1], Saul Jimenez-Ruiz[2], Jose C. Gomez-Villamandos[3] and Antonio Rivero[1,4]

Abstract

Background: It has been shown that wildlife can serve as natural reservoirs of hepatitis E virus (HEV). The wild boar (*Sus scrofa*) is probably the main natural reservoir of HEV and could therefore represent an important route of transmission in Europe, especially in regions where game meat is widely consumed. We evaluated the prevalence of HEV infection in wild boar in the south of Spain, with the aim of identifying associated risk factors. A cross-sectional study that included hunted wild boar was carried out during the 2015/2016 hunting season (October 15 to February 15) in Andalusia (southern Spain). The outcome variable was HEV infection, defined as amplification of HEV RNA in serum by RT-PCR.

Results: A total of 142 animals, selected from 12 hunting areas, were included and formed the study population. Thirty-three wild boars (23.2%; 95% CI: 16.8%–30.7%) were positive for HEV infection. Prevalence peaked in October and November, then gradually declined until the end of December. After multivariate analysis, only hunting date was independently associated with HEV infection across sex and age.

Conclusions: Our study found a relatively high prevalence of HEV infection in wild boar in the south of Spain, suggesting that prevalence may depend on the season when the animal is hunted. In consequence, the potential risk of zoonotic transmission could fluctuate.

Keywords: Hepatitis E, Wild boar, Prevalence, Seasonality, Foodborne

Background

Hepatitis E virus (HEV) is an emerging cause of viral hepatitis in developed countries [1, 2]. The main route of transmission is the consumption of raw or undercooked pork, and pigs have been identified as the main host of HEV [3]. It has been shown that other animals, wildlife in particular, can act as natural reservoirs of HEV [4]. Among wildlife species, the wild boar (*Sus scrofa*) is probably the main reservoir of HEV [5] and could therefore represent an important route of transmission in Europe, especially in regions where game meat is widely consumed. In this respect, we recently described a familial

HEV outbreak in our area that was linked to the consumption of wild boar meat, with a secondary finding in our analysis being a high prevalence of HEV in wild boar [6]. It has been proven experimentally that HEV-infected wild boar can transmit the infection to other animals, such as pigs [4, 7]. This plays an important role in countries where extensive pig farming is widespread, because it facilitates contact between domestic pigs and sympatric species and increases the risk of inter-species transmission. For this reason, the evaluation of HEV infection in wild boar and the identification of risk factors affecting transmission is important in order to determine the zoonotic potential of this emerging viral infection and enable control measures to be established.

Risk factors associated with HEV infection have barely been studied in humans. HEV infection has been associated with older males and certain genetic factors [8–10],

* Correspondence: arjvet@gmail.com
[1]Infectious Diseases Unit. Instituto Maimonides de Investigación Biomédica de Córdoba (IMIBIC), Hospital Universitario Reina Sofía de Córdoba. Universidad de Córdoba, 2° Floor. Box 134.Avenida Menendez Pidal s/n, 14004 Córdoba, Spain
Full list of author information is available at the end of the article

although the reasons remain unknown. At the same time, living in certain regions has also been associated with a higher prevalence of HEV [11, 12]. Here we evaluated the prevalence of HEV infection in wild boar in the south of Spain in order to identify associated risk factors.

Methods
Study design and population
A cross-sectional study that included hunted wild boar was carried out in Andalusia (southern Spain) (36°N–38° 600 N, 1°750 W–7°250 W) during the 2015/2016 hunting season (October 15th to February 15th). Age was determined on the basis of tooth eruption and animals of less than 12 months old were classified as juveniles, those between 12 and 24 months as sub-adults, and those over 2 years old as adults. All animals were classified according to sex. The sample size was calculated on the assumption that 10% of the samples would be positive for HEV. Hence, assuming a confidence interval of 95%, the minimum sample size was estimated at 139 animals.

Variable collection and definition
A whole blood sample was obtained from all hunted animals by puncture of the cavernous sinus of the dura mater [13]. Serum was obtained from whole blood. Epidemiological variables were collected and included age, sex, date of sample collection, and hunting area.

The outcome variable was HEV infection, defined as amplification of HEV RNA in serum by reverse transcription polymerase chain reaction (RT-PCR).

RT-PCR for detection of HEV
Viral RNA was extracted from 200ul of serum using the commercial QIAamp MinElute Virus Spin Kit (QIAgen. Hilden, Germany) and an automated procedure (QIAcube. QIAgen, Hilden, Germany). Samples were frozen at – 80 °C until analysis. For diagnosis of HEV infection, RT-PCR was performed using the Light-Cycler 480 system (Roche. Basel, Switzerland) described elsewhere [14]. For the reaction, the QIAgen One step PCR Kit (QIAgen, Hilden, Germany) was used. The primers (15 µMol) employed were: sense primer HEV5260 (5'-GGTGGTTTCTGGGGTGAC-3') and antisense primer HEV5330 (5'-AGGGGTTGGTTGGAT GAA-3'). The probe employed (20 µMol) was HEV5283 (5'-FAM-TGATTCTCAGCCCTTCGC-TAMRA-3'). The thermal profile was 50 °C for 30 min and 95 °C for 15 min, followed by 45 cycles of 94 °C for 10 s, 55 °C for 20s and 72 °C for 60 s. An external (in-run) standard curve was applied to calculate HEV viral load using a WHO Standard HEV strain supplied by the Paul-Ehrlich-Institute (code 6329/10).

Statistical analysis
HEV prevalence was estimated from the ratio of positive samples to the total number of samples analyzed, with exact binomial confidence intervals of 95%. Throughout the study, we calculated prevalence by age and sex every week in order to evaluate the possible increase or decrease in HEV prevalence over time. Categorical variables were expressed as numbers of cases (percentage). Frequencies were compared using the χ^2 test or Fisher's exact test, and significance was set at a two-tailed p-value of less than 0.05. Bivariate analysis was carried out to discover the variables related to HEV infection, and multivariate logistic regression analysis was also performed. Analyses were carried out using the SPSS statistical software package, version 18.0 (IBM Corporation, Somers, NY, USA), GraphPad Prism, version 6 (Mac OS X version; GraphPad Software; San Diego, California, USA) and Winpepi software, version 11.36 (Brixton Health).

Results
Population
A total of 142 animals were included and constituted the study population. These animals were selected from 12 hunting areas (Fig. 1). Sixty-four animals were male (45.1%) and 78 females (64.9%). Ninety-seven were adults (68.4%) and 45 non-adults (31.6%).

HEV infection prevalence and associated factors
Thirty-three wild boars (23.2%; 95% CI: 16.8%–30.7%) were positive for HEV infection.

When prevalence was compared and analyzed according to sex, 20 males (31.2%; 95% CI: 21.2%–43.4%) and 13 females (16.7%; 95% CI: 9.9%–16.9%) ($p = 0.047$) presented HEV infection. No significant differences in prevalence were found between adults (25 of 97, 25.8%; 95% CI: 18.1%–35.3%) and non-adults (8 of 45, 17.8%; 95% CI: 9%–31.6%) ($p = 0.394$). An analysis of prevalence according to the week when the animals were hunted showed that it was higher in the first weeks of the study than at the end of the hunting season, February 15th (Fig. 2). Prevalence peaked in October and November, then gradually declined until the end of December. The prevalence of HEV infection varied between 60 and 0%, depending on the date of sample collection (Fig. 2).

By multivariate analysis, only hunting date was independently associated with HEV infection across sex and age (Table 1).

Discussion
Our study found a 23.2% prevalence of active HEV infection in wild boars in the south of Spain. Interestingly, the prevalence varied significantly according to the period of the hunting season, with a higher HEV prevalence during the last weeks of October and the first

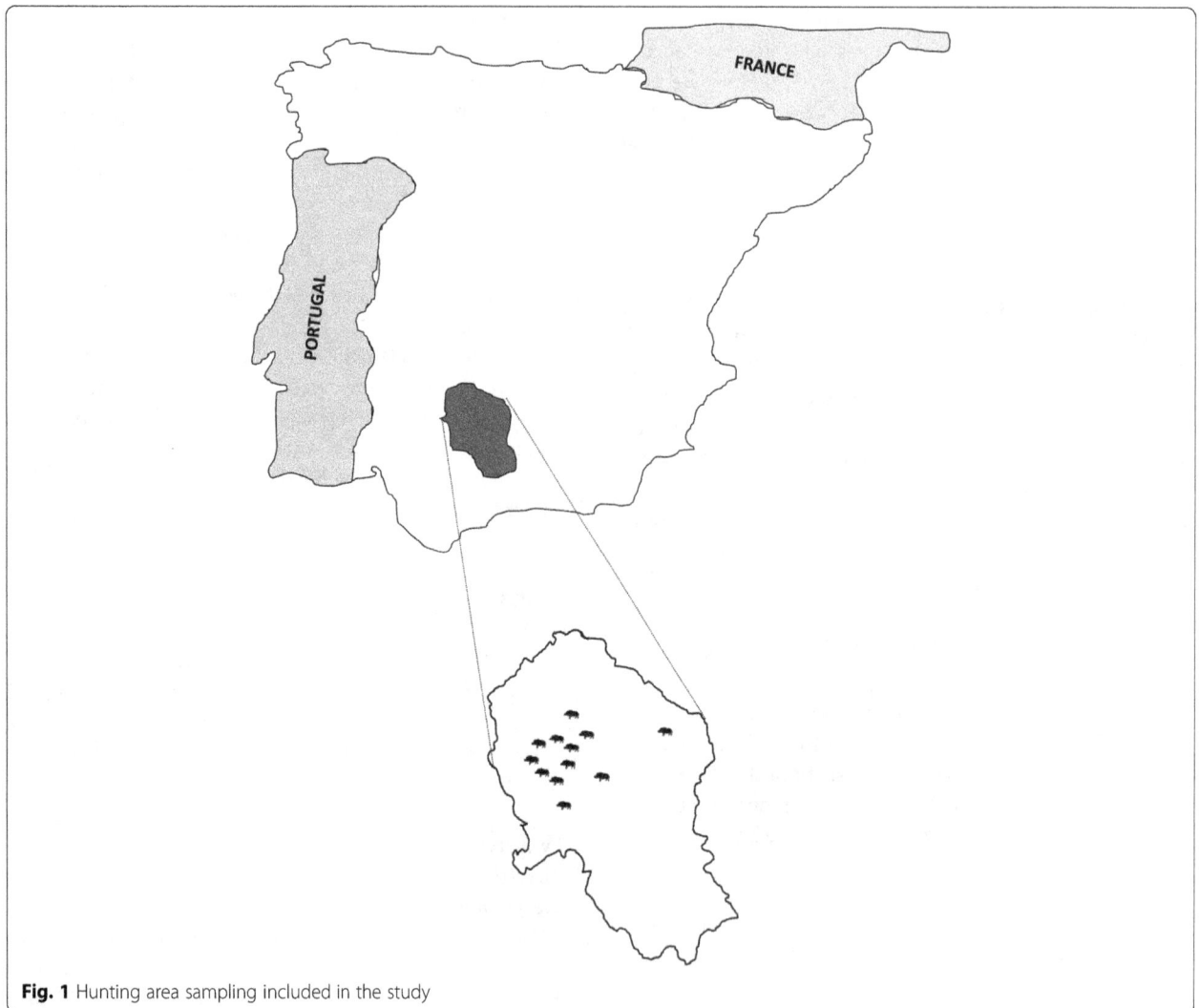

Fig. 1 Hunting area sampling included in the study

weeks of November. This finding suggests a possible seasonal pattern for HEV infection in this species.

Studies performed in Europe have reported a variable prevalence of HEV infection in wild boar, fluctuating between 2 and 68%. A study performed in North Germany showed an HEV prevalence of 5.3% (10/189) [15], while in Central Germany it was 15.2% (7/46) [16] and rose to 68% in other areas [17]. Elsewhere, countries such as Italy and the Netherlands have reported a prevalence of 9.4% (6/64) and 8% (8/106), respectively [18, 19], while in Estonia and Hungary, the reported prevalence was 17.2% (81/471) and 12% (9/74) [20, 21]. Our results are consistent with those previously reported in Central Spain, where 27 out of 138 (19.6%) animals tested positive for HEV [22]. Our study shows an HEV prevalence of 23.2%. Differences between studies could be associated with various factors, including the sensitivity and specificity of the RT-PCR assay employed. These studies however did not consider external factors to explain the differences. Other studies evaluated risk factors

associated with HEV infection in wild boar and identified host and environmental factors. Burri et al. studied 303 serum samples collected from wild boar killed between 2008 and 2012 in 10 different cantons in Switzerland [23]. That study reported a HEV seroprevalence of 12.5% and found that age (adults = 22.5%) and region of origin were factors associated with higher HEV seroprevalence [23]. Likewise, a study carried out in France found that the seroprevalence of HEV IgG antibodies was higher in the south (22.6%) than in the central part (9%) or the north (7.3%) [24]. In another study performed in Corsica, Jori et al. found that hunting season and age were risk factors for HEV seroprevalence [25]. Interestingly, hybrid wild boar showed higher HEV seroprevalence than pure wild boar and domestic farm pigs, suggesting they play an important role in the HEV reservoir [25]. Finally, using RT-PCR, Shielke et al. reported a higher prevalence of HEV infection in wild boars hunted in rural habitats than in urban areas [26], indicating that there may be a more efficient virus

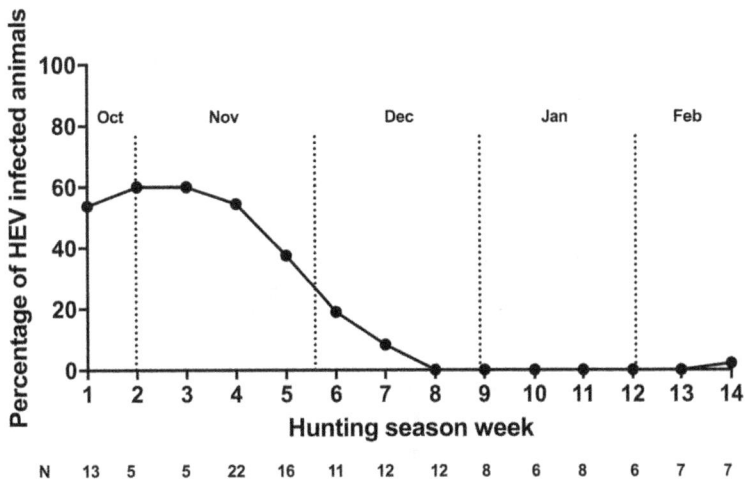

Fig. 2 The prevalence of hepatitis E virus during each week of the hunting season (Oct 15 to Feb 15)

spread in wild boar populations in rural settings. Our study did not find that either age or sex were risk factors for HEV infection, a finding previously reported by others [27]. Our findings did however suggest that other factors, such as the season, could affect the rate of HEV infection in wild boar. In this respect, our study found that the major peak of HEV infection in wild boar was October and November, and then decreased significantly during the rest of the hunting season until February.

Seasonal patterns of HEV infection have previously been described in humans from Asian countries and are clearly linked to environmental factors, such as monsoons and floods, which have a markedly seasonal behavior. In this respect, one study conducted in China showed that the cumulative number of cases of acute HEV is concentrated in the cold season [28]. In studies carried out in India, Pakistan and Nepal, peak HEV infection is linked to floods during the monsoon season [29]. In these countries, the cumulative number of cases can easily be explained as due to the principal route of HEV transmission, which is fecal-oral [8]. Likewise, the authors of a study conducted in China that included

farm pigs also described a seasonal pattern, with a major peak of HEV infection being reported in March–April in Eastern China, and a secondary peak in September–October in Southwest China [30]. Nevertheless, in countries where the main route of transmission in humans is via consumption of contaminated food, the seasonal behavior of the disease has not been well established and remains controversial. A study performed in the Southwest of England found that the highest number of cases occurred in the spring and summer [31], although the reason was unknown. By contrast, in another study carried out in France involving cases collected over a 5-year period, no seasonal variation in the number of cases of HEV infection cases was found at any time [32]. Our study suggests that there is a seasonal component in the prevalence of HEV infection in wild boar, with most cases concentrated in late autumn and gradually decreasing in early winter. This finding is striking and represents the first evidence of an environmental influence on HEV infection in a European country. The explanation for this, bearing in mind current knowledge of the epidemiology and pathogenesis of HEV, is unknown.

Table 1 Multivariate logistic regression model of HEV infection

Variable	Condition	N	HEV-infected	OR (95% CI)	P
Sex	Male	64	20	1.84 (0.69–4.92)	0.22
	Female	78	13	1	
Age	Adult	97	25	1.706 (0.55–5.23)	0.35
	Non-adult	45	8	1	
Hunting date	Oct 15 - Nov 15	23	13	44.73 (8.49–235.53)	< 0.001
	Nov 16 - Dec 15	49	16	28.09 (5.94–132.67)	< 0.001
	Dec 16 - Jan 15	78	3	1.03 (0.13–2.89)	0.99
	Jan 16 – Feb 15	34	1	1	

Legend: *N* number of animals, *HEV* hepatitis E virus, *OR* odds ratio, *95% CI* 95% confidence interval, *Oct* October, *Nov* November, *Dec* December, *Jan* January, *Feb* February

An important point is that the lower prevalence rate in our study coincided with the reproductive season, which is usually between late November and January, when there is extensive contact between animals and the risk of transmission would therefore be expected to be much higher. Our study in fact found the opposite; there was a very low rate of HEV infection in this period compared with the pre-reproductive season. This could be explained by a route of transmission in wild boar that is as yet unknown, as well as by direct contact between the animals, and this may occur in the first weeks of autumn. It should also be mentioned that at the beginning of this period, extensive Iberian pig farming in the southwest of the Iberian Peninsula occupies agricultural land associated with hunting areas, which leads to increased animal population densities, and spaces and resources being shared with other wild animals. This favors an interspecies transmission of pathogens that needs to be elucidated. These points require further investigation.

Several limitations should be noted in this work. Firstly, our study evaluated only the prevalence of HEV infection in a single hunting season and a single region, and we were therefore unable to establish whether the seasonal behavior observed in our study can be extrapolated to other areas or later hunting seasons. Finally, we did not include other environmental or behavioral factors that may have influenced HEV prevalence in these animals, which may explain the seasonality found in our study.

Conclusions

Our study found a relatively high prevalence of HEV infection in wild boar in the south of Spain. This finding suggests that the transmission of HEV infection to humans via meat consumption or contact with infected boar found in the wild may be an important factor. Nevertheless, our study suggests that prevalence may depend on the season when the animal was hunted, and the potential risk of zoonotic transmission may therefore fluctuate.

Abbreviations

95% CI: 95% confidence interval; Dec: December; Feb: February; HEV: Hepatitis E virus; Jan: January; N: Number of animals; Nov: November; Oct: October; OR: Odds ratio

Acknowledgements
The authors are grateful to Ismael Zafra and Laura Ruiz for their technical support.

Funding
This work was supported by the Ministerio de Sanidad (RD12/0017/0012) integrated in the Plan Nacional de I + D + I and cofinanced by the ISCIII-Subdirección General de Evaluación and the Fondo Europeo de Desarrollo Regional (FEDER), and Fundación para la Investigación en Salud (FIS) del Instituto Carlos III (PI16/01297). The funders did not play any role in the design, conclusions or interpretation of the study.

Authors' contributions
Dr. Rivero-Juarez had full access to all of the data in the study and takes responsibility for the integrity of the data and the accuracy of the data analysis. *Study concept and design:* ARJ, MAR, MF, IGB, AR. *Analysis and interpretation of the data:* ARJ, MF, IGB. *Drafting of the manuscript:* ARJ, AR. *Critical revision of the manuscript for important intellectual content:* All authors. *Statistical analysis:* ARJ. *Obtained funding:* ARJ. Acquisition of animal samples and data: DCT, SJR, PLL Perform the experiment and technical procedures: ARJ, DCT, SJR, PLL. *Study supervision:* AR. All authors read and approved the final manuscript.

Ethics approval
This study did not involve purposeful killing of animals. Professional personnel collected blood and liver samples mostly from hunted-harvested wild boar during the hunting season. These animals were legally hunted under Spanish and EU legislation and all hunters had hunting licenses. No ethical approval was deemed necessary; all collection of samples was performed for routine procedures before the design of this study in compliance with the Ethical Principles in Animal Research. Thus, blood or liver samples were not collected specifically for this study. Protocols, amendments and other resources were all done according to the guidelines approved by each Autonomous government following the R.D.1201/2005 of the Ministry of Presidency of Spain.

Consent for publication
Not applicable.

Competing interests
We declare no competing interests. At no time have the authors or their institution received payment or services from a third party for any aspect of the submitted work (data monitoring board, study design, manuscript preparation, statistical analysis, and so on).

Author details
[1]Infectious Diseases Unit. Instituto Maimonides de Investigación Biomédica de Córdoba (IMIBIC), Hospital Universitario Reina Sofía de Córdoba. Universidad de Córdoba, 2° Floor. Box 134.Avenida Menendez Pidal s/n, 14004 Córdoba, Spain. [2]Animal Health Department. Veterinary Science College, Universidad de Córdoba, 14014 Cordoba, Spain. [3]Animal Pathology Department. Veterinary Science College, Universidad de Córdoba, Cordoba, Spain. [4]Unidad de Enfermedades Infecciosas. Hospital Provincial, Complejo Hospitalario reina Sofía de Córdoba, Avenida Menendez Pidal s/n, 14006 Cordoba, Spain.

References
1. Dalton HR, Bendall R, Ijaz S, Banks M. Hepatitis E: an emerging infection in developed countries. Lancet Infect Dis. 2008;8:698–709.
2. Hartl J, Otto B, Madden RG, Webb G, Woolson KL, Kriston L, Vettorazzi E, Lohse AW, Dalton HR, Pischke S. Hepatitis E Seroprevalence in Europe: a meta-analysis. Viruses. 2016;8:E211.
3. Doceul V, Bagdassarian E, Demange A, Pavio N. Zoonotic hepatitis E virus: classification, animal reservoirs and transmission routes. Viruses. 2016;8:E270.
4. Lhomme S, Top S, Bertagnoli S, Dubois M, Guerin JL, Izopet J. Wildlife reservoir for hepatitis E virus, southwestern France. Emerg Infect Dis. 2015;21:1224–6.
5. Pavio N, Meng XJ, Doceul V. Zoonotic origin of hepatitis E. Curr Opin Virol. 2015;10:34–41.
6. Rivero-Juarez A, Frias M, Martinez-Peinado A, Risalde MA, Rodriguez-Cano D, Camacho A, García-Bocanegra I, Cuenca-Lopez F, Gomez-Villamandos JC, Rivero A. Familial hepatitis E outbreak linked to wild boar meat consumption. Zoonoses Public Health. 2017;64:561–5.
7. Schlosser J, Eiden M, Vina-Rodriguez A, Fast C, Dremek P, Lange E, Ulrich RG, Groschup MH. Natural and experimental hepatitis E virus genotype 3-infection in European wild boar is transmissible to domestic pigs. Vet Res. 2014;45:121.
8. Hoofnagle JH, Nelson KE, Purcell RH. Hepatitis E. N Engl J Med. 2012;367:1237–44.
9. Rivero-Juarez A, Martinez-Dueñas L, Martinez-Peinado A, Camacho A, Cifuentes C, Gordon A, Frias M, Torre-Cisneros J, Pineda JA, Rivero A. High hepatitis E virus seroprevalence with absence of chronic infection in HIV-infected patients. J Inf Secur. 2015;70:624–30.

10. Weller R, Todt D, Engelmann M, Friesland M, Wedemeyer H, Pietschmann T, Steinmann E. Apolipoprotein E polymorphisms and their protective effect on hepatitis E virus replication. Hepatology. 2016;64:2274–6.

11. Mansuy JM, Gallian P, Dimeglio C, Saune K, Arnaud C, Pelletier B, Morel P, Legrand D, Tiberghien P, Izopet J. A nationwide survey of hepatitis E viral infection in French blood donors. Hepatology. 2016;63:1145–54.

12. Hunter JG, Madden R, Stone A, Osborne N, Wheeler B, Vine L, Dickson A, Barlow M, Lewis J, Bendall RP. Coastal clustering of HEV; Cornwall, UK. Eur J Gastroenterol Hepatol. 2015;28:323–7.

13. Arenas-Montes A, García-Bocanegra I, Paniagua J, Franco JJ, Miró F, Fernández-Morente M, Carbonero A, Arenas A. Blood sampling by puncture in the cavernous sinus from hunted wild boar. Eur J Wild Res. 2013;59:299–303.

14. Abravanel F, Sandres-Saune K, Lhomme S, Dubois M, Mansuy JM, Izopet J. Genotype 3 diversity and quantification of hepatitis E virus RNA. J Clin Microbiol. 2012;50:897–902.

15. Schielke A, Ibrahim V, Czogiel I, Faber M, Schrader C, Dremsek P, Ulrich RG, Johne R. Hepatitis E virus antibody prevalence in hunters from a district in Central Germany, 2013: a cross-sectional study providing evidence for the benefit of protective gloves during disembowelling of wild boars. BMC Infect Dis. 2015;15:440.

16. Kaci S, Nöckler K, Johne R. Detection of hepatitis E virus in archived German wild boar serum samples. Vet Microbiol. 2008;128:380–5.

17. Adlhoch C, Wolf A, Meisel H, Kaiser M, Ellerbrok H, Pauli G. High HEV presence in four different wild boar populations in east and West Germany. Vet Microbiol. 2009;139:270–8.

18. Mazzei M, Nardini R, Verin R, Forzan M, Poli A, Tolari F. Serologic and molecular survey for hepatitis E virus in wild boar (Sus Scrofa) in Central Italy. New Microbes New Infect. 2015;7:41–7.

19. Rutjes SA, Lodder-Verschoor F, Lodder WJ, van der Giessen J, Reesink H, Bouwknegt M, de Roda Husman AM. Seroprevalence and molecular detection of hepatitis E virus in wild boar and red deer in The Netherlands. J Virol Methods. 2010;168:197–206.

20. Ivanova A, Tefanova V, Reshetnjak I, Kuznetsova T, Geller J, Lundkvist Å, Janson M, Neare K, Velström K, Jokelainen P, Lassen B, Hütt P, Saar T, Viltrop A, Golovljova I. Hepatitis E virus in domestic pigs, wild boars, pig farm workers, and hunters in Estonia. Food Environ Virol. 2015;7:403–12.

21. Reuter G, Fodor D, Forgách P, Kátai A, Szucs G. Characterization and zoonotic potential of endemic hepatitis E virus (HEV) strains in humans and animals in Hungary. J Clin Virol. 2009;44:277–81.

22. de Deus N, Peralta B, Pina S, Allepuz A, Mateu E, Vidal D, Ruiz-Fons F, Martín M, Gortázar C, Segalés J. Epidemiological study of hepatitis E virus infection in European wild boars (Sus scrofa) in Spain. Vet Microbiol. 2008;129:163–70.

23. Burri C, Vial F, Ryser-Degiorgis MP, Schwermer H, Darling K, Reist M, Wu N, Beerli O, Schöning J, Cavassini M, Waldvogel A. Seroprevalence of hepatitis E virus in domestic pigs and wild boars in Switzerland. Zoonoses Public Health. 2014;61:537–44.

24. Carpentier A, Chaussade H, Rigaud E, Rodriguez J, Berthault C, Boué F, Tognon M, Touzé A, Garcia-Bonnet N, Choutet P, Coursaget P. High hepatitis E virus seroprevalence in forestry workers and in wild boars in France. J Clin Microbiol. 2012;50:2888–93.

25. Jori F, Laval M, Maestrini O, Casabianca F, Charrier F, Pavio N. Assessment of domestic pigs, wild boars and feral hybrid pigs as reservoirs of hepatitis E virus in Corsica, France. Viruses. 2016;8(8):236.

26. Schielke A, Sachs K, Lierz M, Appel B, Jansen A, Johne R. Detection of hepatitis E virus in wild boars of rural and urban regions in Germany and whole genome characterization of an endemic strain. Virol J. 2009;6:58.

27. Weigand K, Weigand K, Schemmerer M, Müller M, Wenzel JJ. Hepatitis E Seroprevalence and genotyping in a cohort of wild boars in southern Germany and eastern Alsace. Food Environ Virol. 2017; https://doi.org/10.1007/s12560-017 9329-x.

28. Zhu FC, Huang SJ, Wu T, Zhang XF, Wang ZZ, Ai X, Yan Q, Yang CL, Cai JP, Jiang HM, Wang YJ, Ng MH, Zhang J, Xia NS. Epidemiology of zoonotic hepatitis E: a community-based surveillance study in a rural population in China. PLoS One. 2014;9:e87154.

29. Khuroo MS, Khuroo MS, Khuroo NS. Transmission of hepatitis E virus in developing countries. Viruses. 2016;8:E253.

30. Lu YH, Qian HZ, Hu AQ, Qin X, Jiang QW, Zheng YJ. Seasonal pattern of hepatitis E virus prevalence in swine in two different geographical areas of China. Epidemiol Infect. 2013;141:2403–9.

31. Dalton HR, Stableforth W, Thurairajah P, Hazeldine S, Remnarace R, Usama W, Farrington L, Hamad N, Sieberhagen C, Ellis V, Mitchell J, Hussaini SH, Banks M, Ijaz S, Bendall RP. Autochthonous hepatitis E in Southwest England: natural history, complications and seasonal variation, and hepatitis E virus IgG seroprevalence in blood donors, the elderly and patients with chronic liver disease. Eur J Gastroenterol Hepatol. 2008;20:784–90.

32. Mansuy JM, Abravanel F, Miedouge M, Mengelle C, Merviel C, Dubois M, Kamar N, Rostaing L, Alric L, Moreau J, Peron JM, Izopet J. Acute hepatitis E in south-west France over a 5-year period. J Clin Virol. 2009;44:74–7.

8

Tick-borne encephalitis in a naturally infected sheep

Brigitte Böhm[1]*[iD], Benjamin Schade[1], Benjamin Bauer[1], Bernd Hoffmann[2], Donata Hoffmann[2], Ute Ziegler[3], Martin Beer[2], Christine Klaus[4], Herbert Weissenböck[5] and Jens Böttcher[1]

Abstract

Background: Tick-borne encephalitis (TBE) is the most important viral tick borne zoonosis in Europe. In Germany, about 250 human cases are registered annually, with the highest incidence reported in the last years coming from the federal states Bavaria and Baden-Wuerttemberg. In veterinary medicine, only sporadic cases in wild and domestic animals have been reported; however, a high number of wild and domestic animals have tested positive for the tick-borne encephalitis virus (TBEV) antibody.

Case presentation: In May 2015, a five-month-old lamb from a farm with 15 Merino Land sheep and offspring in Nersingen/Bavaria, a TBEV risk area, showed impaired general health with pyrexia and acute neurological signs. The sheep suffered from ataxia, torticollis, tremor, nystagmus, salivation and finally somnolence with inappetence and recumbency. After euthanasia, pathological, histopathological, immunohistochemical, bacteriological, parasitological and virological analyses were performed. Additionally, blood samples from the remaining, healthy sheep in the herd were taken for detection of TBEV antibody titres. At necropsy and accompanying parasitology, the sheep showed a moderate to severe infection with Trichostrongylids, Moniezia and Eimeria species. Histopathology revealed mild to moderate necrotising, lymphohistiocytic and granulocytic meningoencephalitis with gliosis and neuronophagia. Immunohistochemistry for TBEV was negative. RNA of a TBEV strain, closely related to the Kumlinge A52 strain, was detected in the brain by quantitative reverse transcriptase polymerase chain reaction (RT-qPCR) and subsequent PCR product sequencing. A phylogenetic analysis revealed a close relationship to the TBEV of central Europe. TBEV was cultured from brain tissue. Serologically, one of blood samples from the other sheep in the herd was positive for TBEV in an enzyme-linked immunosorbent assay (ELISA) and in a serum neutralisation test (SNT), and one was borderline in an ELISA.

Conclusion: To the authors' knowledge this is the first report of a natural TBEV infection in a sheep in Europe with clinical manifestation, which describes the clinical presentation and the histopathology of TBEV infection.

Keywords: Tick-borne encephalitis virus, TBEV, Histopathology, RT-qPCR, ELISA, *Ixodes ricinus*

Background

Tick-borne encephalitis (TBE), caused by tick-borne encephalitis virus (TBEV), a *Flavivirus*, has been recognised for decades in Europe and some parts of Asia as an important viral zoonosis with between 5352 (2008) and up to 12,733 (1996) human cases annually [1]. In almost all cases, TBEV is transmitted by two types of ticks: *Ixodes (I.) ricinus*, which are found mainly in Western Europe and in Germany, and *I. persulcatus*, which are found in Eastern Europe and Siberia. However, an alimentary infection via raw milk or raw milk products from ruminants in the viraemic phase may also occur rarely [2, 3]. In Germany, TBE is a notifiable human disease with an average of 250 cases per year reported annually [4, 5]. TBEV circulates between vector ticks and competent hosts in so-called natural foci in a patchy distribution, whose geographic extension as a rule is strictly limited and can be very small [6]. In Germany, TBE risk areas are defined by the incidence of autochthonous human cases as the official statistical tool and are published every year [4, 5]. The TBEV infection risk in humans seems to be influenced by a combination

* Correspondence: brigitte.boehm@tgd-bayern.de
[1]Bavarian Animal Health Service, Senator-Gerauer-Straße 23, 85586 Poing, Germany
Full list of author information is available at the end of the article

of landscape and climatic variables as well as host-species dynamics [7]. Nearly 90% of all human cases occurred in the federal states of Bavaria and Baden-Wuerttemberg.

In veterinary medicine, clinical cases of TBE with neurological symptoms are rarely described, but have been reported in dogs for more than 30 years [8–10], as well as in horses [11–13] and monkey [14, 15]. In ruminants, like cows, goats and sheep as well as wild species such as roe deer and red foxes, TBEV antibody titres are observed without clinical symptoms [3, 16–19]. Very rarely, single clinical cases were registered, for example, in a goat in Switzerland [20] and a mouflon (*Ovis ammon musimon*) in Austria [21].

In this report, we describe the clinical manifestation of TBEV infection in a sheep with severe neurological symptoms, raised in a pasture in a district in Bavaria (GMS N 48°; O 10°; 456 m) defined as a TBE risk area [5]. Potential differential etiologies were ruled out (*Listeria monocytogenes*, *Borna disease virus*, *West Nile virus*, *Louping ill virus*).

Case presentation

In May 2015, a five-month-old lamb from a farm with 15 Merino Land sheep showed impaired general health with pyrexia (41.2 °C) and acute neurological signs. The sheep suffered from ataxia, torticollis, tremor, nystagmus, salivation and finally somnolence with inappetence and recumbency. The veterinarian suspected listeriosis. The remaining sheep were healthy. The animal was euthanised and submitted for pathological examination. Samples of the brain stem, cerebellum, medulla oblongata, pons, cerebral cortex, hippocampus and thalamus were fixed in 10% buffered formalin for 48 h and subsequently embedded in paraffin wax. Sections of 4 μm thickness were mounted on glass slides and stained with haematoxylin and eosin (HE) and examined histologically.

A swab was taken from the brain stem and cultured for the suspected infection with *Listeria monocytogenes*. Twenty-four hours of aerobic incubation at 37 °C on Columbia Agar supplemented with 5% sheep's blood as well as enrichment on Listeria enrichment broth and Fraser enrichment broth with subsequent use of Oxford Listeria agar and Brilliance™ Listeria agar (Oxoid, Wesel, Germany) showed no evidence of a Listeria infection.

Immunohistochemistry for TBEV antigen with a polyclonal Western-TBEV antibody (Dilution: 1:3000; source: Department of Virology; Medical University of Vienna) was performed with an automated Immunostainer (Thermo Autostainer 360-2D; Thermo-Fisher Scientific, Fremont; CA) using the UltraVision LP detection system (Thermo-Fisher) and DAB Plus as chromogen (Thermo-Fisher). As positive control brain of a TBEV-infected dog was used.

For parasitological analysis of the faeces, a flotation method with a zinc-chloride solution (specific gravity of 1.35) was used.

For detection of TBEV-specific RNA, two independent RT-qPCR protocols according to Schwaiger and Cassinotti [22], and adapted by Klaus et al. [23, 24], were applied. The TBEV strain Leila BH 95–15 was isolated from the brain material using BHK21 cells (Collection of Cell Lines in Veterinary Medicine (CCLV) 179, Friedrich-Loeffler-Institute, Greifswald-Insel Riems, Germany). The RNA of the brain sample was used for direct sequencing of the TBEV strain Leila BH 95–15. The complete coding sequence was generated by the use of primer-based Sanger sequencing (primer sequences available upon request). For phylogenetic analysis multiple alignments of 23 selected TBEV strains, Langat virus and the TBEV sequence from sheep Leila_BH95–15 were performed using the MAFFT method [25]. The evolutionary history was inferred using the neighbour-joining method [26]. The optimal tree with the sum of branch length = 1.19366931 is shown. The percentage of replicate trees in which the associated taxa clustered together in the bootstrap test (500 replicates) is shown next to the branches [27]. The tree is drawn to scale, with branch lengths in the same units as those of the evolutionary distances used to infer the phylogenetic tree. The evolutionary distances were computed using the Kimura 2-parameter method [28] and are in the units of the numbers of base substitutions per site. The rate variation among sites was modelled with a gamma distribution (shape parameter = 1). The analysis involved 25 nucleotide sequences. All positions containing gaps and missing data were eliminated. There were a total of 10,242 positions in the final dataset. Evolutionary analyses were conducted in MEGA6 [29]. Blood samples from the whole flock of sheep ($n = 21$) were collected and tested for TBEV antibodies by ELISA and serum neutralization test (SNT). A commercially available ELISA (Immunozym® FSME IgG all species, Progen Biotechnik GmbH, Heidelberg, Germany) was used, and sera were analysed according to the manufacturer's protocol. Data were expressed as Vienna units (VIEU/ml). Values between 63 and 126 VIEU/ml were interpreted as borderline; lower and higher values were negative and positive, respectively.

In order to avoid false positive results of TBEV antibody titres detected in ELISA, all positive ELISA results of field collected samples were confirmed by the SNT as a gold standard according to a modified version of that established by Holzmann et al. [30]. For the SNT, the low pathogenic strain Langat was used with 100 $TCID_{50}$/well. The virus titre used was confirmed by re-titrations. Serum samples were titrated in triplicates starting at a dilution of 1:5 in Minimum Essential Medium (MEM) Earle's medium, and after a 24-h incubation time (37 °C), BHK-21 cell suspension was added to the virus-serum-sample and incubated at 37 °C for an additional 4 days. Subsequently, virus replication was detected by immunofluorescence analysis using a TBEV specific rabbit-antiserum. Titres were expressed as the reciprocal of dilutions that caused 50% neutralisation (ND50).

At necropsy and accompanying parasitology, the sheep showed a moderate to severe infestation with Trichostrongylids, Moniezia and Eimeria species. Histopathology of the brain revealed a mild to moderate, necrotising, lymphohistiocytic and granulocytic meningoencephalitis. The most severe inflammatory changes were noticed in the thalamus, hypothalamus and mesencephalon, followed by less severe lesions in the cerebral cortex, cerebral white matter and hippocampus as well as in the medulla oblongata, pons, cerebellum and leptomeninges. Frequent features were degeneration and necrosis of neurons surrounded by neuronophagic nodules as well as multifocal foci of gliosis. Inflammatory perivascular cuffs in the neuroparenchyma and diffuse inflammatory infiltrates in the leptomeninges were composed of mononuclear, lymphohistiocytic cells and a few neutrophils (Figs. 1a–d). Cytoplasmic or intranuclear inclusion bodies were not found. The cultured swab of the brain, especially for *L. monocytogenes*, was negative. Immunohistochemistry results for TBEV antigen detection were negative. The positive control (brain of a TBEV-infected dog) showed the expected specific staining. By TBEV quantitative RT-PCR, viral RNA was detected in brain samples of different locations and ascertained Cq values of 21 to 25. PCR analysis for Borna disease virus, West Nile virus and Louping ill virus scored negative. The brain sample was used for nearly complete sequencing of the TBEV genome (GenBank accession number KU884607). BlastN search revealed the highest nucleotide similarities of 98.63% for TBEV strain Kumlinge A52 (10,897 of 11,048 nt identical). The phylogenetic analysis confirmed the closest relation of the TBEV genome of the lamb to the classical European TBEV strains (Fig. 2). Virus isolation using BHK21 cells finally resulted, in the third passage, in a high-titre stock (10^7 $TCID_{50}$ ml^{-1}) of the sheep isolate Leila BH95–15. Serum samples from individual sheep of the herd were tested for TBEV specific antibodies. One sheep serum of the flock tested positive for TBEV antibodies, and one tested as borderline for TBEV antibodies. The positive one of these two sera was confirmed as specific TBEV positive in SNT (ND50 1:30).

TBEV infection was diagnosed based on the clinical symptoms, the histopathology of the brain and the detection of TBEV in the brain via RT-qPCR and subsequent sequencing.

Discussion and conclusions

To the authors' knowledge, this is the first report of a natural TBEV infection in a sheep in Europe with

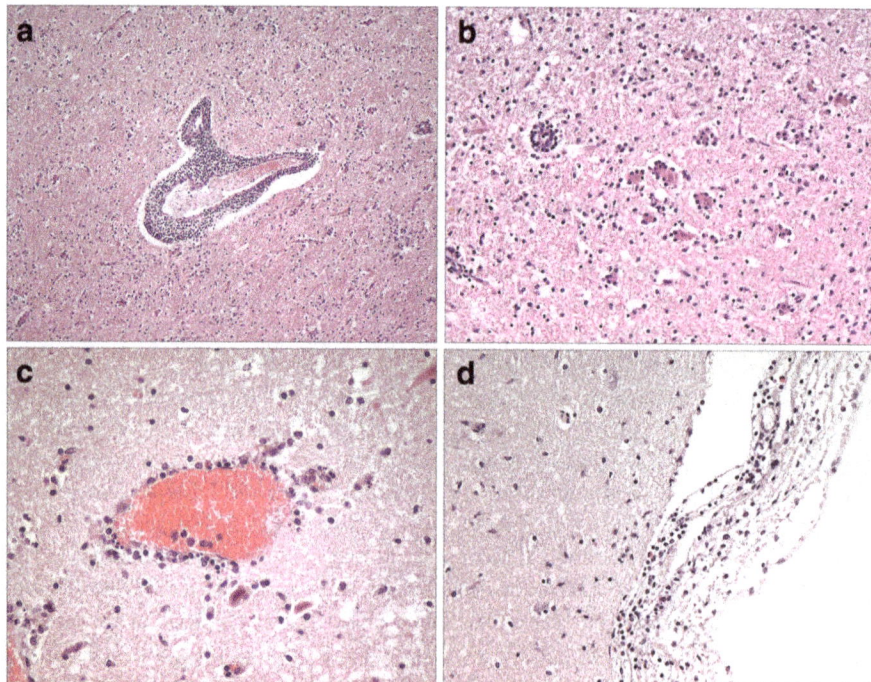

Fig. 1 a Brain, thalamus, severe lymphohistiocytic perivascular infiltration of neuroparenchyma and gliosis, HE, 100 x. **b** Brain, thalamus, necrosis of neurons surrounded by neuronophagic nodules, HE, 200 x. **c** Brain, cerebral cortex, moderate lymphohistiocytic and neutrophil granulocytic perivascular infiltration of neuroparenchyma, HE, 400 x. **d** Leptomeninx, moderate, diffuse, lymphohistiocytic and neutrophil granulocytic infiltration, HE, 200 x

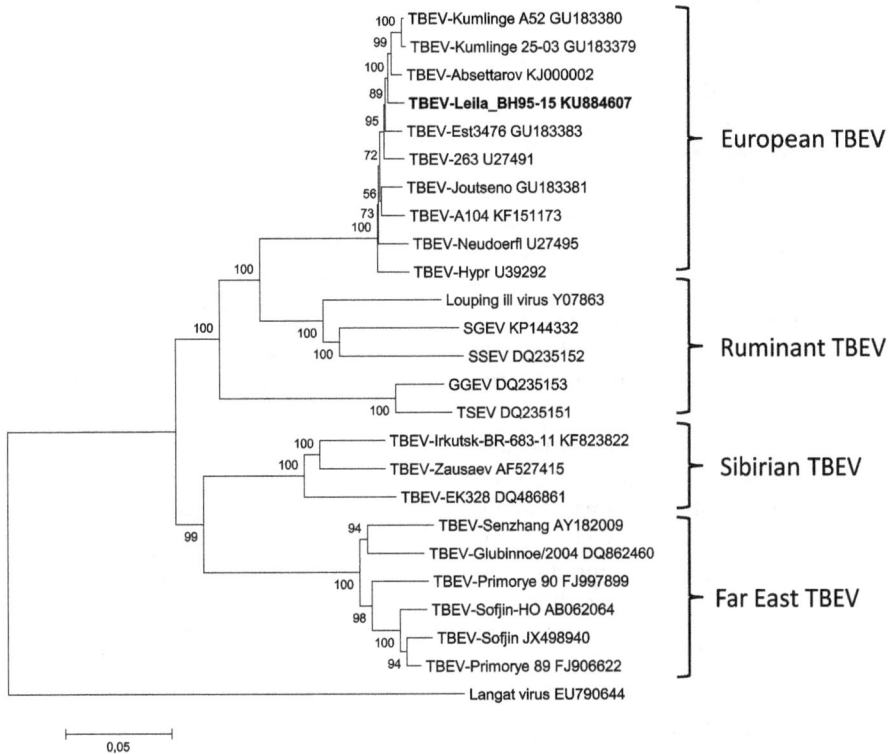

Fig. 2 Phylogenetic analysis was performed with whole genome sequences and showed that TBEV Leila_BH95–15 belongs to the Eurasian TBEV clade. The optimal tree with the sum of branch length = 1.19366931 is shown

clinical manifestation. Single cases of TBEV infection have been described in dogs [8–10], horses [11–13], small ruminants [20, 21, 31, 32] and a monkey [14, 15]. The number of unreported/undetected cases remains unclear as the clinical signs, albeit neurological, are non-specific and maybe interpreted by owners and veterinarians most likely as a *L.monocytogenes* infection. Since several animals of the flock had ticks, it can be assumed that the sheep was infected at the pasture by a tick bite. The species in this specific case was not determined, but large domestic animals, such as goats, sheep and cattle, are potential hosts for *I. ricinus*. Possible TBEV prevalence in this area was also confirmed by the TBEV antibody titre of one further sheep in the flock. Interestingly, during experimental inoculation using TBEV sheep and goats showed almost no clinical symptoms except fever [33, 34]. Only a minority of TBEV-positive small ruminants displayed prominent neurological signs [20, 31, 32]. Additionally, Bagó et al. [21] reported only a moribund condition of the TBEV-positive mouflon. Further investigations are necessary to evaluate the relevant factors, such as host immunity, co-infections (e.g. parasitic infestation), age, TBEV strain and virus load, causing clinical signs in small ruminants.

Neuropathological lesions of this case resemble these in human beings [35] and in wild and domestic animals

[10, 11, 14, 21] after TBEV infection. A glial shrubbery as described by Bagó et al. [21] and Weissenböck et al. [10] was not observed. The presence of a pronounced neuronophagia and numerous resulting glial nodules corresponds well with a protracted disease course. This may explain the negative outcome of immunohisto-chemical antigen detection because rapid virus clearance is a typical feature of TBE [10]. Borna disease-, rabies-, scrapie-, listeriosis associated histopathological lesions were not found. Other potential flavivirus infections with similar histopathological lesions for example Louping ill disease (dramatic clinical course, up to now not yet reported in Germany) and West Nile Encephalitis (not yet reported in Germany) were ruled out by PCR.

Sheep or goat flocks, especially those that are used for the production of raw milk or raw milk products and live on pastures with possible tick contact, should be tested for TBEV, because these can be the source of transmission of so-called alimentary TBE to humans. Experimental investigations showed that during the viraemic stage, the virus is excreted in milk for 3–7 days and consumption of non-pasteurised milk or milk products might lead to infection [16, 33, 36, 37]. In the past few years, alimentary TBE cases in humans were reported from Slovakia, Poland, Estonia, Austria and Hungary [2, 3, 38–41]. In Germany the first case of an alimentary TBEV infection was reported in June 2016 [42]. Two persons in Reutlingen, a TBE risk area of Baden-Wuerttemberg, which is 68 km away from the location of the reported case, were affected by consumption of raw goat milk products. However, individual cases or small group outbreaks in humans may be possible. Therefore, consumption of raw milk should be avoided to reduce the risk of human infection. In Hungary, after an outbreak of TBEV in a milking goat, flock animals were successfully immunised against TBEV to allow further raw milk production connected with grazing on pastures [34]. Besides the very rare clinical symptoms in sheep and goats, serological investigations can provide information about possible infection with TBEV in an area. It is an inexpensive method to improve consumer's protection against TBE in those cases when milking goats and sheep (very seldom cows), especially in TBE risk areas, but also in areas with only single human TBE cases [43]. As the viraemic period is very short and clinical symptoms might be absent or misinterpreted, testing of TBEV specific antibodies could highlight the TBEV specific epidemiological situation of a flock.

For neurological symptoms in sheep, especially in TBE risk areas, TBEV infection should be taken into consideration, especially in cases where other diseases with neurological symptoms were excluded, for example Louping ill disease, Borna disease, rabies, scrapie, listeriosis, tetanus, malnutrition and intoxications.

Abbreviations
DAB: Diaminobenzidin; ELISA: Enzyme-linked Immunosorbent assay; HE: Hematoxylin and eosin; I.: Ixodes; L.: Listeria; RNA: Ribonucleic acid; RT-qPCR: Quantitative reverse transcriptase Polymerase chain reaction; SNT: Serum neutralisation test; TBE: Tick-borne encephalitis; TBEV: Tick-borne encephalitis virus; VIEU: Vienna units

Acknowledgements
The authors thank Astrid Nagel, Birgit Isele-Rüegg, Fritz Fassler, Thomas Wagner, Regine Forster, Christian Korthase and Mareen Lange for excellent technical assistance.

Funding
Parts of this work were financially supported by the Free State of Bavaria and the Bavarian Joint Founding Scheme for the Control and Eradication of Contagious Lifestock Diseases. The funding body had no influence on the design of the study and collection, analysis and interpretation of data and on writing the manuscript.

Authors' contributions
BB examined the lamb pathologically and histopathologically and drafted the manuscript. BS participated in the histopathological examination of the lamb. B. Bauer examined the lamb clinically. BH, DH and MB performed the virological analysis of the lamb. UZ performed the WNV PCR. CK performed the molecular biological analysis and examined the flat serologically. HW performed the immunohistochemistry. JB examined the flat serologically. B. Böhm and JB led the coordination of the examinations. All authors revised the manuscript, read and approved the final version of the manuscript.

Consent for publication
Not applicable.

Competing interests
The authors confirm that no competing financial interests exist.

Author details
[1]Bavarian Animal Health Service, Senator-Gerauer-Straße 23, 85586 Poing, Germany. [2]Institute of Diagnostic Virology, Friedrich-Loeffler-Institute, Südufer 10, 17493 Greifswald-Insel Riems, Germany. [3]Institute of Novel and Emerging Infectious Diseases, Friedrich-Loeffler-Institute, Südufer 10, 17493 Greifswald-Insel Riems, Germany. [4]Institute of bacterial Zoonoses and Infections, Friedrich-Loeffler-Institute, Naumburger Straße 96 a, 07743 Jena, Germany. [5]Institute of Pathology and Forensic Veterinary Medicine, University of Veterinary Medicine, Veterinärplatz 1, 1210 Vienna, Austria.

References
1. Süss J. Tick-borne encephalitis 2010: epidemiology, risk areas, and virus strains in Europe and Asia – an overview. Ticks Tick Borne Dis. 2011;2:2–15.
2. Holzmann H, Aberle SW, Stiasny K, Werner P, Mischak A, Zainer B, Netzer M, Koppi S, Bechter E, Heinz FX. Tick-borne encephalitis from eating goatcheese in a mountain region of Austria. Emerg Infect Dis. 2009;15:1671–3.
3. Balogh Z, Ferenczi E, Szeles K, Stefanoff P, Gut W, Szomor K, Takacs M, Berencsi G. Tick-borne encephalitis outbreak in Hungary due to consumption of raw goat milk. J Virol Methods. 2010;163:481–5.
4. Robert Koch-Institute. FSME: Risikogebiete in Deutschland. Epid Bull. 2007; 15:129–35.
5. Robert Koch-Institute. FSME: Risikogebiete in Deutschland. Epid Bull. 2016;2:151–62.
6. Kupča AM, Essbauer S, Zoeller G, de Mendonca PG, Brey R, Rinder M, Pfister K, Spiegel M, Doerrbecker B, Pfeffer M, Dobler G. Isolation and molecular characterization of a tick-borne encephalitis virus strain from a new tick-borne encephalitis focus with severe cases in Bavaria,Germany. Ticks Tick Borne Dis. 2010;1:44–51.
7. Kiffner C, Zucchini W, Schomaker P, Vor T, Hagedorn P, Niedrig M, Rühe F. Determinants of tick-borne encephalitis in counties of southern Germany, 2001-2008. Int J Health Geogr. 2010;9:42–52.
8. Leschnik MW, Kirtz GC, Thalhammer JG. Tick-borne encephalitis (TBE) in dogs. Int J Med Microbiol. 2002; 291Suppl 33:66–9.
9. Tipold A, Fatzer R, Holzmann H. Zentraleuropäische Zeckenencephalitis beim Hund. Kleintierpraxis. 1993;38:619–28.
10. Weissenböck H, Suchy A, Holzmann H. Tick-borne encephalitis in dogs: neuropathological findings and distribution of antigen. Acta Neuropathol. 1998;95:361–6.
11. Waldvogel K, Matile H, Wegmann C, Wyler R, Kunz C. Zeckenenzephalitis beim Pferd. Schweiz Arch Acta Neuropathol Tierheilkd. 1981;123:227–33.
12. Grabner A. Klinische Differentialdiagnose infektiös bedingter Krankheiten des ZNS beim Pferd. Collegium Veterinarium. 1993;24:27–31.
13. Luckschander N, Kölbl S, Enzesberger O, Zipko HT, Thalhammer JG. Frühsommermeningoenzephalitis-(FSME)Infektion in einer österreichischen Pferdepopulation. Tierarztl Prax Aus G Grosstiere Nutztiere. 1999;27:235–8.
14. Süss J, Gelpi E, Klaus C, Bagon A, Liebler-Tenorio EM, Budka H, Stark B, Müller W, Hotzel H. Tickborne encephalitis in naturally exposed monkey (Macaca sylvanus). Emerg Infect Dis. 2007;13:905–7.
15. Süss J, Dobler G, Zöller G, Essbauer S, Pfeffer M, Klaus C, Liebler-Tenorio EM, Gelpi E, Stark B, Hotzel H. Genetic characterisation of a tick-borne encephalitis virus isolated from the brain of a naturally exposed monkey (Macaca sylvanus). Int J Med Microbiol. 2008;298(Suppl 1):295–300.
16. Grešíková M. Excretion of the tick-borne encephalitis virus in the milk of subcutaneously infected cows. Acta Virol. 1958;2:188–92.
17. Nosek J, Kožuch O, Ernek E, Lichard M. The importance of goats in the maintenance tick-borne encephalitis virus in nature. Acta Virol. 1967;11:470–2.
18. Kiffner C, Vor T, Hagedorn P, Niedrig M, Rühe F. Determinants of tcik-borne encephalitis virus antibody presence in roe deer (Capreolus Capreolus) sera. Med Vet Entomol. 2012;26:18–25.
19. Rieger MA, Nübling M, Müller W, Hasselhorn H-M, Hoffmann F. TBE in foxes

study group. Foxes as an indicator for TBE endemicity – a comparative serological investigation. Zentralblatt für Bakteriologie. 1999;289:610–8.

20. Zindel W, Wyler R. Zeckenenzephalitis bei einer Ziege im Prättigau. Schweiz Arch Tierheilkd. 1983;125:383–6.

21. Bagó Z, Bauder B, Kolodziejek J, Nowotny N, Weissenböck H. Tickborne encephalitis in a mouflon (Ovis Ammon Musimon). Vet Rec. 2002;150: 218–20.

22. Schwaiger M, Cassinotti P. Development of a quantitative real-time RT-PCR assay with internal control for the laboratory detection of tick borne encephalitis virus (TBEV) RNA, 2003. J Clin Virol. 2003;27:136–45.

23. Klaus C, Hoffmann B, Hering U, Mielke B, Sachse K, Beer M, Süss J. Tick-borne encephalitis (TBE) virus prevalence and virus genome characterization in field-collected ticks (Ixodes ricinus) in risk, non-risk, and former risk areas of TBE, and in ticks removed from humans in Germany. Clin Microbiol Infect. 2010;16:238–44.

24. Klaus C, Hoffmann B, Beer M, Müller W, Stark B, Bader W, Stiasny K, Heinz FX, Süss J. Seroprevalence of tick-borne encephalitis (TBE) in naturally exposed monkeys (Macaca Sylvanus) and sheep and prevalence of TBE virus in ticks in a TBE endemic area in Germany. Ticks Tickborne Dis. 2010;1:141–4.

25. Katoh K, Misawa K, Kuma KL, Miyata T. MAFFT: a novel method for rapid multiple sequence alignment based on fast Fourier transform. Nucleic Acids Res. 2002;30:3059–66.

26. Saitou N, Nei M. The neighbor-joining method: a new method for reconstructing phylogenetic trees. Mol Biol Evol. 1987;4:406–25.

27. Felsenstein J. Confidence limits on phylogenies: an approach using the bootstrap. Evolution. 1985;39:783–91.

28. Kimura M. A simple method for estimating evolutionary rate of base substitutions through comparative studies of nucleotide sequences. J Mol Evol. 1980;16:111–20.

29. Tamura K, Stecher G, Peterson D, Filipski A, Kumar S. MEGA6: molecular evolutionary genetics analysis version 6.0. Mol Biol Evol. 2013;30:2725–9.

30. Holzmann H, Kundi M, Stiasny K, Clement J, McKenna P, Kunz C, Heinz FX. Correlation between ELISA, hemagglutination inhibition, and neutralization tests after vaccination against tick-borne encephalitis. J Med Virol. 1996;48:102–7.

31. Protas II, Votiakov VI, Bortkevich VS. Spontaneous morbidity in sheep from tick-borne encephalitis in white Russia. Med Parazitol. 1973;42:293–8.

32. Gray D, Webster K, Berry JE. Evidence of louping ill and tick-borne fever in goats. Vet Rec. 1988;122:66.

33. Grešíková M. Recovery of the tick-borne encephalitis virus from the blood and milk of subcutaneously infected sheep. Acta Virol. 1958;2:113–9.

34. Balogh Z, Egyed L, Ferenczi E, Bán E, Szomor K, Takács M, Berencsi G. Experimental infection of goats with tick-borne encephalitis virus and the possibilities to prevent virus transmission by raw goat milk. Intervirology. 2012;55:194–200.

35. Jellinger K. The histopathology of tick-borne encephalitis. In: Kunz C, editor. Tick-borne encephalitis. Wien: Facultas; 1979. p. 59–75.

36. Van Tongeren HA. Encephalitis in Austria. IV. Excretion of virus by milk of the experimentally infected goat. Arch gesamte Virusforsch. 1955;6:158–62.

37. Grešíková M, Rehacek J. Isolation of the tick-borne encephalitis virus from the blood and milk of domestic animals (sheep and cow) after infection by ticks of the family Ixodes ricinus. Arch Gesamte Virusforsch. 1959;9:360–4.

38. Kohl I, Kozuch O, Elecková E, Labuda M, Zaludko J. Family outbreak of alimentary tick-borne encephalitis in Slovakia associated with a natural focus of infection. Eur J Epidemiol. 1996;12:373–5.

39. Matuszczyk I, Tarnowska H, Zabicka J, Gut W. The outbreak of an epidemic of tick-borne encephalitis in Kielce province induced by milk ingestion. Przegl Epidemiol. 1997;51:381–8.

40. Labuda M, Elecková E, Licková M, Sabó A. Tick-borne encephalitis virus foci in Slovakia. Int J Med Microbiol. 2002;291(Suppl 33):43–7.

41. Kerbo N, Donchenko I, Kutsar K, Vasilenko V. Tick-borne encephalitis outbreak in Estonia linked to raw goat milk. Euro Surveill. 2005;10:2–4.

42. District office, Reutlingen. Zwei Fälle von seltener Frühsommer-Meningoenzephalitis-Virus-Übertragung durch Ziegenrohmilchprodukte. Pressemitteilung Nr.95/2016, District Office, Reutlingen. 2016.

43. Klaus C, Hoffmann B, Moog U, Schau U, Beer M, Süss J. Can goats be used as sentinels for tick-borne encephalitis (TBE) in non-endemic areas? Experimental studies and epizootological observations. Berl Munch Tierärztl Wochenschr. 2010;123:10–4.

9

First molecular detection and characterization of Marek's disease virus in red-crowned cranes (*Grus japonensis*)

Xue Lian[1†], Xin Ming[1†], Jiarong Xu[1], Wangkun Cheng[2], Xunhai Zhang[3], Hongjun Chen[4], Chan Ding[4], Yong-Sam Jung[1*] (ID) and Yingjuan Qian[1*]

Abstract

Background: Marek's disease virus (MDV) resides in the genus *Mardivirus* in the family *Herpesviridae*. MDV is a highly contagious virus that can cause neurological lesions, lymphocytic proliferation, immune suppression, and death in avian species, including Galliformes (chickens, quails, partridges, and pheasants), Strigiformes (owls), Anseriformes (ducks, geese, and swans), and Falconiformes (kestrels).

Case presentation: In 2015, two red-crowned cranes died in Nanjing (Jiangsu, China). It was determined that the birds were infected with Marek's disease virus by histopathological examination, polymerase chain reaction (PCR), gene sequencing and sequence analysis of tissue samples from two cranes. Gross lesions included diffuse nodules in the skin, muscle, liver, spleen, kidney, gizzard and heart, along with liver enlargement and gizzard mucosa hemorrhage. Histopathological assay showed that infiltrative lymphocytes and mitotic figures existed in liver and heart. The presence of MDV was confirmed by PCR. The sequence analysis of the Meq gene showed 100% identity with Md5, while the VP22 gene showed the highest homology with CVI988. Furthermore, the phylogenetic analysis of the VP22 and Meq genes suggested that the MDV (from cranes) belongs to MDV serotype 1.

Conclusion: We describe the first molecular detection of Marek's disease in red-crowned cranes based on the findings previously described. To our knowledge, this is also the first molecular identification of Marek's disease virus in the order Gruiformes and represents detection of a novel MDV strain.

Keywords: Marek's disease virus, Red-crowned crane, Clinical necropsy, PCR, Homology

Background

The red-crowned crane (*Grus japonensis*) is classified as an endangered species with a small global population of 1830 mature individuals [1]. The red-crowned crane breeds in south-eastern Russia, north-east China, Mongolia, and eastern Hokkaido, Japan [2]. The Russian and Chinese populations mainly migrate to the Yellow River Delta and the coast of Jiangsu province, China, and the demilitarized zone of North Korea/South Korea in winter [3]. The number of over wintering cranes in China is now only 8% of what it was in the 1980s due to habitat degradation [4]. This also leads to the overconcentration of cranes at a few sites, which therefore decrease the genetic diversity [5]. The small population and low genetic diversity make this species especially vulnerable to epidemic diseases.

Marek's disease (MD) is a highly contagious disease, characterized by immunosuppression, neurological disorder, CD4$^+$ T cells transformation and eventual tumor formation in peripheral nerves and visceral organs [6]. The causative agent of MD is Marek's disease virus (MDV) which is a member of the genus *Mardivirus* belonging to the subfamily *Alphaherpesvirinae* of the

* Correspondence: ysjung@njau.edu.cn; yqian@njau.edu.cn
†Equal contributors
[1]MOE Joint International Research Laboratory of Animal Health and Food Safety, College of Veterinary Medicine, Nanjing Agricultural University, Nanjing, China
Full list of author information is available at the end of the article

family *Herpesviridae*. MD was first reported in chicken by Josef Marek in 1907 [7], and is prevalent throughout the world. Later, it was also reported in Galliformes (such as turkeys [8], quails [9] and pheasants [10]), Strigiformes (owls), Anseriformes (ducks, geese and swans) and Falconiformes (kestrels). The most important natural hosts for MDV are domestic and wild chicken, including game fowl [11], native breeds [12] and jungle fowl [13]. MDV contains three serotypes: MDV-1 (GaHV-2), MDV-2 (GaHV-3) and MDV-3 (MeHV-1) [14]. However, only MDV-1 can induce disease in chicken, whereas MDV-2 and MDV-3 are avirulent and are used as vaccines. Symptoms of MD depend on the age of the bird [15, 16]. For example, young chickens infected with virulent MDV strains may exhibit high mortality in 8–16 days post infection with early mortality syndrome [17]. MDV can be transmitted horizontally through aerosols and enter the host through the respiratory tract, but it cannot be transmitted vertically from chicken to eggs [18]. The infectious virus particles replicate in epithelial cells of the feather follicles, and then are shed to the environment with the skin dander [19]. MDV in the dust survives for at least several months at room temperature [20]. Thus, skin dander, house dust, feces, and saliva can be a source of infection. Because cross-species spreading of the virus is possible, this mode of transmission enables it to infect an even broader spectrum of hosts [21]. Despite several cases of herpesvirus infection in cranes that were previously reported [22, 23], there is a lack of molecular evidence on MDV infection and knowledge of which species are infected.

This article reports the molecular detection of a novel MDV isolate in the red-crowned crane. It is also the first molecular determination of MDV from individual red-crowned cranes (*Grus japonensis*), indicating that red-crowned cranes can be infected by MDV.

Case presentation

In August 2015, reduced feed consumption was observed in a 2-month-old red-crowned crane (crane A) that was bred and incubated in Hongshan Forest Zoo in Nanjing, China. After segregation and palliative treatment, it exhibited astasia and then died. Subcutaneous palpation examination revealed many diffuse soybean-sized nodules (Table 1). Red-crowned crane B showed obvious swelling of its left knee joint. After antibiotic treatment, although it showed an improved appetite and a decrease in symptoms, the crane died unexpectedly ten days later (Table 1). The dead cranes were not immunized with the MDV vaccine. Although there had been no previous reports of MDV-infected cranes, the cranes were fed in mesh cages that were accessible to wild birds.

Clinical necropsy revealed diffuse yellow-white nodules from millet to soybean size in the skin, muscle, trachea, liver, gizzard, and heart in both cranes (Fig. 1a-f).

Table 1 Timeline table of the information in the case report

Date	Clinical Observation	Intervention	Samples Collected
Aug 19-20	Red-crowned crane A[a]: emaciation, reduced appetite. Diagnostic check: Heart rate: 106 bp/min Body temperature: 39.7°C Weight: 1.6kg; Crane A[a]: astasia.	Rocephin (100 mg/kg) + Dexamethasone (0.25 mg/kg) + 0.9% NaCl 15 ml (intravenous infusion once daily for 3 days) Vitamin C (100 mg/kg) + Vitamin B$_1$(15 mg/kg) + ATP (5 mg/kg) + Inosine (15 mg/kg) + coenzyme A (15 IU/kg) + 10% glucose 15 ml (intravenous infusion once daily for 3 days) 10% glucose + mixture of fish and corn, orally, for 3 days Calcium gluconate (25 mg/kg) + 10% glucose 15 ml (intravenous infusion once daily), for 1 day	Crane A[a]: •Histopathological examination of liver, heart, and spleen. Three repeated samples were collected from each tissue. The ratio of MDV-positive tissues: Heart: 67%, liver: 100%, spleen: 67%. •RNA extraction of feather, liver, spleen, kidney and muscle. Three of each. Positive ratio of all samples was 100%. •Genome extraction of muscle, feather, feather-pulp and spleen; three of each. The positive ratio of all samples was 100%. Crane B[b]: same as crane A.
Aug 21	Diagnostic check: Heart rate: 156 bp/min Body temperature: 39.8°C Crane A[a] died.		
Aug 22	Clinical necropsy: Nodules found in skin, muscle, liver, heart, trachea, and gizzard. (MD suspected).		
Sept 11	Crane B[b]: obvious swelling of its left knee joint.	Musk analgesic aerosol, external use	
Sept 12-17	Clinical improvement: Recovered appetite, left knee joint swelled	Musk analgesic aerosol + ichthammol ointment, external use Calcium tablets (10 mg/kg) + Vitamin D$_2$ (500 IU/kg) + Vitamin A (1500 IU/kg), orally	
Sept 18-19	Relapse, anorexia, depression		
Sept 20	Crane B[b] died. Clinical necropsy: Inflammatory exudates; diffuse yellow-white nodules in muscle, liver, spleen, kidney, pancreas, gizzard, and heart.		

[a]Crane A: hatched on June 10, died on Aug 22, 2015. Body weight: 1.6 kg
[b]Crane B: hatched on June 9, died on Sep 22. Body weight: 2.4 kg

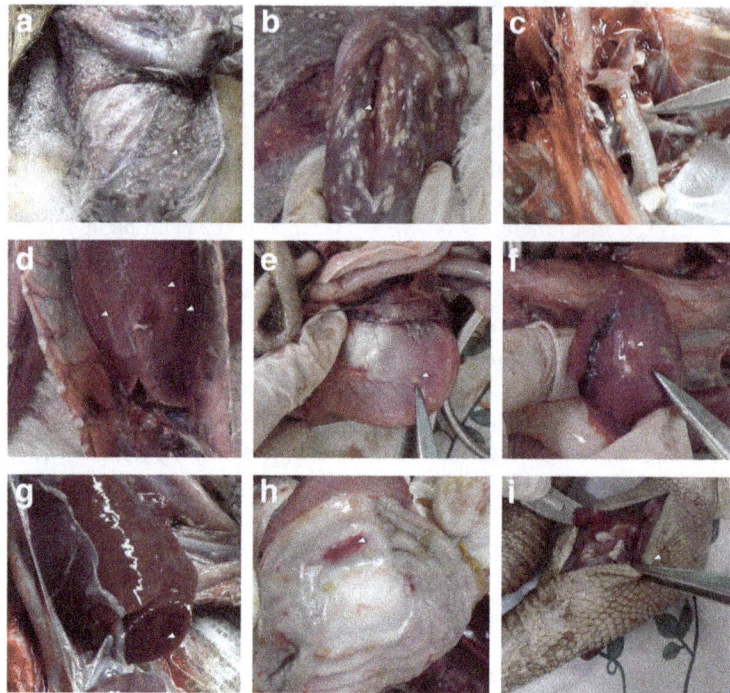

Fig. 1 Clinical symptoms and pathological lesions. **a-f**, Nodules in the (**a**) skin, (**b**) muscle, (**c**) trachea, (**d**) liver, (**e**) gizzard, (**f**) heart; **g**, nodules in the vertical section of the liver; **h**, hemorrhage sites in the gizzard mucosa; **i**, left knee joint swelling

Diffused yellow-white nodules were also observed in the vertical section of the liver (Fig. 1g). Several hemorrhage sites were observed in the gizzard mucosa (Fig. 1h). A large amount of inflammatory exudates oozed from a surface cut of the swollen joint (Fig. 1i). No other significant gross lesions were observed.

Representative tissue samples were collected during necropsy from the liver, spleen, kidney, muscle, heart, feather follicles, and skin, and these were used for pathological and virological investigations (Table 1). Two sets of tissue samples were collected for different purposes, and were processed in different ways: one set was used for virology analysis and stored at − 80 °C and the other was used for conventional histopathology analysis and fixed in 10% neutral formalin. Formalin-fixed samples were subsequently dehydrated through graded alcohols before being embedded in paraffin wax. Several 4 μm-thick sections were cut from each sample and stained using hematoxylin and eosin (H&E). The heart histology showed that numerous lymphoid cells had infiltrated between adjacent cardiac muscle fibers. We also found that there were various polymorphic lymphoid cells with mitotic figures (Fig. 2a-b). Similarly, the liver structure was disrupted, and a large number of variably sized lymphoblastic cells was dispersed or exhibited mass distribution in the hepatic lobules (Fig. 2c-d). These findings were consistent with the symptoms of MDV infection. However, bacterial infection cannot be excluded from this case, as swollen joints could be a consequence of bacterial infection. It is possible that the birds were infected with MDV, which is immunosuppressive, but succumbed to other conditions such as bacterial superinfection.

For virological detection, frozen tissue samples of feather, liver, spleen, kidney, and muscle were homogenized using a dispersing homogenizer (IKA Ultra-Turrax, T10, Germany). Total RNA was extracted using TRIzol reagent (Sigma, USA) according to the manufacturer's protocol. The cDNA was prepared using an iScript™ Reverse Transcription Supermix kit (Bio-Rad, USA). The PCR was carried out with 2× Taq Master Mix (Vazyme, China) with the following conditions: denaturation at 94 °C for 5 min; 28 cycles of denaturation at 94 °C for 30 s, annealing at 55 °C for 30 s, and extension at 72 °C for 30s; final extension at 72 °C for 7 min; and maintenance at 4 °C. To optimize the PCR conditions, primers were modified on the basis of Yamaguchi et al. [24]. The Meq primer used herein amplified the DNA 10 bp longer than published primer, due to the forward primer being located 10 bp upstream from the published forward primer. For example, the sequence of the published primer is 5'-GGTCGACTTCGAGACGGAAA-3', while here, we used 5'-CTTTCTCTCGGGTCGACTTC-3'.

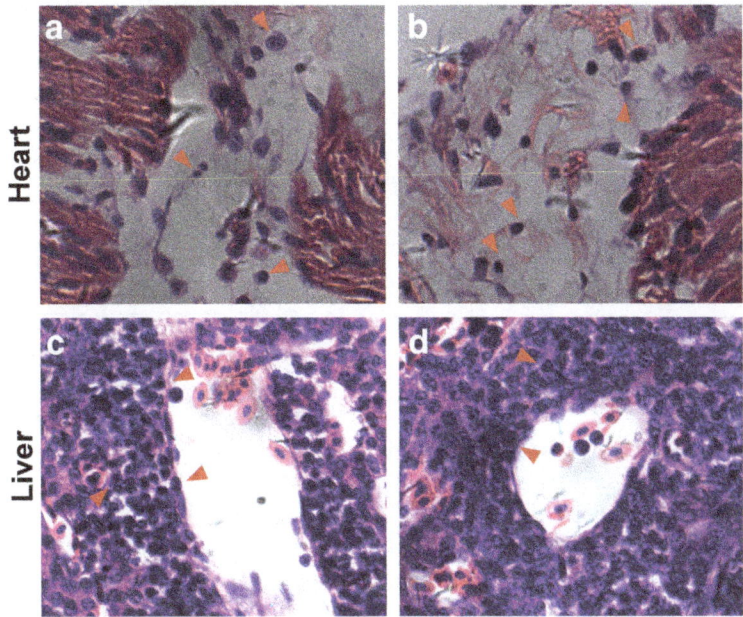

Fig. 2 Histopathological section (H&E staining) of the liver and heart. **a** and **b**, lymphomatous infiltration in the liver; **c** and **d**, lymphomatous infiltration in the heart. The arrow shows a range of leukocytes, including large lymphocytes, small lymphocytes lymphoblasts, and malignant cells with mitotic figure.

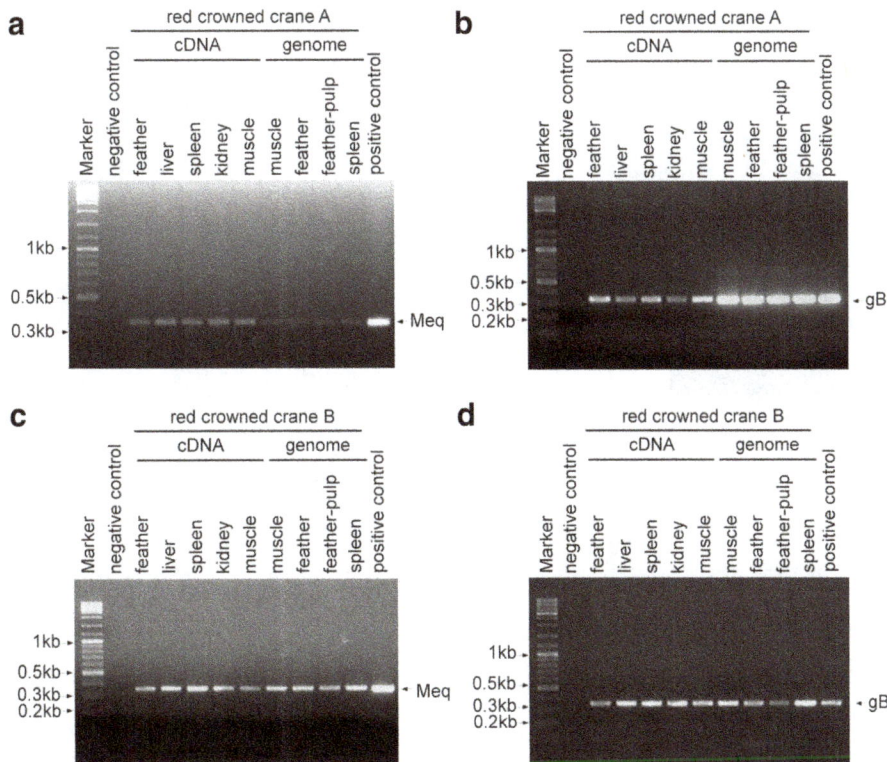

Fig. 3 Agarose gel electrophoresis of MDV from RT-PCR of the Meq and gB genes. **a** and **c**, Meq (347 bp, partial) was amplified from cDNAs and genomes of different tissues. **b** and **d**, gB (338 bp, partial) was amplified from cDNAs and genomes of different tissues. The feather sample was a homogenate of ground feathers and skin that contained feather follicle epithelium. The entire cell genome from Md5-infected CEF cells was used as a positive control. The cDNA from CEF cells was used as a negative control

Table 2 Primers used for amplification of MDV viral genes

Name	Sequence	Length
Meq-RT-F	CTTTCTCTCGGGTCGACTTC	347 bp
Meq-RT-R	GTAAGCAGTCCAAGGGTCAC	
gB-RT-F	CTTCACAGTTGGGTGGGAC	338 bp
gB-RT-R	GAGCCAGGGATTTGGATAG	
MDV-Meq-F	AGAGATGTCTCAGGAGCCAGAGCC	1020 bp
MDV-Meq-R	ATCATCAGGGTCTCCCGTCACCTG	
MDV-VP22-F	ATCGGATCCATGGGGGATTCTGAAAG	750 bp
MDV-VP22-R	ACACTCGAGTTATTCGCTATCACTGC	

The underlined parts are the overlapped sequence. Our gB primers can amplify shorter sequences, but still within the same region compared with the published one. The amplified fragments were resolved by electrophoresis in 1.5% agarose gels (Sigma, USA). Here, RT-PCR was used to determine the viral gene expression. Meq mRNA was examined to measure general mRNA expression regardless of the phase of infection, and gB mRNA was examined to indicate virus lytic replication. Samples were analyzed for the presence of MDV and the reactions were positive for all sampled tissues when amplified with both Meq (Fig. 3a) and gB (Fig. 3b) primers [24] in both crane A (Fig. 3a, b) and crane B (Fig. 3c, d).

To further characterize the virus, samples from spleen, skin, muscle, feather and feather pulp were cut and digested overnight at 55 °C in lysis buffer (10 mM Tris, 100 mM EDTA, 0.5% SDS, and 0.2 mg/ml Proteinase K), and total cellular DNA was extracted with phenol-chloroform-isoamylalcohol (25:24:1), then precipitated with ethanol, and dissolved in TE buffer (10 mM Tris-HCl 1 mM EDTA pH 8.0). Total cellular DNA samples were used as templates for amplifying viral genes. The PCR was carried out with LA Taq DNA polymerase (Takara, Japan). Primers were designed to target the entire-length of the Meq and VP22 genes of Md5 (Table 2). The Meq gene was successfully amplified from the spleen tissue (Fig. 4a). The sequence alignment revealed that the analyzed fragment presented 100% sequence identity with reference MDV strains such as Md5 and Md11 (Fig. 4b and c). It also shared a high homology (97.9%) to the very virulent strain LMS from China (Fig. 4b). An additional PCR was carried out to amplify the VP22 gene, which is a major structural component of the virion [25] (Fig. 5a). The obtained band was sequenced and aligned using DNAMAN software.

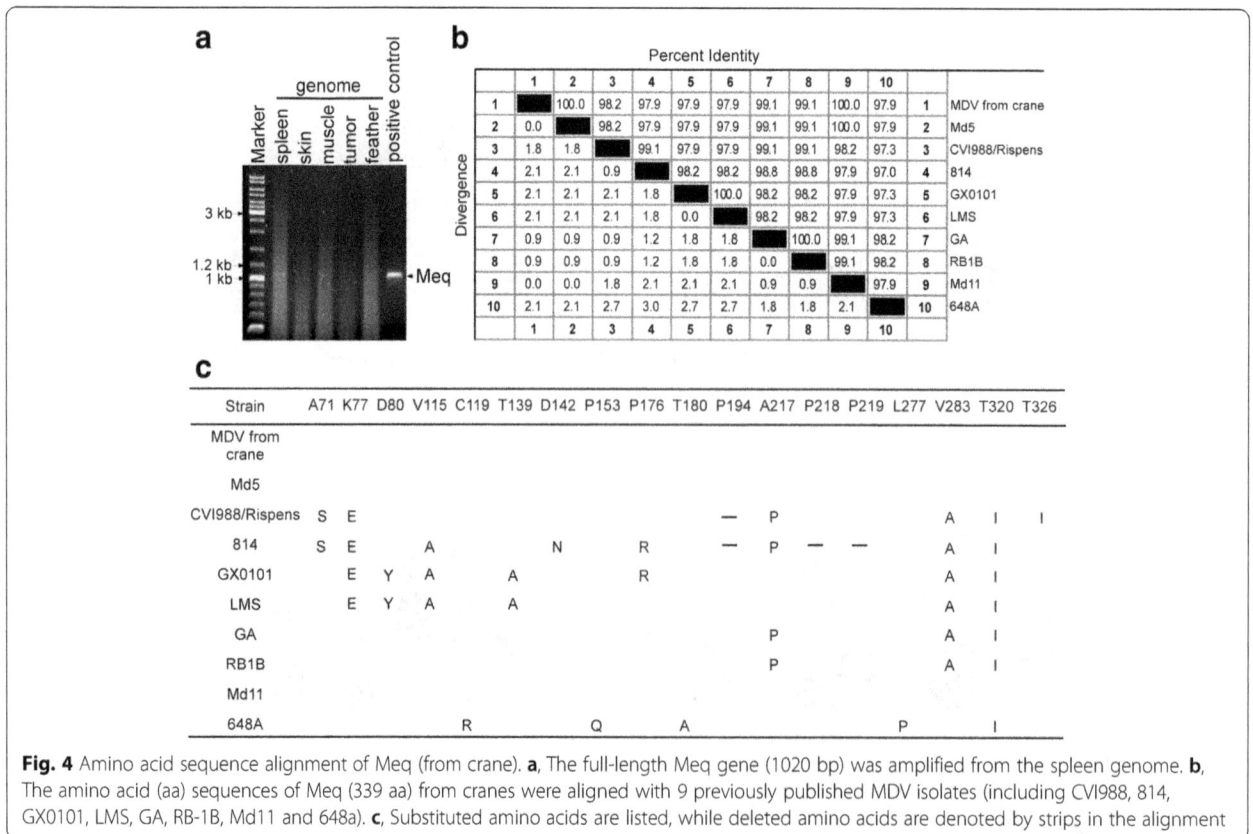

b Percent Identity

	1	2	3	4	5	6	7	8	9	10		
1		100.0	98.2	97.9	97.9	97.9	99.1	99.1	100.0	97.9	1	MDV from crane
2	0.0		98.2	97.9	97.9	97.9	99.1	99.1	100.0	97.9	2	Md5
3	1.8	1.8		99.1	97.9	97.9	99.1	99.1	98.2	97.3	3	CVI988/Rispens
4	2.1	2.1	0.9		98.2	98.2	98.8	98.8	97.9	97.0	4	814
5	2.1	2.1	2.1	1.8		100.0	98.2	98.2	97.9	97.3	5	GX0101
6	2.1	2.1	2.1	1.8	0.0		98.2	98.2	97.9	97.3	6	LMS
7	0.9	0.9	0.9	1.2	1.8	1.8		100.0	99.1	98.2	7	GA
8	0.9	0.9	0.9	1.2	1.8	1.8	0.0		99.1	98.2	8	RB1B
9	0.0	0.0	1.8	2.1	2.1	2.1	0.9	0.9		97.9	9	Md11
10	2.1	2.1	2.7	3.0	2.7	2.7	1.8	1.8	2.1		10	648A
	1	2	3	4	5	6	7	8	9	10		

Divergence (lower-left), Percent Identity (upper-right)

c

Strain	A71	K77	D80	V115	C119	T139	D142	P153	P176	T180	P194	A217	P218	P219	L277	V283	T320	T326
MDV from crane																		
Md5																		
CVI988/Rispens	S	E									—	P				A	I	I
814	S	E		A			N		R		—	P	—	—		A	I	
GX0101		E	Y	A		A			R							A	I	
LMS		E	Y	A		A										A	I	
GA												P				A	I	
RB1B												P				A	I	
Md11																		
648A						R		Q	A						P		I	

Fig. 4 Amino acid sequence alignment of Meq (from crane). **a**, The full-length Meq gene (1020 bp) was amplified from the spleen genome. **b**, The amino acid (aa) sequences of Meq (339 aa) from cranes were aligned with 9 previously published MDV isolates (including CVI988, 814, GX0101, LMS, GA, RB-1B, Md11 and 648a). **c**, Substituted amino acids are listed, while deleted amino acids are denoted by strips in the alignment

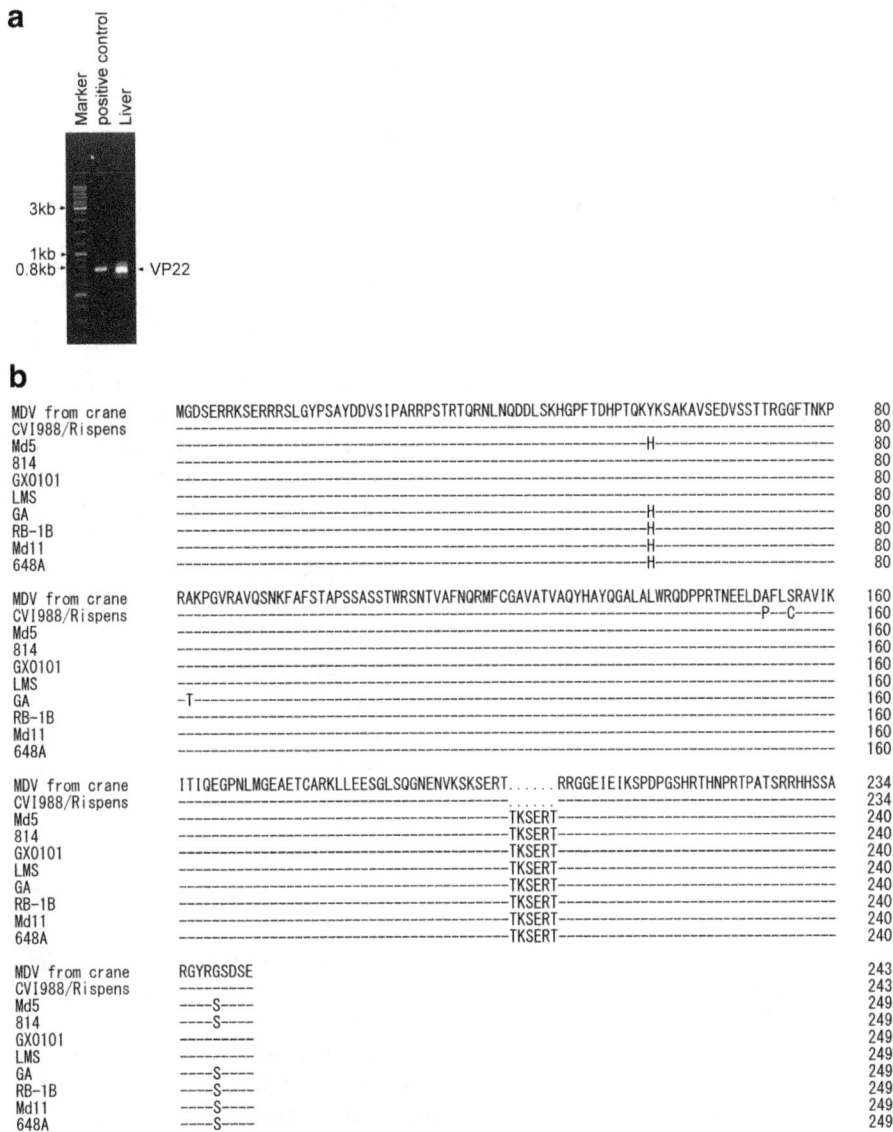

Fig. 5 Amino acid sequence alignment of VP22 (from crane). **a**, The full-length VP22 gene (750 bp) was amplified from the liver genome. **b**, The amino acid sequence of VP22 (243 aa) from cranes was aligned with 9 previously published MDV isolates (including CVI988, 814, GX0101, LMS, GA, RB-1B, Md11 and 648a). Identical amino acids are denoted by strips in the alignment, and deleted amino acids are denoted by dots in the alignment

The sequence alignment showed that the VP22 protein shared the highest identity with the vaccine strain CVI988 as compared to other strains (Fig. 5b). Two substitutions at residues 152 (A → P), 155 (S → C) were found in CVI988, while 2–3 substitutions were found when aligned with other eight isolates. The same as CVI988, six amino acids (TKSERT) were deleted from residues 201 to 206 compared with other isolates (Fig. 5b). The deleted site was located in the domain (SKSERTTK-SERT), which contained two copies of motif (KSERT) in virulent MDV strains. It has been shown that regardless of the deletion, the VP22 of the CVI988/Rispens vaccine strain still maintains an intercellular-trafficking function [26].

Phylogenetic analysis of the Meq gene showed that our MDV clustered with very virulent strain and was most closely related to Md5, Md11 and W (Fig. 6a). However, phylogenetic analysis of the VP22 gene showed that our MDV was closely related with attenuated virulent strain CVI988 (Fig. 6b).

The Meq gene is only present in MDV1m and its gene product, MEQ, which is an MDV1-specific 339 amino

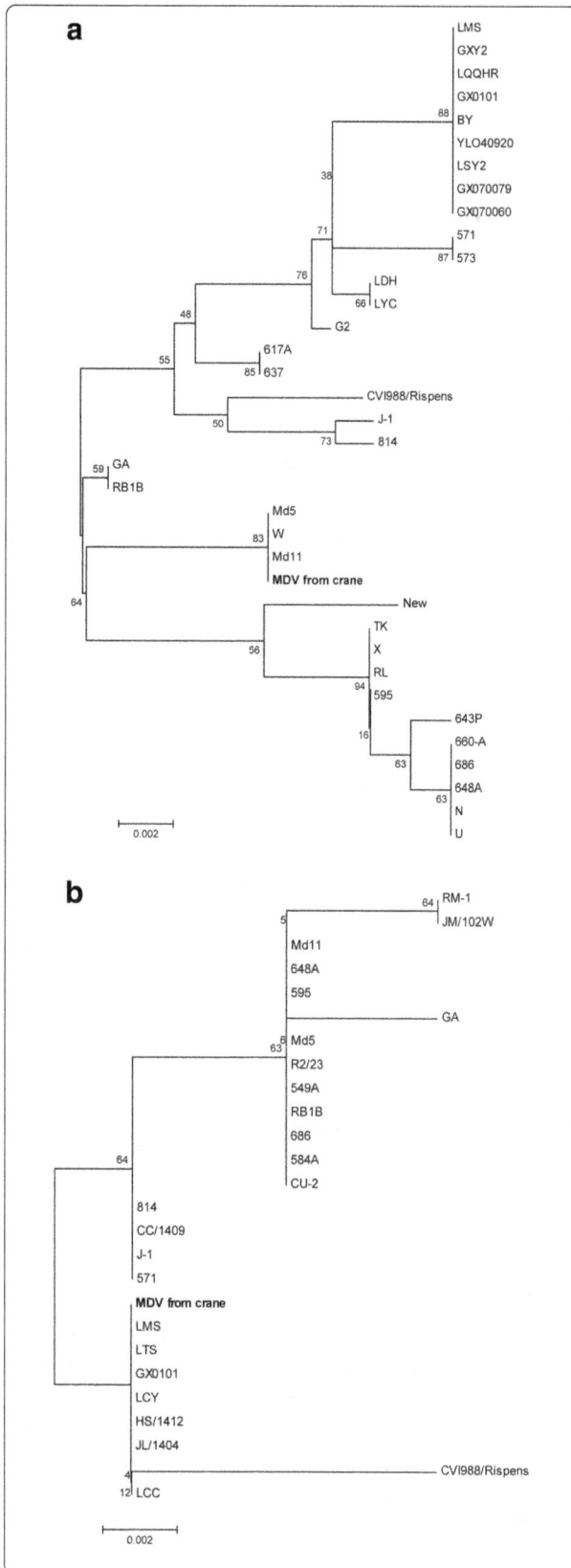

Fig. 6 Phylogenetic profile showing the relationships among MDV isolates based on a comparison of the (**a**) Meq gene and (**b**) VP22 gene. Phylogenetic analysis of MDV based on (**a**) Meq and (**b**) VP22 amino acid sequences. The tree was constructed using the neighbor-joining (N-J) analysis method in the MEGA 5.0 program with bootstrapping (1000)

acids protein with an N-terminal basic leucine zipper (bZIP) [27, 28]. L-Meq is a longer isoform of Meq that contains a 180 bp nucleotide insertion in the transactivation domain of Meq and is normally detected in non-oncogenic MDV1 vaccine strains such as CVI988 [29]. Interestingly, L-Meq was also amplified from crane spleen cDNA (Fig. 7a). Sequence alignment indicated that it shared 100% identity with the BC-1 strain (Fig. 7b). One substitution was found in the American strains GA and JM10 (Fig. 7c), and four substitutions were found in the JM102 strain (Fig. 7c).

Discussion and conclusions

This article describes the first reported case of MD in red-crowned cranes and confirms that red-crowned cranes are susceptible to MDV infection. Previous research has suggested that multiple avian species can be infected by MDV such as Galliformes (turkeys [8], quails [9] and pheasants [10]), Strigiformes (owls), Anseriformes (ducks, geese, swans), and Falconiformes (kestrels). Therefore, this is also the first molecular detection of Marek's disease virus in the order Gruiformes. One possible sources of infection for red-crowned cranes might be from migrating wild birds. Murata et al. tested feather-tip samples of wild geese from Japan and the Far East region of Russia by nested PCR, and 30% of analyzed white-fronted geese contained MDV [30], which suggested that migratory birds such as white-fronted geese could be MDV carriers during their migration. Previous research showed that MDV can be transmitted horizontally and across-species. Kenzy and Cho reported that a MDV-positive quail can transmit MD to monitor chicks by contact exposure [31]. Additionally, MDV is fairly resistant to environmental factors. Indeed, MDV can survive at least a few months in the dust at room temperature [32]. The infection source is also diverse, including chicken feather follicle, scurf, house dust, feces, and saliva. Many chickens with a normal appearance can be MDV carriers and spread infection. All of these provide a conducive environment for the spreading of MDV. Considering that the red-crowned cranes were fed in mesh cages that were accessible to wild birds and were never immunized with MDV vaccine before, it is possible that the MDV infection originated from migratory birds.

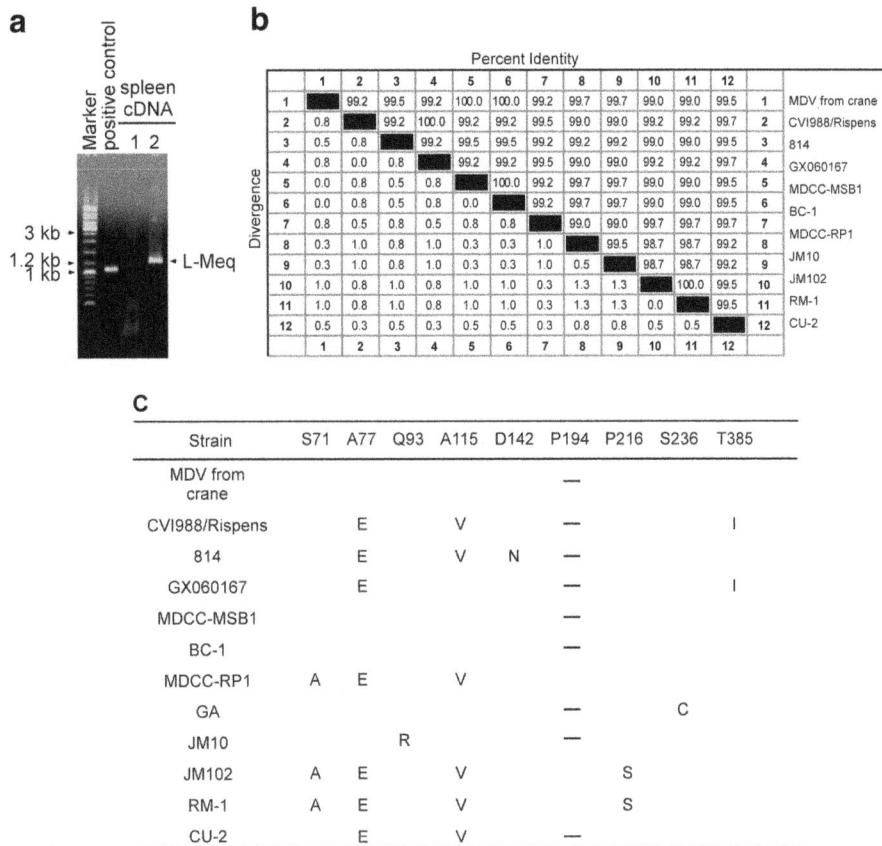

Fig. 7 Amino acid sequence alignment of L-Meq (from crane). **a**, The full-length L-Meq gene (1197 bp) was amplified from spleen cDNA. **b**, The amino acid sequence of L-Meq (398 aa) from cranes was aligned with 11 previously published MDV isolates (including CVI988, 814, GX060167, MDCC-RP1, MDCC-MSB1, BC-1, GA, JM10, JM102, RM-1 and CU-2). Sequences were aligned using the DNAMAN software. **c**, Substituted amino acids are listed, while deleted amino acids are denoted by strips in the alignment

This case report describes the first detection and characterization of MDV from red-crowned cranes in a Chinese zoo. The molecular analysis of MDV genes suggested that the MDV (from crane) belongs to MDV serotype 1. The red-crowned cranes described in this study showed clinical signs of disease including decreased feed consumption, depression, paralysis, joint swelling, and eventual death. The results of necropsies of dead cranes revealed diffuse nodules in skin, muscle, liver, gizzard and heart. Histological investigation demonstrated lymphomatous infiltration into the affected tissue and neoplastic changes with evidence of mitosis. Additionally, a range of leukocytes was observed, including small/large lymphocytes, lymphoblasts, and occasional plasma cells. The demonstration of polymorphic leukocytes along with molecular detection is highly suggestive of Marek's disease. RT-PCR targeting oncogenic gene Meq and glycoprotein B (gB) was performed to confirm the presence of MDV, with observation of expected amplification products throughout all tissue samples. Further phylogenetic analysis of the Meq and VP22 genes suggested that our MDV could belong to MDV-1, but it is a novel MDV that contains both Meq and L-Meq gene in its genome.

The detection of MDV in red-crowned cranes suggests that the virus may have a huge impact on the endangered crane's survival in the wild. However, because this is the first verified instance of crane mortality resulting from MD, its true impact on the population is unknown. A possible solution for management of this species is to vaccinate all the red-crowed cranes to provide immunity against this disease. There are several available MDV vaccines including serotypes MDV-2 (SB-1), HVT, and attenuated MDV-1 (CVI988/Rispens), which are used singly or jointly [33]. MDV vaccination began in the late 1960s and made great contributions to protecting the poultry industry for more than 50 years. However, the current problem involves attenuated MDV vaccine that cannot provide effective protection against new virulent MDV strains. For example, previous reports have shown

Table 3 GenBank accession numbers of the Meq amino acid sequences used in this study

Isolate	Country of Origin	Accession Number	Pathotype
Md5	USA	AAG14255.1	vvMDV
Md11	USA	AAS01627.1	vvMDV
GA	USA	AAF66798.1	vMDV
RB1B	USA	AY243332	vvMDV
RL	USA	AAR13332.1	vv + MDV
TK	USA	AAR13333.1	vv + MDV
N	USA	AAR13330.1	vv + MDV
New	USA	AAR13331.1	vv + MDV
U	USA	AAR13334.1	vv + MDV
W	USA	AAR13335.1	vv + MDV
X	USA	AAR13336.1	vv + MDV
571	USA	AAR13322.1	vMDV
573	USA	AAR13323.1	vMDV
595	USA	AAR13327.1	vvMDV
617A	USA	AAR13324.1	vMDV
637	USA	AAR13325.1	vMDV
643P	USA	AAR13328.1	vvMDV
660-A	USA	AAR13338.1	vv + MDV
648A	USA	AAR13337.1	vv + MDV
686	USA	AAR13339.1	vv + MDV
CVI988/Rispens	Netherlands	AAP06940.1	attMDV
814	China	AAL99997.1	attMDV
J-1	China	AEA06596.1	N.A.
LQQHR	China	AEM63533.1	N.A.
LSY2	China	AEM63534.1	N.A.
LMS	China	AEZ51745.1	vvMDV
GX0101	China	AFX97850.1	vvMDV
GX070060	China	ACA13267.1	vvMDV
GX070079	China	ACA13268.1	N.A.
BY	China/Tibet	AEA06593.1	N.A.
GXY2	China	ABQ23868.1	N.A.
LDH	China	AEM63523.1	N.A.
LYC	China	AEM63536.1	N.A.
YLO40920	China	ABA54944.1	N.A.
G2	China	AAM00003.1	vvMDV

Abbreviation: *Att* attenuated, *N.A* not available, *v* virulent, *vv* very virulent, *vv + * very virulent plus

Table 4 GenBank accession numbers of the L-meq amino acid sequences used in this study

Isolate	Country of Origin	Accession Number	Pathotype
CVI988/Rispens	Netherlands	ABF72204.1	attMDV
814	China	ADA83412.1	attMDV
GX060167	China	ACD75766.1	N.A.
HNGS201	China	CCO02736.1	N.A.
MDCC-MSB1	Japan	BAC57989.1	N.A.
MDCC-RP1	Japan	BAC57991.1	N.A.
BC-1	USA	AAR13319.1	vMDV
GA	USA	AAC54628.1	vMDV
JM10	USA	AAP06937.1	vMDV
JM/102 W	USA	ABG22905.1	vMDV
RM-1	USA	ABG22996.1	attMDV
CU-2	USA	ACF94907.1	mMDV
3004	Russia	ABS84657.1	attMDV

Abbreviation: *att* attenuated; *m* mild, *N.A* not available, *v* virulent, *vv* very virulent, *vv +* very virulent plus

Table 5 GenBank accession numbers of the vp22 amino acid sequences used in this study

Isolate	Country of Origin	Accession Number	Pathotype
Md5	USA	YP-001033978.1	vvMDV
CVI988/Rispens	Netherlands	ABF72291.1	attMDV
GA	USA	AAF66784.1	vMDV
RB1B	USA	ABR13137.1	vvMDV
Md11	USA	AAS01692.1	vvMDV
648A	USA	AFM7426.1	vv + MDV
549A	USA	ABV31181.1	vvMDV
571	USA	ABV31182.1	vMDV
584A	USA	ABV31183.1	vv + MDV
595	USA	ABV31184.1	vvMDV
686	USA	ABV31179.1	vv + MDV
JW/102 W	USA	ABV31188.1	vMDV
CU-2	USA	ABV31189.1	mMDV
R2/23	USA	ABV31191.1	N.A.
RM-1	USA	ABV31180.2	N.A.
CC/1409	China	AQN78018.1	N.A.
HS/1412	China	AQN78190.1	N.A.
J-1	China	AQN77146.1	N.A.
814	China	AEV55030.1	attMDV
GX0101	China	AFX97970.1	vvMDV
LMS	China	AEZ51712.1	vvMDV
JL/1404	China	AQN77845.1	N.A.
LCC	China	AQN77320.1	N.A.
LCY	China	ARE59101.1	N.A.
LTS	China	AQN77496.1	N.A.

Abbreviation: *att* attenuated, *N.A* not available, *v* virulent, *vv* very virulent, *vv +* very virulent plus

that molecular evolution of MDV genes could increase the pathogenicity of MDV virulent strains [34]. Infection of very virulent MDV (vvMDV) could also break post-vaccinal protection and cause the outbreak of MDV [35]. Therefore, the safety of this measurement needs to be studied in depth.

Abbreviations
gB: Glycoprotein B; MDV: Marek's disease virus; PCR: Polymerase chain reaction

Acknowledgements
We thank Dr. Yingjun Lv for review of histological slides and constructive input for data analysis strategies.

Funding
This work was financially supported by the National Natural Science Foundation of China (Grant No. 31472218), the Natural Science Foundation of Jiangsu Province (Grant No. BK20140711), the Priority Academic Program Development of Jiangsu Higher Education Institutions (PAPD), SRF for ROCS, SEM([2014]1685), and the Fundamental Research Funds for the Central Universities (Y0201300526; Y0201300527). The funders did not play any role in the design, conclusions or interpretation of the study.

Authors' contributions
XL, XM, YSJ, and YJQ designed the research. XL, WC and XM performed laboratory experiments. XL, XM, JX, WC, XZ, HC, CD, YSJ, and YJQ analyzed the data. WC initiated the study and organized samples. XL, XM, YSJ, and YJQ wrote the paper. All authors read and approved the final manuscript.

Consent for publication
Consent was obtained from the owner of the animal for publication of this case report.

Competing interests
The authors declare that they have no competing interests.

Author details
[1]MOE Joint International Research Laboratory of Animal Health and Food Safety, College of Veterinary Medicine, Nanjing Agricultural University, Nanjing, China. [2]Nanjing Hongshan Forest Zoo, Nanjing, China. [3]Anhui Provincial Key Laboratory for Control and Monitoring of Poultry Diseases, Anhui Science and Technology University, Fengyang, China. [4]Shanghai Veterinary Research Institute, Chinese Academy of Agricultural Sciences, Shanghai, China.

References
1. Grus japonensis:The IUCN Red List of Threatened Species 2016. http://www.iucnredlist.org/details/full/22692167/0.
2. Collar NJ, Andreev A, Chan S, Crosby M, Subramanya S, Tobias J, Srinivasan C, Subramanya HS, Awatramani N, Subramanya HM. Threatened birds of Asia: the BirdLife international red data book. Cambridge (RU): Birdlife international; 2001.
3. Kanai Y, Ueta M, Germogenov N, Nagendran M, Mita N, Higuchi H. Migration routes and important resting areas of Siberian cranes (Grus leucogeranus) between northeastern Siberia and China as revealed by satellite tracking. Biol Conserv. 2002;106(3):339–46.
4. Su L, Zou H. Status, threats and conservation needs for the continental population of the red-crowned crane. Chinese Birds. 2012;3(3):147 64.
5. Wang Q. Threats for red-crowned crane. China Crane News. 2008;12(2):7–12.
6. Calnek BW. Pathogenesis of Marek's disease virus infection. Curr Top Microbiol Immunol. 2001;255:25–55.
7. Osterrieder N, Kamil JP, Schumacher D, Tischer BK, Trapp S. Marek's disease virus: from miasma to model. Nat Rev Microbiol. 2006;4(4):283–94.
8. Davidson I, Borenstein R. Multiple infection of chickens and turkeys with avian oncogenic viruses: prevalence and molecular analysis. Acta Virol. 1999;43(2–3):136–42.
9. Kobayashi S, Kobayashi K, Mikami T. A study of Marek's disease in Japanese quails vaccinated with herpesvirus of turkeys. Avian Dis. 1986;30(4):816–9.
10. Lesnik F, Pauer T, Vrtiak OJ, Danihel M, Gdovinova A, Gergely M. transmission of Marek's disease to wild feathered game. Veterinarni medicina. 1981;26(10):623–30.
11. Kenzy S, Mclean G, Mathey W, Lee H. Preliminary observations of gamefowl neurolymphomatosis. J Nat Cancer Inst. 1964;17:121–30.
12. Grewal G, Singh B, Singh H. Epidemiology of Marek's disease: incidence of viral specific antigen in feather follicle epithelium of domestic fowl of Punjab. Indian journal of poultry science. 1977;
13. Weiss RA, Biggs PM. Leukosis and Marek's disease viruses of feral red jungle flow and domestic fowl in Malaya. J Natl Cancer Inst. 1972;49(6):1713–25.
14. Bulow VV, Biggs PM. Differentiation between strains of Marek's disease virus and Turkey herpesvirus by immunofluorescence assays. Avian pathology : journal of the WVPA. 1975;4(2):133–46.
15. Swayne DE, Fletcher OJ, Schierman LW. Marek's disease virus-induced transient paralysis in chickens. 1. Time course association between clinical signs and histological brain lesions. Avian pathology : journal of the WVPA. 1989;18(3):385–96.
16. Witter RL, Gimeno IM, Reed WM, Bacon LD. An acute form of transient paralysis induced by highly virulent strains of Marek's disease virus. Avian Dis. 1999;43(4):704–20.
17. Stephens EA, Witter RL, Nazerian K, Sharma JM. Development and characterization of a Marek's disease transplantable tumor in inbred line 72 chickens homozygous at the major (B) histocompatibility locus. Avian Dis. 1980;24(2):358–74.
18. Boodhoo N, Gurung A, Sharif S, Behboudi S. Marek's disease in chickens: a review with focus on immunology. Vet Res. 2016;47(1):119.
19. Nazerian K, Witter RL. Cell-free transmission and in vivo replication of Marek's disease virus. J Virol. 1970;5(3):388–97.
20. Calnek BW. Marek's disease–a model for herpesvirus oncology. Crit Rev Microbiol. 1986;12(4):293–320.
21. Imai K, Yuasa N, Kobayashp S, Nakamura K, Tsukamoto K, Hihara H. Isolation of Marek's disease virus from Japanese quail with lymphoproliferative disease. Avian pathology : journal of the WVPA. 1990;19(1):119–29.
22. Docherty DE, Henning DJ. The isolation of a herpesvirus from captive cranes with an inclusion body disease. Avian Dis. 1980;24(1):278–83.
23. Foerster S, Chastel C, Kaleta EF. Crane hepatitis herpesviruses. Zentralbl Veterinarmed B. 1989;36(6):433–41.
24. Yamaguchi T, Kaplan SL, Wakenell P, Schat KA. Transactivation of latent Marek's disease herpesvirus genes in QT35, a quail fibroblast cell line, by herpesvirus of turkeys. J Virol. 2000;74(21):10176–86.
25. Dorange F, Tischer BK, Vautherot JF, Osterrieder N. Characterization of Marek's disease virus serotype 1 (MDV-1) deletion mutants that lack UL46 to UL49 genes: MDV-1 UL49, encoding VP22, is indispensable for virus growth. J Virol. 2002;76(4):1959–70.
26. Chen H, Song C, Qin A, Zhang C. Expression and intercellular trafficking of the VP22 protein of CVI988/Rispens vaccine strain of Marek's disease virus. Sci China Ser C Life Sci. 2007;50(1):75–9.
27. Liu JL, Ye Y, Lee LF, Kung HJ. Transforming potential of the herpesvirus oncoprotein MEQ: morphological transformation, serum-independent growth, and inhibition of apoptosis. J Virol. 1998;72(1):388–95.
28. Liu JL, Kung HJ. Marek's disease herpesvirus transforming protein MEQ: a c-Jun analogue with an alternative life style. Virus Genes. 2000;21(1–2):51–64.
29. Lee SI, Takagi M, Ohashi K, Sugimoto C, Onuma M. Difference in the meq gene between oncogenic and attenuated strains of Marek's disease virus serotype 1. J Vet Med Sci. 2000;62(3):287–92.
30. Murata S, Hayashi Y, Kato A, Isezaki M, Takasaki S, Onuma M, Osa Y, Asakawa M, Konnai S, Ohashi K. Surveillance of Marek's disease virus in migratory and sedentary birds in Hokkaido, Japan. Vet J. 2012;192(3):538–40.
31. Kenzy S, Cho B. Transmission of classical Marek's disease by affected and carrier birds. Avian Dis. 1969;211–4.

32. Calnek BW. Marek's disease virus and lymphoma. In: Rapp F, editor.
     Oncogenic Herpesviruses. Boca Raton: CRC Press; 1980. p. 103-143.
33. Witter RL. Increased virulence of Marek's disease virus field isolates. Avian
     Dis. 1997;41(1):149–63.
34. Wozniakowski G, Samorek-Salamonowicz AE. Molecular evolution of Marek's
     disease virus (MDV) field strains in a 40-year time period. Avian Dis. 2014;
     58(4):550–7.
35. Madej JP, Wozniakowski G, Gawel A. Morphology of immune organs after
     very virulent plus strain of Marek's disease virus infection in vaccinated
     hens. Pol J Vet Sci. 2016;19(2):325–35.

10

Phylogenetic characterization of genes encoding for viral envelope glycoprotein (ORF5) and nucleocapsid protein (ORF7) of porcine reproductive & respiratory syndrome virus found in Malaysia in 2013 and 2014

Seetha Jaganathan King[1,3], Peck Toung Ooi[1*], Lai Yee Phang[2], Zeenathul Nazariah Binti Allaudin[1], Wei Hoong Loh[5], Chiou Yan Tee[5], Shiao Pau How[4], Lai Siong Yip[5], Pow Yoon Choo[5] and Ban Keong Lim[5]

Abstract

Background: Porcine reproductive and respiratory syndrome (PRRS) is one of the most expensive diseases of modern swine production & results in annual economic losses and cost the industry over 600 million USD in U.S. alone and billions of dollars worldwide. Two atypical PRRS cases were observed in 2013 and 2014 characterized by late-term abortion, fever and sudden increase in sow mortality which persisted for a prolonged period of time.

Methods: Lungs, lymph nodes and other samples were collected for disease investigation. Sequencing of the viral envelope glycoprotein (ORF5) and nucleocapsid protein (ORF7) of PRRSV was done using the BigDye Terminator v3. 1 cycle sequencing kit chemistry. The phylogenetic tree was constructed by using the Maximum Likelihood method, generated by Mega 6.06®.

Results: Analysis of the ORF5 and ORF7 showed high degree of sequence homology to PRRSV parent vaccine strain VR-2332, RespPRRSV and other mutant/chimeric virus strains.

Conclusions: Our study suggests that recombination events between vaccine strains and field isolates may contribute to PRRSV virulence in the field.

Keywords: Porcine reproductive and respiratory syndrome virus, PRRSV, Genetic characterization, ORF5 gene, ORF7 gene

Background

Porcine reproductive and respiratory Syndrome (PRRS) is an economically important viral disease that is easily transmitted through direct contact to susceptible pigs and vertically to foetuses. The disease is also known as Mystery Swine Disease, Blue Ear Disease, Porcine Endemic Abortion & Respiratory Syndrome (PEARS) and Swine Infertility Respiratory Syndrome (SIRS) [1, 2]. It is known as one of the most expensive disease of modern swine production. A porcine reproductive and respiratory syndrome virus (PRRSV) outbreak can devastate a herd and determining the origin of the virus can be impossible. PRRS is characterized by an acute viral infection of the porcine macrophage that leads to an immunologically altered state. In extreme cases, respiratory distress, metabolic dysregulation and neuronal involvement result in significant mortality within days to weeks of experimental inoculation with highly pathogenic isolates [3–5]. The virus can also reappear in farms that have taken great lengths to eliminate the virus.

* Correspondence: ooi@upm.edu.my; ooivetupm@gmail.com
[1]Department of Clinical Studies, Faculty of Veterinary Medicine, Universiti Putra Malaysia, UPM, Serdang, Selangor 43400, Malaysia
Full list of author information is available at the end of the article

Endemic disease from emerging and re-emerging PRRSV results in estimated annual economic losses and the virus is estimated to cost the industry over 600 million USD, or 1.5 million USD per day in the U.S. economy alone [6].

Since its emergence, much has been studied and learned about the virus. It was first detected in North America and reported in 1987 [1]. The virus was then subsequently isolated in Europe in 1990 [7]. Since then it has spread rapidly to Asia and throughout the world. The porcine reproductive & respiratory syndrome virus (PRRSV), the causative agent for the syndrome, is a positive-sense single stranded RNA virus, belonging to the family *Arteriviridae* of the order *Nidovirales*, and genus *Arterivirus* [8, 9].

The PRRSV genome organization is similar to other arterivirus and is approximately 15 kilobases in length. There are 10 open reading frames (ORFs), ORF1a and ORF1b encoding polyproteins that are processed into 14 non-structural proteins (nsp) by viral proteases within the virus genome [10]. The glycosylated membrane associated minor structural proteins GP (2a), GP3 and GP4, respectively are encoded by ORF2a, ORF3 and ORF4 [11]. ORF2b encodes 2b protein, a non-glycosylated structural protein which is virion associated and the principal product of ORF2 [12]. Three major structural proteins, GP5, M and N protein within the virus genome are encoded by ORF5, ORF6 and ORF7, respectively. GP5a, which is referred to as ORF5a protein, is a novel structural protein encoded by an alternative ORF of the subgenomic mRNA encoding GP5 and is incorporated into the virion [13, 14].

Based on genetic characterization, there exist two related but antigenically and genetically distinguishable major genotypes with over 50% RNA sequence variation; the European strain (EU genotype, Type 1, with Lelystad virus as the prototype) representing the viruses predominating in Europe and the North American strain (NA genotype, Type 2, with VR-2332 as the prototype) originally and mostly found in North America [15]. Both genotypes have been described to be evolving independently in Europe and North America and the co-existence of both genotypes has been increasingly evident in several countries, including Malaysia, Thailand, Korea and China [16–20]. Most recently, a variant of genotype 2 also known as highly pathogenic strain of PRRSV, genetically characterized by a unique discontinuous deletion of 30 amino acids (aa) in the non structural protein (Nsp2) of the North American strains was confirmed by the Office International Des Epizooties (OIE) and the Food and Agricultural Organization (FAO) as the causative agent for the severe "high fever" disease designated as the highly pathogenic strain of PRRSV in Asia. Because of its economic significance, a great deal of resource has been invested to research the virus and in developing effective prevention and control strategies. But protocols providing consistent success have been elusive due to the high rate of genetic change and antigenic variability [2, 21–24].

Situation in South East Asia

Throughout Asia, PRRS outbreaks were reported in many countries between the late 1980s and early 1990s [9]. The highly pathogenic PRRS (HP-PRRS) which emerged in China in 2006 has spread to South East Asian countries since 2007 [25]. The highly pathogenic PRRS was reported in Vietnam in March 2007 [26], Laos in June 2010 [27], Thailand in 2010 [28], Myanmar in February 2011 [29], Cambodia in August 2010 [30], Philippines in August 2010 [30]. Transboundary spread of HP-PRRSV from southern China to South East Asia suggests that biosecurity failures have occurred, including failure to control animal movements and trading among neighboring countries at borders [31, 32] (Fig. 1).

Situation in Malaysia

In Malaysia, a syndrome very similar to PRRSV has been recorded in various pig farms as early as 1995 [33]. A serological survey done in about 100 farms in major pig rearing states in the country showed that the pigs found in 93 out of the 100 farms had serological evidence of infection. Subsequently a study done in 2008 showed that 94% of the farms and 83.4% of the pigs were tested positive for PRRSV antibodies [34]. In 2012, another local study documented 89.2% sero-prevalence out of 120 sera collected from 12 non-PRRS vaccinated farms in 6 states. In the same study, 27 tissue samples were collected from 11 farms [35]. Twelve of the tissue samples were positive for PRRSV with all positive for US strains in the selected pig farms. The study concluded that there are more US strains in the selected pig farms, however, it should be noted that there were limitations as the number of samples studied were too small. In 2013, another study conducted in Malaysia concluded that both EU and NA strains are present in Malaysian [36].

Farming situation in Malaysia

There is approximately 772 farms in Malaysia with 565 farms in West Malaysia with an approximate total no of sows of 0.17 million of which 0.14 million is located in West Malaysia. The estimated total ex farm value of the swine industry in Malaysia is around 2.2 billion ringgit. Similar to the farming situation in other countries, the farming industry in Malaysia has moved on from backyard to industrialized farming systems, with many small farms that has shut down; merged or bought over by bigger players in the industry which explains the reason for the number of farms that have reduced but the farm sizes that are increasing. The farms are mostly open

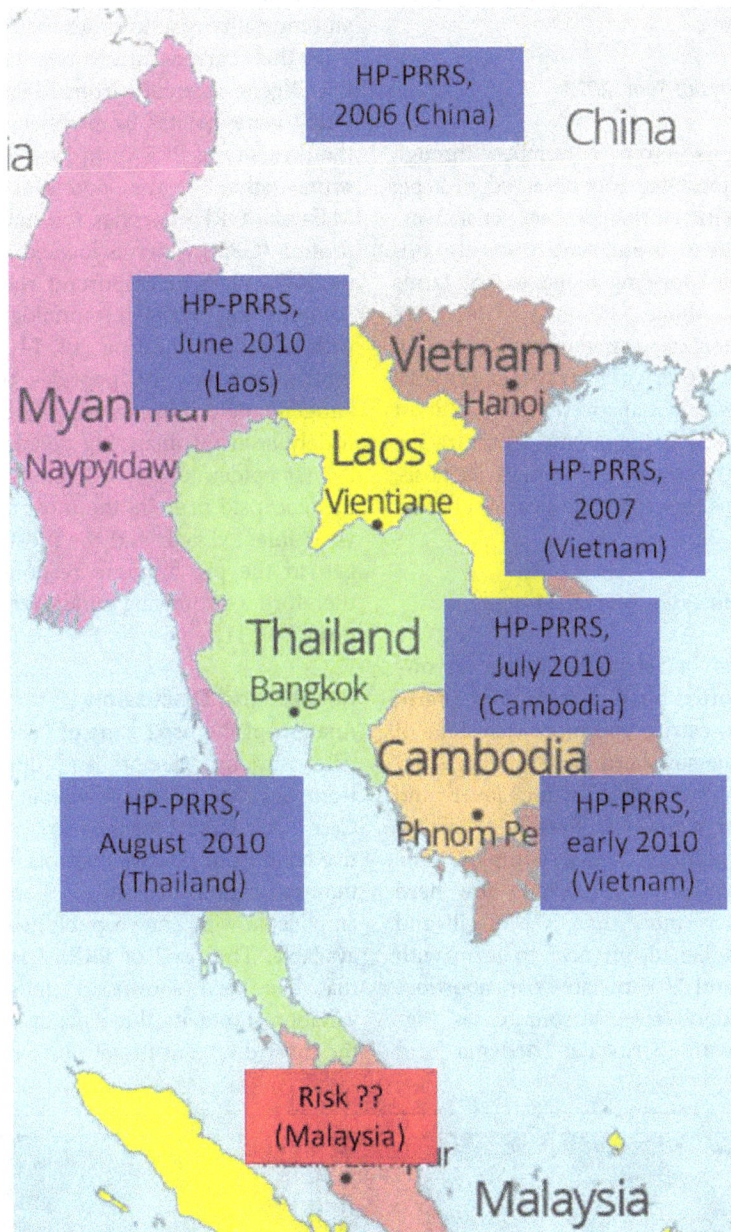

Fig. 1 Is Malaysia at risk of HP-PRRSV? Since the disease started in China, it has quickly spread to the surrounding countries in South East Asia. Malaysia is constantly at risk and threat after Thailand reported its first HP-PRRS in 2010

house, farrow to finish and single site, about 10% of the farms use closed house system. The swine respiratory health status varies from state to state and from farm to farm as well. Selangor (located in Central Malaysia) and Penang (located in North Malaysia) are highly urbanized states with limited land space; therefore the farms are close to each other. Johor (located in South Malaysia and borders with Singapore)—these farms used to export to Singapore and are more established. Perak is located between central and north Malaysia which used to be a mining area; the pig farms are located next to lakes and

ponds. Sabah and Sarawak which is located in East Malaysia have more land and space, thus, in theory there are less issues.

Current study

Between 2013–2014, 22 diagnostic cases were received for diagnostic investigation. Seven out of 22 cases received were positive. Six out of the seven cases were positive for NA strain while one case was positive for both EU and NA strain. Among all those cases that were received, there were two atypical cases observed in 2013 and subsequently in 2014 in which high mortality rates were observed in the farm.

Case 1: Location: East Malaysia; Year: 2013

History & clinical signs From November through December of 2012, high mortality was observed in a pig farm in East Malaysia with its return/repeat service increasing by 30%. The piglets were weak and runt. It was also evident that the disease was spreading along to pig farms located in close proximity. About 2–3 months later, there were occurrence of problems in weaners showing signs of Edema and Classical Swine Fever virus and other bacterial infections. At that point, veterinarians were called to assist to collect samples and investigate the disease. Upon tracing back the case history and through differential diagnosis, there were adequate reasons to suspect the case as a potential high fever PRRSV (Fig. 2).

Case 2: Location: Central Malaysia, Year: 2014

History & Clinical Signs In February 2014, a farrow-finish 300 sow herd reported an outbreak of late-term abortion, 30% of repeat to estrus, and more than 50% of pre-weaning mortality. The sow herd had been regularly vaccinated with Aujeszky's vaccine, swine fever and PRRS MLV. Mortality of grower & finisher were increased with porcine respiratory disease complex. One week after a schedule blanket vaccination in sow herd with PRRS MLV, there were more than 50% of gilt and first parity sow where sudden death near to term with hyperemia and pyrexia, and 20% of late-term abortion. Umbilical hemorrhage was observed in some cases. Piglet showed ill thrift with periocular oedema and

conjunctivitis. Weaners exhibited dyspnea and lethargy and mortality rose to more than 80% (Fig. 3).

In this current study, we compare the ORF5 and ORF7 gene sequence from Malaysia with the ORF5 and ORF7 gene isolates from other Asian countries to study the diversity of PRRSV in Malaysia which may help shed some light on the potential origin of PRRSVs in Malaysia. ORF5 encodes the major viral envelope glyco-protein (GP5), which is located on the surface of the virion. GP5 plays an important role in viral infectivity and contains important immunological domains associated with viral neutralization [13, 14]. Several peptide/protein motifs, such as signal peptides, trans-membrane regions, antigenic determinants and glycosylation sites have been widely used for analyzing genetic variation and the molecular epidemiology of PRRSV. The ORF7 encodes the nucleocapsid protein, the most abundant viral protein in virus-infected cells and the most immunodominant antigen in the pig immune response to PRRSV. ORF7 is, therefore a promising candidate for detection and diagnosis [10, 11].

Results and Discussion
Analysis of the nsp2 gene of Case 1 and Case 2
After various attempts and optimization strategies, the team did not manage to obtain any positive bands from Case 1 and Case 2 for the nsp2 gene despite getting a positive band from positive controls. It is highly suspected that the nsp2 regions of both cases are highly mutated and not amplifiable with known published primers that are readily available. The nsp2 of PRRSV is a multi-domain protein that has been shown to undergo remarkable genetic variation. From its three major domains [37–39], nsp2 is the most divergent protein between PRRSV types 1 and 2

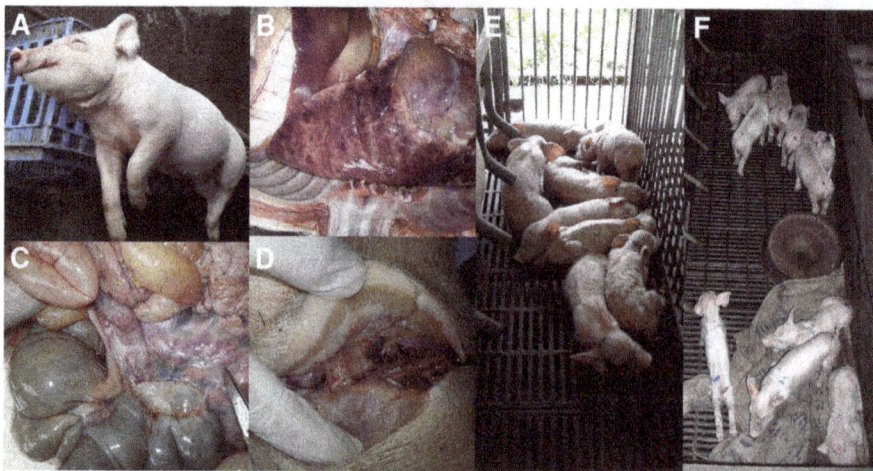

Fig. 2 Case 1 (Sar01/2013) East Malaysia. **a**, **b**, **c**, **d**: At the time of sample collection, the case looked like Edema and the animals had secondary bacterial infections. **e**, **f**: The virus had spread to the neighbouring farms and showed similar clinical signs. 3 days old piglets looked weak, runt and emaciated

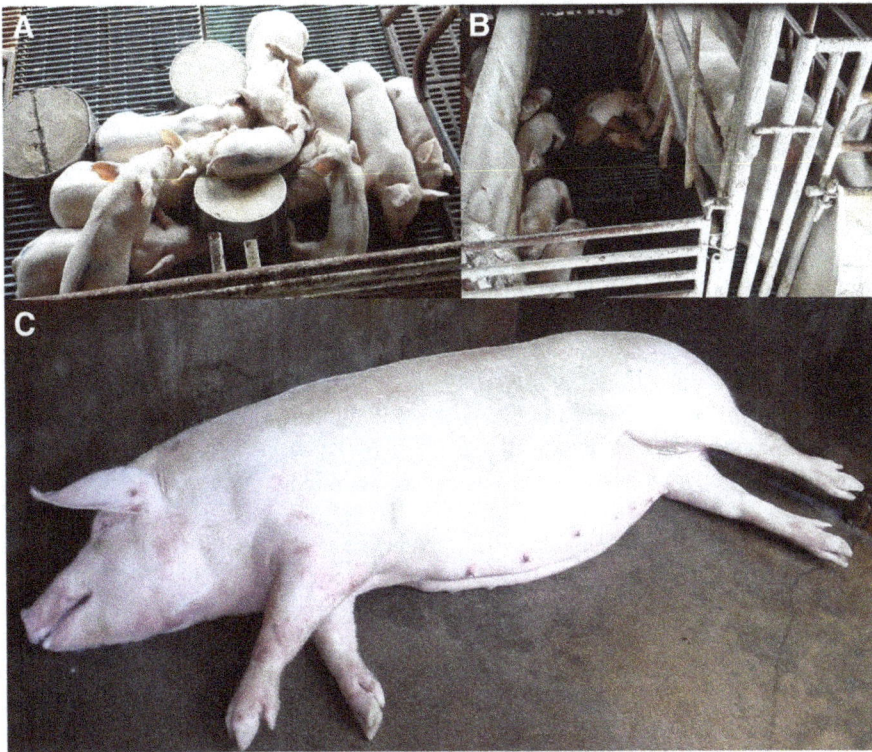

Fig. 3 Case 2 (Sel01/2014) Central Malaysia. **a** and **b**: The piglets are weak, runt and chilled. They are clustering together to stay warm. **c**: Sudden death detected in sow with hyperemia and pyrexic

[40, 41] and also between the pathogenic PRRSV 16244B and an attenuated vaccine and its parental strain VR2332 [40–42]. A large amount of data supports the theory that the middle section of the nsp2 protein is highly susceptible to mutation and tolerant to insertions and deletions, regardless of the pathogenic phenotype of the isolates [39–50]. In fact, reverse genetics experiments have mapped several non-essential viral replication regions in nsp2, including a large deletion of 402 nucleotides in the middle region of the gene [39, 51].

Analysis of the ORF5 gene (KU512850) of Case 1 (Sar01/2013) from East Malaysia

BLAST analysis of the ORF5 sequence derived from Case 1 (Sar01/2013) from East Malaysia (KU512850) on Genbank showed that the sequence had high similarities to PRRSV sequences from China (KR612142; KR018787), South Korea (KP317086) and USA (KT905092; KT904941). Based on the phylogenetic study, the ORF5 gene sequence derived from Case 1 (Sar01/2013) from East Malaysia (KU512850) clustered together with the ORF5 gene sequences PL97-1 from South Korea (AY585241), 4034-2-v-2008 from South Korea (FJ972733), MD-001 from Taiwan (AF121131) and RespPRRS MLV (AF066183), MLV RespPRRS AF159149 and ATCC VR-2332 U87392.3, the parent strain of the vaccine Ingelvac PRRS MLV. The sequence comparison

showed that the nucleotide sequence of the ORF5 gene derived from Case 1 (Sar01/2013) from East Malaysia (KU512850) had 99.3% nucleotide sequence homology with RespPRRS MLV (AF066183) and RespPRRS MLV (AF159149), 99% nucleotide sequence homology with ATCC VR-2332 (U87392.3) the parent strain of the vaccine Ingelvac PRRS MLV and 98% nucleotide sequence homology with 4034-2-V-2008 (FJ972733) from South Korea.

Three mismatched amino acid changes were observed in position 3, 4 and 13 throughout the ORF5 gene in comparison to ATCC VR-2332 (U87392.3). Amino acid substitutions at positions 3 (Glutamate, E) was substituted with (Glycine, G); amino acid substitution at position 4 (Lysine, K) was substituted with (Glutamate, E) and amino acid substitution at position 13 (Arginine, R) was substituted with (Glutamine, Q) was observed in Case 1 from East Malaysia (KU512850) (Fig. 6).

Analysis of the ORF7 gene (KU512849) of Case 1 (Sar01/2013) from East Malaysia

BLAST analysis on Genbank shows that the ORF7 gene sequence (KU512849) of Case 1 from East Malaysia derived in this study is highly similar to PRRSV strain JN-HS from Shandong, China, 2008 (Accession No: HM016158). Further analysis by phylogenetic tree studying only selected sequences of the ORF7 gene (KU512849) of Case 1 (Sar01/2013) from East Malaysia with other highly similar sequences from Genbank showed that it clustered with sequences from China (HM016158, KM453699, KM453698), Vietnam (JQ860406, JQ860392, JQ860419, JQ860410) and Laos (JN626287). The phylogenetic tree also suggest that the ORF7 gene (KU512849) of Case 1 (Sar01/2013) from East Malaysia is a derivative of JX-AI from Jiangxi, China (EF112445) which is a representative strain of highly pathogenic PRRSV in China since 2006. The sequence comparison of the nucleotide sequence of ORF7 gene (KU512849) of Case 1 (Sar01/2013) from East Malaysia showed 97.8% sequence homology to to JN-HS from China, 96.7% sequence homology to JX-A1 from Jiangxi, China and 97.5% sequence homology to these sequences from China (HM016158, KM453699, KM453698), Vietnam (JQ860406, JQ860392, JQ860419, JQ860410) and Laos (JN626287). The high genetic similarity between Case 1 (Sar01/2013) and JX-A1 (EF112445) suggest that Case 1 (Sar01/2013) from East Malaysia may be potentially a first HP-PRRSV found in Malaysia. Multiple mismatched amino acid changes were observed in position 15, 46, 91, 109, 117 and 122 throughout the ORF7 gene of Case 1 (Sar01/2013) from East Malaysia in (KU512849) comparison to ATCC VR-2332 (U87392.3).

Analysis of the ORF5 gene (KU512851) of Case 2 (Sel01/2014) from Central Malaysia

BLAST analysis on Genbank showed that the ORF5 gene sequence derived in this study was highly similar to PRRSV isolate Shizuoka from Japan (Accession No: AB175704.1) and other isolates from America (KT894735.1, U34298.1, DQ477864.1, DQ477718.1). One of the sequence that is highly similar to the ORF5 gene of Case 2 (Sel01/2014) from Central Malaysia is a synthetic contruct clone of PRRSV. Based on the phylogenetic study, the ORF5 gene sequence derived from this study clustered together with the ORF5 gene sequences from Taiwan (AF121131), South Korea (FJ972733; AY585241); RespPRRS MLV (AF066183), MLV RespPRRS AF159149 and ATCC VR-2332 U87392.3, the parent strain of the vaccine Ingelvac PRRS MLV. Sequences for constructing the phylogenetic tree were selected based on genetic relatedness and completeness of sequences that were available on Genbank. The sequence comparison demonstrated that the nucleotide sequence of ORF5 gene derived from Case 2 (Sel01/2014), Central Malaysia (KU512851) has 89.5% sequence homology with sequence MD-001 (AF121131) from Taiwan. The sequence homology of the ORF5 gene from Case 2 (Sel01/2014), Central Malaysia (KU512851) with the vaccine strains were relatively lower than expected with 85.5% with RespPRRS MLV (AF066183), 85.5% with MLV RespPRRS AF159149 and 86% with ATCC VR-2332 U87392.3, the parent strain of the vaccine Ingelvac PRRS MLV. This is not surprising

as it has been documented that PRRSV strains differ in virulence [22, 52–55] and vary genetically [56–59] suggesting that the ORF5 gene derived from Case 2 (Sel01/2014), Central Malaysia (KU512851) may be a derivative of the Ingelvac PRRS MLV vaccine, a possible recombinant of the vaccine virus and a wild-type virus, or a truly wild-type virus that is partially homologous to the original parent vaccine strain, VR2332, which may be still circulating in the field [55].

As expected that the ORF5 that encodes for the major envelope glycoprotein would have a high degree of mutation as it is known to be the most variable region [55], multiple mismatched amino acids were observed throughout the ORF5 gene of Case 2 (Sel01/2014), Central Malaysia (KU512851) in comparison to ATCC VR-2332 (U87392.3) (Fig. 4).

Analysis of the ORF7 gene (KU512848) of Case 2 (Sel01/2014) from Central Malaysia

BLAST analysis on Genbank of the ORF7 gene sequence from Case 2 (Sel01/2014), Central Malaysia (KU512848) derived in this study demonstrates high similarity to PRRSV strains from South Korea, China, Denmark and USA. Further analysis by phylogenetic tree studying only selected sequences of the ORF7 gene with other highly similar sequences from Genbank suggest that the ancestor for the sequence from this study is ATCCVR2332 which is the parent strain to Ingelvac PRRS MLV vaccine (Fig. 5). The phylogenetic tree also suggest that the ORF7 gene sequence derived from this study groups closely with other sequences from Genbank such as V7-HA-myc (FJ524376), RespPRRS MLV (AF066183), RVRp22 (KM386622), FJSD (KP998474), DK-2004-1-7-PI (KC862578), DK-1997-19407B (KC862576), YN-2011 (JX857698), pMLV/MN184-3UTR (FJ629371) and pMLV/MN184ORF5-6 (FJ629369); two of which are chimeric infectious Porcine reproductive and respiratory syndrome virus type 2 clone representing a background of strain Ingelvac PRRS MLV and 3′ UTR of MN184. The sequence comparison confirmed that the nucleotide sequence of ORF7 gene derived from this study is 98.3% similar to ATCC VR-2332 which is the parent strain to Ingelvac PRRS MLV vaccine and has 100% similarity to other sequences as listed in Table 1; two of which are chimeric infectious PRRSV pMLV/MN184-3′UTR FJ629371 & pMLV/MN184-3′UTR FJ629371. Three mismatched amino acid changes were observed in position 49, 54 and 56 throughout the ORF7 gene of Case 2 (Sel01/2014), Central Malaysia (KU512848) in comparison to ATCC VR-2332 (U87392.3) (Fig. 7). At first glance, the four sequences derived from these two cases display a vast diversity in terms of

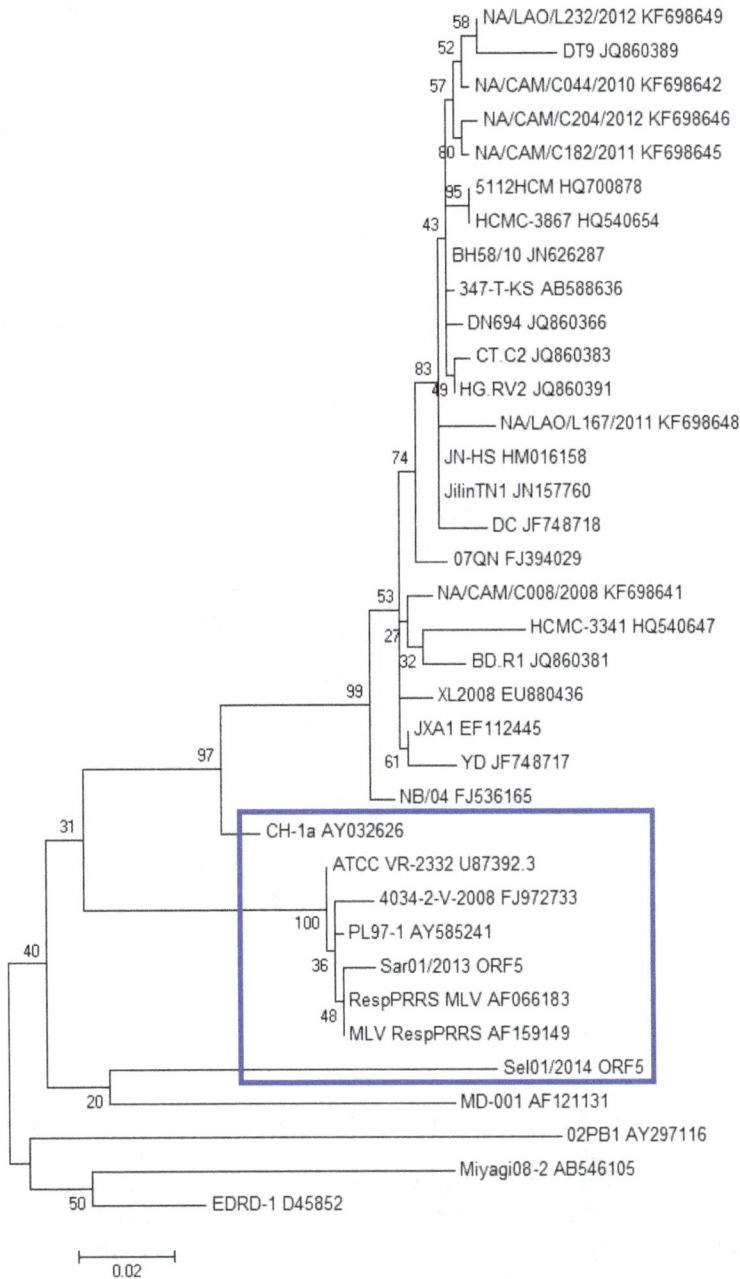

Fig. 4 Molecular Phylogenetic analysis of the ORF5 gene of Case 1 (Sar01/2013) & Case 2 (Sel01/2014) from East Malaysia. (Source of Map: http://aseanup.com/free-maps-asean-southeast-asia/)

geographical origin of the PRRS virus found in Malaysia. However, when looked closely, all four sequences seem to have one thing in common, they display high sequence homology to the modified live vaccine virus strain regardless of which geographical location the sequences are highly similar to.

Both cases displayed a high degree of mutation in both genes, ORF5 and ORF7. There are several possibilities that the ORF5 and ORF7 genes from Case 1 (Sar01/2013) and Case 2 (Sel01/2014) are derivatives of the

Ingelvac PRRS MLV vaccine, a possible recombinant of the vaccine virus and a wild-type virus, or a truly wild-type virus that is highly homologous to the original parent vaccine strain, VR 2332. This theory was discussed by Opriessnig in 2002 [55].

Further to that, it has been reported that the use of synthetic porcine reproductive and respiratory syndrome virus strain confers unprecedented levels of heterologous protection because current vaccines do not provide sufficient levels of protection against divergent PRRSV

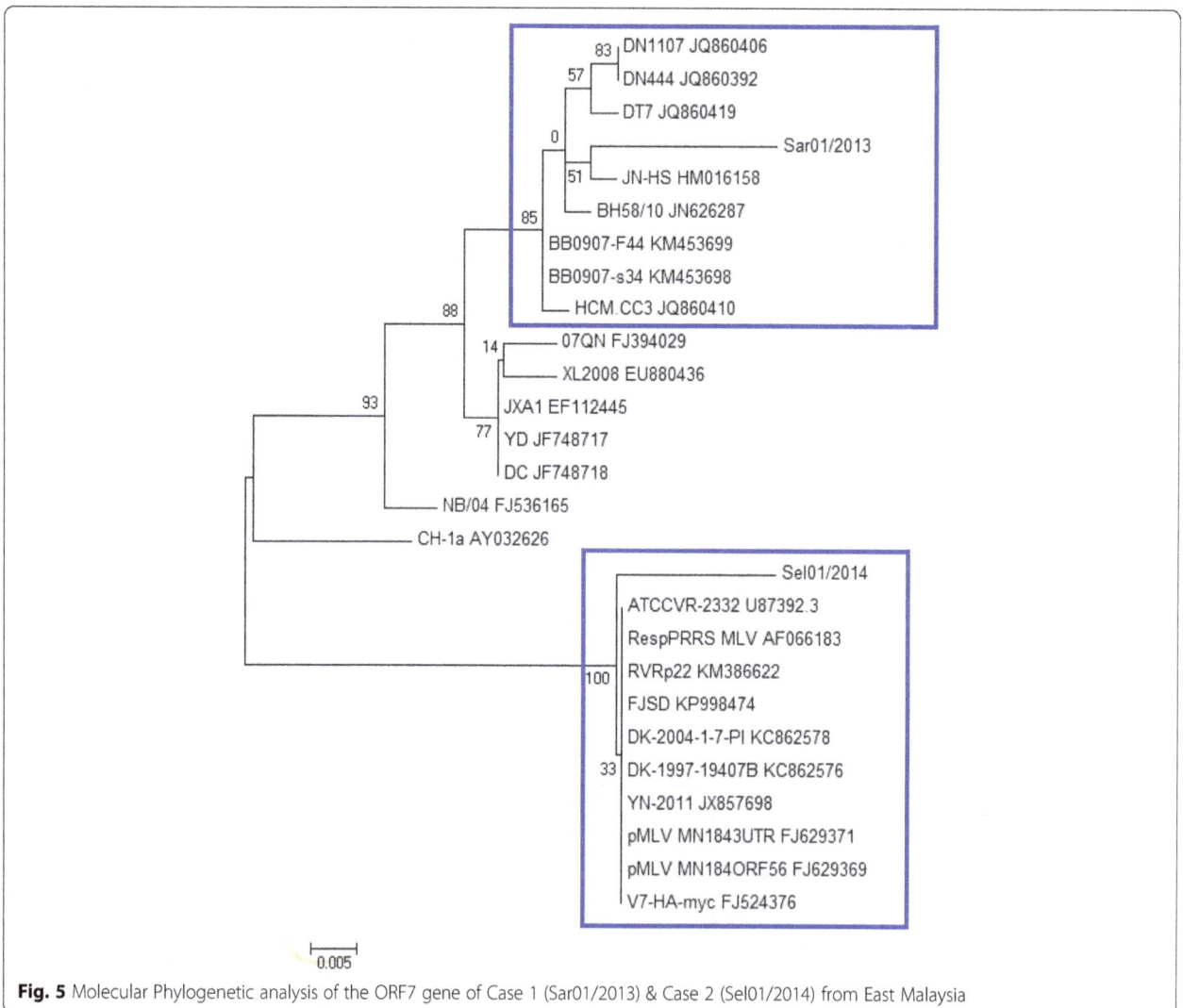

Fig. 5 Molecular Phylogenetic analysis of the ORF7 gene of Case 1 (Sar01/2013) & Case 2 (Sel01/2014) from East Malaysia

strains circulating in the field, mainly due to the substantial variation of the viral genome [60]. However, based on the phylogenetic study and sequence comparison, our study suggest that these synthetic chimeric infectious PRRSV constructs are highly similar to the PRRSV that are causing high mortality rate in the farm in Case 2 (Sel01/2014) in Central Malaysia. Therefore, it is apparent that the rapid rate of PRRSV gene mutation remains a huge challenge for the practicality of such synthetic clone vaccines.

Conclusion

Over the years, there have been biased and mixed reports of the presence of PRRSV in Malaysia. Based on a collection of studies, it can be confirmed that both European and North American strains are present in Malaysia. However, over the years there have been more North American strains reported compared to European strains. Despite global and intensive use of MLV vaccine, repeated

PRRS outbreaks continue due to constant genetic changes in field isolates. Rapid evolution due to high mutation rate has led to new generations of genetically and antigenically variable virus strains in the field. Vaccine strains or derivatives of vaccine strains may also induce disease in the field and persist in vaccinated pigs and spread to non-vaccinated pigs contributing to PRRSV virulence in the field and the inability for effective vaccination. As reported by many other researchers in other countries [55], the results from this study suggest that the virus that persisted in both farms is a product of a recombination event between vaccine strains and field isolates. For future studies, genetic and evolutionary analyses of full length genomes are important to delineate the degree of homology among PRRSVs and for effective vaccine design. We are unable to confirm the presence of highly pathogenic PRRSV in Malaysia. The high degree of nucleotide homology between the local Malaysian isolates with representative strains of highly pathogenic PRRSV strains from China

Table 1 PRRSV isolates derived from this study and other isolates reported previously used for comparison and constructing the phylogenetic tree in this study

No	Isolate Name	Genbank® Accession Number	Gene region	Year	Country	Reference
1	Sar01/2013	KU512850	ORF5	2013	Malaysia	This study
2	Sar01/2013	KU512849	ORF7	2013	Malaysia	This study
3	Sel01/2014	KU512851	ORF5	2014	Malaysia	This study
4	Sel01/2014	KU512848	ORF7	2014	Malaysia	This study
5	NA/LAO/L232/2012	KF698649	ORF5	2012	Laos	Genbank®
6	DT9	JQ860389	ORF5	2012	Vietnam	Genbank®
7	NA/CAM/C044/2010	KF698642	ORF5	2010	Cambodia	Genbank®
8	NA/CAM/C204/2012	KF698646	ORF5	2012	Cambodia	Genbank®
9	NA/CAM/C182/2011	KF698645	ORF5	2011	Cambodia	Genbank®
10	5112HCM	HQ700878	ORF5	2010	Vietnam	Genbank®
11	HCMC-3867	HQ540654	ORF5	2010	Vietnam	Genbank®
12	BH58/10	JN626287	ORF5/ORF7	2010	Laos	Genbank®
13	347-T-KS	AB588636	ORF5	2010	Vietnam	Genbank®
14	DN694	JQ860366	ORF5	2008	Vietnam	Genbank®
15	CT.C2	JQ860383	ORF5	2012	Vietnam	Genbank®
16	HG.RV2	JQ860391	ORF5	2012	Vietnam	Genbank®
17	NA/LAO/L167/2011	KF698648	ORF5	2011	Laos	Genbank®
18	JN-HS	HM016158	ORF5/ORF7	2008	China	Genbank®
19	JilinTN1	JN157760	ORF5	2011	China	Genbank®
20	DC	JF748718	ORF5/ORF7	2010	China	Genbank®
21	07QN	FJ394029	ORF5/ORF7	2007	Vietnam	Genbank®
22	NA/CAM/C008/2008	KF698641	ORF5	2008	Cambodia	Genbank®
23	HCMC-3341	HQ540647	ORF5	2010	Vietnam	Genbank®
24	BD.R1	JQ860381	ORF5	2010	Vietnam	Genbank®
25	XL2008	EU880436	ORF5/ORF7	2008	China	Genbank®
26	JXA1	EF112445	ORF5/ORF7	2006	China	Genbank®
27	YD	JF748717	ORF5/ORF7	2009	China	Genbank®
28	NB/04	FJ536165	ORF5/ORF7	2004	China	Genbank®
29	CH-1a	AY032626	ORF5/ORF7	1996	China	Genbank®
30	ATCC VR-2332	U87392.3	ORF5/ORF7	1995	USA	Genbank®
31	PL97-1	AY585241	ORF5	1997	Korea	Genbank®
32	RespPRRS MLV	AF066183	ORF5/ORF7	2005	USA	Genbank®
33	MLV RespPRRS	AF159149	ORF5	1999	USA	Genbank®
34	4034-2-V-2008	FJ972733	ORF5	2008	Korea	Genbank®
35	MD-001	AF121131	ORF5	1997	Taiwan	Genbank®
36	02 PB1	AY297116	ORF5	2002	Thailand	Genbank®
37	Miyagi08-2	AB546105	ORF5	2008	Japan	Genbank®
38	EDRD-1	D45852	ORF5	1992	Japan	Genbank®
39	DN1107	JQ860406	ORF7	2009	Vietnam	Genbank®
40	DN444	JQ860392	ORF7	2008	Vietnam	Genbank®
41	DT7	JQ860419	ORF7	2012	Vietnam	Genbank®
42	BB0907-F44	KM453699	ORF7	2009	China	Genbank®
43	BB0907-s34	KM453698	ORF7	2014	China	Genbank®

Table 1 PRRSV isolates derived from this study and other isolates reported previously used for comparison and constructing the phylogenetic tree in this study *(Continued)*

44	HCM.CC3	JQ860410	ORF7	2010	Vietnam	Genbank®
45	V7-HA-myc	FJ524376	ORF7	2010	USA	Genbank®
46	RVRp22	KM386622	ORF7	2014	Korea	Genbank®
47	FJSD	KP998474	ORF7	2015	China	Genbank®
48	DK-2004-1-7-PI	KC862578	ORF7	2004	Denmark	Genbank®
49	DK-1997-19407B	KC862576	ORF7	1997	Denmark	Genbank®
50	YN-2011	JX857698	ORF7	2011	China	Genbank®
51	pMLV/MN184-3UTR	FJ629371	ORF7	2010	USA	Genbank®
52	pMLV/MN184ORF5-6	FJ629369	ORF7	2010	USA	Genbank®

suggest that a derivative of the highly pathogenic virus strain may be present in Malaysia.

Methods
Sampling
Two sets of samples from two atypical PRRS cases were sent for disease investigation. One of the sample was from Central Malaysia, Selangor (lungs and lymp nodes) and another from East Malaysia, Sarawak (brains and lungs). The organ samples were pooled for diagnostic testing.

Animals were humanely slaughtered for disease investigation purpose by professionally trained veterinarians. There was no experimental research done on the animals.

Nucleic acid extraction
Nucleic acid extraction was carried out on the pooled organ samples by using TRIsure® (Bioline®). 100–200 mg of tissue was collected and homogenized using a mortar and pestle and 1 mL of Phosphate Buffered Saline (PBS). 750 uL of cold TRIsure® was added to 250 uL of the homogenized samples in a microcentrifuge tube. The samples were then vortexed vigorously and incubated at room temperature for 5 min. 200 uL of cold chloroform was then added into the same microcentrifuge tube, vortexed vigorously and incubated at room temperature for 5 min. The microcentrifuge tube is then centrifuged at 12, 000 rpm for 15 min at 4 °C. The upper aqueous solution was then transferred to a new and clean microcentrifuge tube. 5 uL polyacryl carrier (Molecular Research Centre Inc) and 500 uL of cold isopropanol are then added and the microcentrifuges containing these reagents were incubated at room temperature for 10 min. After 10 min, centrifuge at 12, 000 rpm for 10 min at 4 °C. Decant the supernatant and wash pellet with 1 mL of 75% cold ethanol. Centrifuge at 12, 000 rpm for 5 min at 4 °C. The pellet was resuspended in 50 uL of TE buffer.

PCR amplification of PRRSV—ORF5 and ORF7 gene
The presence of PRRSV in the samples were assessed using a previously described reverse transcriptase nested PCR assay that amplifies a 241 bp nucleotide (European strain) and 337 bp nucleotide (North American strains) respectively [61, 62]. Three sets of primers were used. PLS: 5'-ATG GCC AGC CAG TCA ATC-3'; PLR: 5-TCG CCC TAA TTG AAT AGG TG-3' [62–64] to reverse transcribes and amplifies a common site in the ORF 7 region of both strains. The nested primer sets for the North American and European Strains were P-US-s: 5'-AGT CCA GAG GCA AGG GAC CG-3'; P-US-as:5'-TCA ATC AGT GCC ATT CAC CAC-3' and P-EU-s:5'-ATG ATA AAG TCC CAG CGC CAG CGC CAG-3'; P-EU-as:5'-CTG TAT GAG CAA CCG GCA GCA T-3'.

PCR amplification with ORF5 gene
The primer pairs used were ORF5-F: 5'-ATGTTGGG-GAAGTGCTTGACC-3' and ORF5-R: 5'-CTAGAGAC-GACCCCATTGTTCCGC-3' [65].

PCR amplification with nsp2 gene
The primer pairs used were 2492-F: 5'-GRACTTCCT-CARCTTCTTGC-3' and 3160-R: 5'-TCGACGAGCT-TAAAGACCAGA-3' [51].

Sequence alignment & phylogenetic analysis of the ORF5 and ORF7 region
Virus sequences were derived from clinical samples sent to the laboratory for diagnostic investigation, with the permission of local veterinary authorities. Virus RNA was extracted using TRIzol reagent according to the manufacturer's instructions. RNA was quantified by using spectrophotometer (SpectraMax® Plus 384, Molecular Devices). A 603 bp (ORF5) and 337 bp (ORF7) region of the was amplified by reverse transcriptase PCR. The PCR products were purified using the PCR clean-up gel extraction kit according to the manufacturer's protocol (Macherey-Nagel, Germany). Sequencing

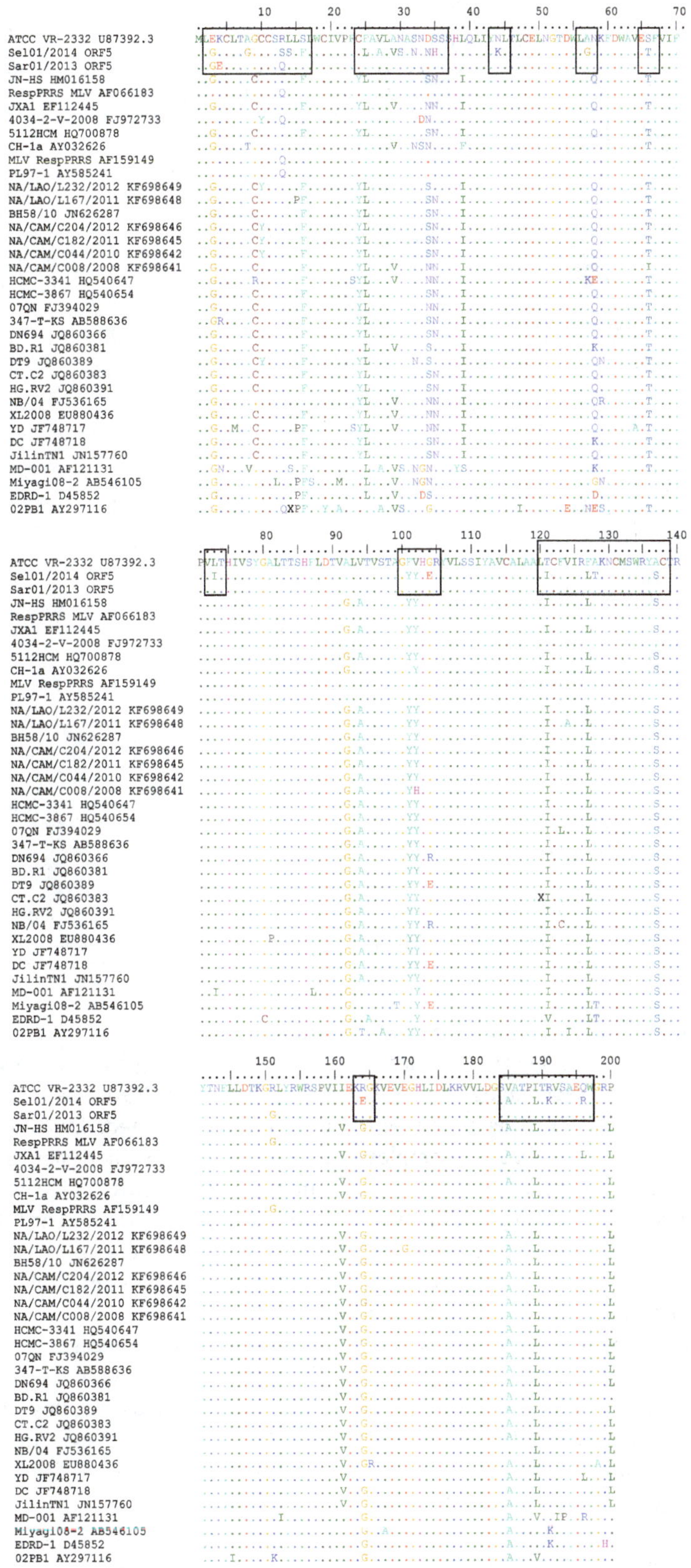

Fig. 6 Amino acid sequence alignment of the ORF5 gene of Case 1 (Sar01/2013) and Case 2 (Sel01/2014) from East and Central Malaysia. Identical sequences are *doted*. *Boxed* portions denote mutations/changes in the amino acid

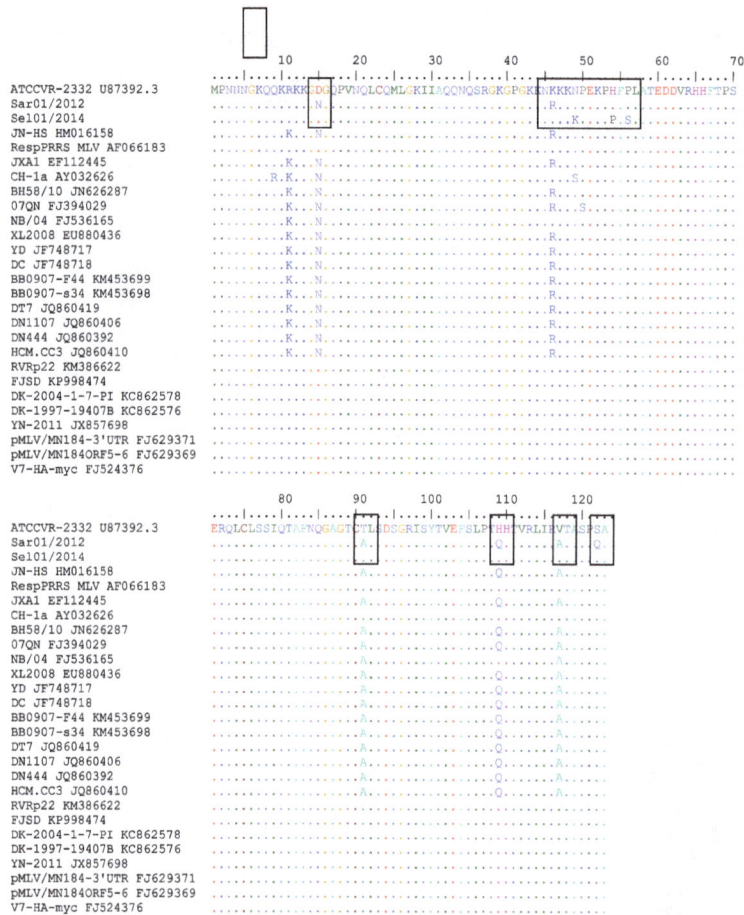

Fig. 7 Amino acid sequence alignment of the nucleocapsid protein (ORF7) gene of Case 1 (Sar01/2013) and Case 2 (Sel01/2014) from East and Central Malaysia. Identical sequences are *doted*. *Boxed* portions denote mutations/changes in the amino acid

of the complete genome of PCV2 was done in a commercial sequencing facility using the BigDye Terminator v3.1 cycle sequencing kit. After sequencing, a Basic Local Alignment Search Tool (BLAST) was performed as a preliminary measure to confirm that all samples were true PRRSV when compared with other sequences deposited in Genbank. The sequence editing and assembly were done by using CLC Workbench. At least one forward and one reverse primer were used to generate a consensus sequence for each gene, which were visually checked for errors prior to alignment using CLC Workbench. Multiple sequence alignments were done by using ClustalW. The phylogenetic tree was constructed by using the Maximum Likelihood method based on the Tamura-Nei model. The percentage of trees in which the associated taxa clustered together is shown next to the branches. Initial tree (s) for the heuristic search were obtained automatically by applying Neighbor-Join and BioNJ algoritms to a matrix of pairwise distances estimated using the Maximum Composite Likelihood (MCL) approach, and then selecting the topology with

superior log likelihood value. The tree is drawn to scale, with branches lengths measures in the number of substitutions per site. Positions containing gaps and missing data were eliminated. Evolutionary analyses were conducted in Mega 6 (Biodesign Institute, Tempe, Arizona). The sequence identity matrix data was generated with BioEdit Sequence Alignment Editor version 7.0.5.2 (Tom Hall, US). Sequences used for constructing the phylogenetic tree are listed in Table 1.

Nucleotide sequence accession numbers
The complete genomic sequences of the ORF5 and ORF7 gene reported in this paper were deposited into the GenBank database under accession numbers KU512848, KU512849, KU512850 and KU512851.

Abbreviations
ORF: Open reading frame; PCR: Polymerase chain reaction; PRRSV: Porcine reproductive and respiratory syndrome virus

Acknowledgements
The authors would like to thank Vet Food Agro Diagnostics (M) Sdn. Bhd for providing the samples for this study and Rhone Ma Malaysia Sdn. Bhd. for

funding the research. The authors would also like to thank Prof. Dr. Henry Too and Dr. Francois Joisel for their advice & contribution.

Funding
The study was funded by Rhone Ma Malaysia Sdn. Bhd.

Authors' contributions
SJ participated in the conceptual aspect of the work, conceived the research, performed the experiments and wrote the manuscript. TCY and LWH collected the samples, documented the clinical signs and assisted in the research design. OPT, PLY, ZNA, HSP, YLS, CPY, LBK provided consultation, research advise and coordination. All authors read and approved the final manuscript.

Competing interest
The authors declare that they have no competing interests.

Consent for publication
Not applicable.

Ethics approval
Organ samples were sent to the laboratory for diagnostic investigation. These diagnostic samples were collected under the supervision of institution veterinarians. The study was conducted following the guidelines as stated in the Code of Practice for Care and use of Animals for Scientific Purposes as stipulated by Universiti Putra Malaysia and complied with the current guidelines for the care and use of animals and was approved by the Animal Care and Use Committee (ACUC), Faculty of Veterinary Medicine, Universiti Putra Malaysia. There was no experimental research done on the animals. No animals were deliberately sacrificed or injured. Every effort was made to minimize any distress or unnecessary culling.

Disclaimer
This document is provided for scientific purposes only. Any reference to a brand or trademark herein is for informational purposes only and for providing specific information and is not intended for commercial purpose, to dilute the rights of the respective owner (s) of the brand (s) or trademark (s) and does not imply recommendation or endorsement by the authors.

Author details
[1]Department of Clinical Studies, Faculty of Veterinary Medicine, Universiti Putra Malaysia, UPM, Serdang, Selangor 43400, Malaysia. [2]Department of Biotechnology, Faculty of Biotechnology & Molecular Science, Universiti Putra Malaysia, UPM, Serdang, Selangor 43400, Malaysia. [3]Asia-Pacific Special Nutrients Sdn. Bhd, Lot 18B, Jalan 241, Section 51A, Petaling Jaya, Selangor 46100, Malaysia. [4]Vet Food Agro Diagnostic Sdn. Bhd, Lot 18B, Jalan 241, Section 51A, Petaling Jaya, Selangor 46100, Malaysia. [5]Vet Food Agro Diagnostic (M) Sdn. Bhd, Lot 18B, Jalan 241, Section 51A, Petaling Jaya, Selangor 46100, Malaysia.

References
1. Keffaber KK. Reproductive failure of unknown etiology. Am Assoc Swine Vet. 1989;1(2):1–9.
2. Xiao XL, Wu H, Yu YG, Cheng BZ, Yang XQ, Chen G, Liu DM, Li XF. Rapid detection of a highly virulent Chinese-type isolate of Porcine Reproductive and Respiratory Syndrome Virus by real-time reverse transcriptase PCR. J Virol Methods. 2008;149(1):49–55.
3. Tian K, Yu X, Zhao T, Feng Y, Cao Z, Wang C, Hu Y, Chen X, Hu D, Tian X, Liu D, Zhang S, Deng X, Ding Y, Yang L, Zhang Y, Xiao H, Qiao M, Wang B, Hou L, Wang X, Yang X, Kang L, Sun M, Jin P, Wang S, Kitamura Y, Yan J, Gao GF. Emergence of fatal PRRSV variants: unparalleled outbreaks of atypical PRRS in China and molecular dissection of the unique hallmark. PLoS One. 2007;2:e526.
4. Guo B, Lager KM, Henningson JN, Miller LC, Schlink SN, Kappes MA, Kehrli Jr ME, Brockmeier SL, Nicholson TL, Yang HC, Faaberg KS. Experimental infection of United States swine with a Chinese highly pathogenic strain of porcine reproductive and respiratory syndrome virus. Virology. 2013;435:372–84.
5. Kappes MA, Miller CL, Faaberg KS. Highly Divergent Strains of Porcine Reproductive and Respiratory Syndrome Virus Incorporate Multiple Isoforms of Nonstructural Protein 2 into Virions. J Virol. 2013;87(24):13456–65.
6. Holtkamp DJ, Kliebenstein JB, Neumann EJ, Zimmerman JJ, Rotto HF, Yoder TK, Wang C, Yeske PE, Mowrer CL, Haley CA. Assessment of the economic impact of porcine reproductive and respiratory syndrome virus on United States pork producers. J Swine Health Prod. 2013;21(2):72–84.
7. Wensvoort G, Terpstra C, Pol JM, ter Laak EA, Bloemraad M, de Kluyver EP, Kragten C, van Buiten L, den Besten A, Wagenaar F, Broekhuijsen JM, Moonen PLJM, Zetstra T, de Boer EA, Tibben HJ, de Jong MF, van Veld P, Greenland GJR, Van't Gennep JA, Voets M, Verheijden JHM, Braamskamp J. Mystery swine disease in the Netherlands: the isolation of Lelystad virus. Vet Q. 1991;13:121–30.
8. Cavanagh D. Nidovirales: a new order comprising Coronaviridae and Arteriviridae. Arch Virol. 1997;142:629–33.
9. Zimmerman JJ, Yoon KJ, Pirtle EC, Wills RW, Sanderson TJ, McGinley MJ. Studies of porcine reproductive and respiratory syndrome (PRRS) virus infection in avian species. Vet Microbiol. 1997;55:329–36.
10. Fang Y, Snijder EJ. The PRRSV replicase: exploring the multifunctionality of an intriguing set of nonstructural proteins. Virus Res. 2010;154:61–76.
11. Music N, Gagnon CA. The role of porcine reproductive and respiratory syndrome (PRRS) virus structural and non-structural proteins in virus pathogenesis Anim. Health Res Rev. 2010;11:135–63.
12. Wu WH, Fang Y, Farwell R, Steffen-Bien M, Rowland RR, Christopher-Hennings J, Nelson EA. A 10-kDa structural protein of porcine reproductive and respiratory syndrome virus encoded by ORF2b. Virology. 2001;287:183–91.
13. Firth AE, Zevenhoven-Dobbe JC, Wills NM, Go YY, Balasuriya UB, Atkins JF, Snijder EJ, Posthuma CC. Discovery of a small arterivirus gene that overlaps the GP5 coding sequence and is important for virus production. J Gen Virol. 2011;92:1097–106.
14. Johnson CR, Griggs TF, Gnanandarajah J, Murtaugh MP. Novel structural protein in porcine reproductive and respiratory syndrome virus encoded by an alternative ORF5 present in all arteriviruses. J Gen Virol. 2011;92:1107–16.
15. Allende R, Lewis TL, Lu Z, Rock DL, Kutish GF, Ali A, Doster AR, Osorio FA. North American and European porcine reproductive and respiratory syndrome viruses differ in nonstructural protein coding regions. J Gen Virol. 1999;80(Pt 2):307–15.
16. Indik S, Schmoll F, Sipos W, Klein D. Genetic variability of PRRS virus in Austria: consequences for molecular diagnostics and viral quantification. Vet Microbiol. 2005;107:171–8.
17. Kim SH, Roh IS, Choi EJ, Lee C, Lee CH, Lee KH, Lee KK, Song YK, Lee OS, Park CK. A molecular analysis of European porcine reproductive and respiratory syndrome virus isolated in South Korea. Vet Microbiol. 2010;143:394–400.
18. Lee C, Kim H, Kang B, Yeom M, Han S, Moon H, Park S, Song D, Park B. Prevalence and phylogenetic analysis of the isolated type I porcine reproductive and respiratory syndrome virus from 2007 to 2008 in Korea. Virus Genes. 2010;40:225–30.
19. Thanawongnuwech R, Amonsin A, Tatsanakit A, Damrongwatanapokin S. Genetics and geographical variation of porcine reproductive and respiratory syndrome virus (PRRSV) in Thailand. Vet Microbiol. 2004;101:9–21.
20. Tun HM, Shi M, Wong CL, Ayudhya SN, Amonsin A, Thanawonguwech R, Leung FC. Genetic diversity and multiple introductions of porcine reproductive and respiratory syndrome viruses in Thailand. Virol J. 2011;8:164.
21. Christopher-Hennings J, Nelson E, Hines R, Nelson J, Swenson S, Zimmerman J, Chase C, Yaeger M, Benfield D. Persistence of porcine reproductive and respiratory syndrome virus in serum and semen of adult boars. J Vet Diagn Invest. 1995;7:456–64.
22. Kapur V, Elam MR, Pawlovich TM, Murtaugh MP. Genetic variation in porcine reproductive and respiratory syndrome virus isolates in the midwestern United States. J Gen Virol. 1996;77:1271–76.
23. Halbur PG, Paul PS, Meng XJ, Lum MA, Andrews JJ, Rathje JA. Comparative pathogenicity of nine US porcine reproductive and respiratory syndrome virus (PRRSV) isolates in a five-week-old cesarean-derived, colostrum-deprived pig model. J Vet Diagn Investig. 1996;8:11–20.
24. Meng XJ. Heterogeneity of porcine reproductive and respiratory syndrome virus: Implications for current vaccine efficacy and future vaccine development. Vet Microbiol. 2000;74:309–29.

25. OIE. Miscellaneous: "Swine high fever disease" in pigs in China (People's Rep. of), Disease Information (Weekly info), 11 November 2005. 2006;18(45). http://web.oie.int/eng/info/hebdo/AIS_78.HTM#Sec15.

26. Youjun F, Zhao T, Nguyen T, Inui K, Ma Y, Nguyen TH, Nguyen VC, Liu D, Bui QA, To LT, Wang C, Tian K, Gao GF. Porcine Respiratory and Reproductive Syndrome Virus Variants, Vietnam and China, 2007. Emerg Infect Dis. 2008;14(11):1774–6.

27. Ni J, Yang S, Bounlom D, Yu X, Zhou Z, Song J, Khamphouth V, Vatthana T, Tian K. Emergence and pathogenicity of highly pathogenic Porcine reproductive and respiratory syndrome virus in Vientiane, Lao People's Democratic Republic. J Vet Diagn Invest. 2012;24(2):349–54.

28. ProMED-mail. Porcine reprod. & resp. syndrome—Thailand. OIE; 2010. www.promedmail.org/direct.php?id=20101201.4320. Accessed 12 Oct 2010.

29. ProMED-mail. Porcine reprod. & resp. syndrome—Myanmar. Naypyidaw; 2011. 20110427.1307 http://www.promedmail.org/post/20110427.1307. Accessed 17 Nov 2011.

30. ProMED-mail. Undiagnosed disease, porcine—Cambodia; 2010. RFI 20100803.2618. www.promedmail.org/direct.php?id=20100803.2618. Accessed 3 Aug 2010.

31. ProMED-mail. Classical swine fever & PRRS, Philippines: (KA), susp. 2010. RFI 20100901.3123. https://www.promedmail.org/post/20100901.3123. Accessed 10 Feb 2011.

32. Nguyen T. PRRS Control in the region. OIE; 2013.

33. FAO. Porcine Reproductive and Respiratory Syndrome (PRRS); EMPRES Bulletin (2–2007). 2007. ftp://ftp.fao.org/docrep/fao/011/ai340e/ai340e00.pdf. Accessed 17 Nov 2010.

34. Too HL. In: A Guide to Pig Diseases in Malaysia. Faculty of Veterinary Medicine. Malaysia: Universiti Putra Malaysia; 1997. pp. 158–161.

35. Jasbir S, Kamaruddin MI, Latiffah H. Update on Porcine reproductive and respiratory syndrome (PRRS) seroprevalence in Malaysia. EMPRES Transboundary Animal Diseases Bulletin, FAO. 13th International Symposium for the World Association of Veterinary Laboratory Diagnosticians (WAVLD). 2008;30:38–9. ftp://ftp.fao.org/docrep/fao/010/i0059e/i0059e00.pdf.

36. Vania KTL, Ooi PT. Characterization of Porcine Reproductive and Respiratory Syndrome Virus (PRRSV) strains at selected farms in Malaysia. The 24th Veterinary Association Malausia Congress. 21-23th September 2012. Malaysia: Marriot Putrajaya; 2012.

37. Jaganathan S, Ooi PT, Phang LY, Zeenathul NA, Tuam SM, Ong LP, Kalaiwaney M, How SP, Tee CY, Lim HC, Choo PY, Lim BK. An update on the status of Porcine Reproductive and Respiratory Syndrome Virus (PRRSV) isolated in Malaysia. Ho Chi Minh: The 6th Asian Pig Veterinary Society Congress; 2013. September 23–25, 2013.

38. den Boon JA, Faaberg KS, Meulenberg JJ, Wassenaar AL, Plagemann PG, Gorbalenya AE, Snijder EJ. Processing and evolution of the N-terminal region of the arterivirus replicase ORF1a protein: identification of two papain-like cysteine proteases. J Virol. 1995;69:4500–5.

39. Han J, Burkhart KM, Vaughn EM, Roof MB, Faaberg KS. Replication and expression analysis of PRRSV defective RNA. In: The Nidoviruses. US: Springer; 2006. p. 445–8.

40. Han J, Liu G, Wang Y, Faaberg KS. Identification of nonessential regions of the nsp2 replicase protein of porcine reproductive and respiratory syndrome virus strain VR-2332 for replication in cell culture. J Virol. 2007;81:9878–90.

41. Allende R, Lewis TL, Lu Z, Rock DL, Kutish GF, Ali A, Doster AR, Osorio FA. North American and European porcine reproductive and respiratory syndrome viruses differ in non-structural protein coding regions. J Gen Virol. 1999;80:307–15.

42. Nelsen CJ, Murtaugh MP, Faaberg KS. Porcine reproductive and respiratory syndrome virus comparison: Divergent evolution on two continents. J Virol. 1999;73:270–80.

43. Allende R, Kutish GF, Laegreid W, Lu Z, Lewis TL, Rock DL, Friesen J, Galeota JA, Doster AR, Osorio FA. Mutations in the genome of porcine reproductive and respiratory syndrome virus responsible for the attenuation phenotype. Arch Virol. 2000;145:1149–61.

44. Shen S, Kwang J, Liu W, Liu DX. Determination of the complete nucleotide sequence of a vaccine strain of porcine reproductive and respiratory syndrome virus and identification of the Nsp2 gene with a unique insertion. Arch Virol. 2000;145:871–83.

45. Fang Y, Kim DY, Ropp S, Steen P, Christopher-Hennings J, Nelson EA, Rowland RR. Heterogeneity in Nsp2 of European-like porcine reproductive and respiratory syndrome viruses isolated in the United States. Virus Res. 2004;100:229–35.

46. Fang Y, Rowland RR, Roof M, Lunney JK, Christopher-Hennings J, Nelson EA. A full-length cDNA infectious clone of North American type 1 porcine reproductive and respiratory syndrome virus: expression of green fluorescent protein in the Nsp2 region. J Virol. 2006;80:11447–55.

47. Gao ZQ, Guo X, Yang HC. Genomic characterization of two Chinese isolates of porcine respiratory and reproductive syndrome virus. Arch Virol. 2004;149:1341–51.

48. Ropp SL, Wees CE, Fang Y, Nelson EA, Rossow KD, Bien M, Arndt B, Preszler S, Steen P, Christopher-Hennings J, Collins JE, Benfield DA, Faaberg KS. Characterization of emerging European-like porcine reproductive and respiratory syndrome virus isolates in the United States. J Virol. 2004;78:3684–703.

49. Han J, Wang Y, Faaberg KS. Complete genome analysis of RFLP 184 isolates of porcine reproductive and respiratory syndrome virus. Virus Res. 2006;122:175–82.

50. Kim WI, Lee DS, Johnson W, Roof M, Cha SH, Yoon KJ. Effect of genotypic and biotypic differences among PRRS viruses on the serologic assessment of pigs for virus infection. Vet Microbiol. 2007;123:1–14.

51. Ran ZG, Chen XY, Guo X, Ge XN, Yoon KJ, Yang HC. Recovery of viable porcine reproductive and respiratory syndrome virus from an infectious clone containing a partial deletion within the Nsp2-encoding region. Arch Virol. 2008;153(5):899–907.

52. Metwally S, Mohamed F, Faaberg K, Burrage T, Prarat M, Moran K, Bracht A, Mayr G, Berninger M, Koster L, To TL, Nguyen VL, Reising M, Landgraf J, Cox L, Lubroth J, Carrillo C. Pathogenicity and Molecular Characterization of Emerging Porcine Reproductive and Respiratory Syndrome Virus in Vietnam in 2007. Transbound Emerg Dis. 2010;57:315–29.

53. Halbur PG, Paul PS, Frey ML, Landgraf J, Eernisse K, Meng XJ, Lum MA, Andrews JJ, Rathje JA. Comparison of the pathogenicity of two U.S. porcine reproductive and respiratory syndrome virus isolates with that of the Lelystad virus. Vet Pathol. 1995;34:648–60.

54. Halbur PG, Paul PS, Frey ML, Landgraf J, Eernisse K, Meng XJ, Andrews JJ, Lum MA, Rathje JA. Comparison of the antigen distribution of two U.S. porcine reproductive and respiratory syndrome virus isolates with that of the Lelystad virus. Vet Pathol. 1996;33:159–70.

55. Mengeling WL, Lager KM, Vorwald AC. Clinical consequences of exposing pregnant gilts to strains of porcine reproductive and respiratory syndrome (PRRS) virus isolated from field cases of "atypical" PRRS. Am J Vet Res. 1998;59:1540–4.

56. Opriessnig T, Halbur PG, Yoon KJ, Pogranichniy RM, Harmon KM, Evans R, Key KF, Pallares FJ, Thomas P, Meng XJ. Comparison of Molecular and Biological Characteristics of a Modified Live Porcine Reproductive and Respiratory Syndrome Virus (PRRSV) Vaccine (Ingelvac PRRS MLV), the Parent Strain of the Vaccine (ATCC VR2332), ATCC VR2385, and Two Recent Field Isolates of PRRSV. J Virol. 2002;76(23):11837–44.

57. Meng XJ, Paul PS, Halbur PG. Molecular cloning and nucleotide sequencing of the 3-terminal genomic RNA of porcine reproductive and respiratory syndrome virus. J Gen Virol. 1994;75:1795–801.

58. Meng XJ, Paul PS, Halbur PG, Lum MA. Phylogenic analyses of the putative M (ORF 6) and N (ORF 7) genes of porcine reproductive and respiratory syndrome virus (PRRSV): implication for the existence of two genotypes of PRRSV in the USA and Europe. Arch Virol. 1995;140:745–55.

59. Meng XJ, Paul PS, Halbur PG, Morozov I. Sequence comparison of open reading frames 2 to 5 of low and high virulence United States isolates of porcine reproductive and respiratory syndrome virus. J Gen Virol. 1995;76:3181–8.

60. Meng XJ, Paul PS, Halbur PG, Lum MA. Characterization of a high-virulence US isolate of porcine reproductive and respiratory syndrome virus in a continuous cell line, ATCC CRL11171. J Vet Diagn Investig. 1996;8:374–81.

61. Vu HLX, Ma F, Laegreid WW, Pattnaik AK, Steffen D, Doster AR, Osorio FA. A synthetic porcine reproductive and respiratory syndrome virus strain confers unprecedented levels of heterologous protection. J Virol. 2015;89:12070–83.

62. Pesch S. Etablierung einer Nachweismethode fur die zwei Genotypen von PRRSV und ein Beitrag zu seiner molekularen Epidemiologie. Doctoral Vet. Med. Leipzig: Thesis, Veterinary Faculty Universitat Leipzig; 2003.

63. Truyen U, Wilhelm S, Genzow M, Schagemann G. Porcine Reproductive and Respiratory Syndrome Virus (PRRSV): A Ring Test Performed in Germany to Assess RT-PCR Detection Methods. J Vet Med. 2006;53(2):68–74.

64. Mardassi H, Mounir S, Dea S. Identification of major differences in the nucleocapsid protein genes of a Quebec strain and European strains of porcine reproductive and respiratory syndrome virus. J Gen Virol. 1994;75:681–5.

65. Feng Y, Zhao T, Nguyen T, Inui K, Ma Y, Nguyen TH, Nguyen VC, Liu D, Bui QA, To LT, Wang C, Tian K, Gao GF. Porcine respiratory and reproductive syndrome virus variants, Vietnam and China, 2007. Emerg Infect Dis. 2008; 14(11):1774–6.

11

Nonstructural protein 11 (nsp11) of porcine reproductive and respiratory syndrome virus (PRRSV) promotes PRRSV infection in MARC-145 cells

Xibao Shi[1,2*†], Xiaozhuan Zhang[1,3†], Yongzhe Chang[2], Bo Jiang[5], Ruiguang Deng[2], Aiping Wang[4] and Gaiping Zhang[2,3*]

Abstract

Background: Porcine reproductive and respiratory syndrome virus (PRRSV) induces one of most important devastating disease of swine worldwide, and the current methods poorly control it. Previous studies have indicated that the nonstructural protein 11 (nsp11) of PRRSV may be an important protein for the immune escape of PRRSV.

Results: Here, we firstly explored the effect of over-expression of nsp11 on PRRSV infection and found that over-expression of nsp11 enhanced the PRRSV titers while the small interfering RNA (siRNAs) specifically targeting nsp11 could reduce the PRRSV titers in MARC-145 cells.

Conclusion: In conclusion, PRRSV nsp11 promotes PRRSV infection in MARC-145 cells and siRNAs targeting nsp11 may be a potential therapeutic strategy to control PRRSV in future.

Keywords: PRRSV, Small interfering RNA, nsp11

Background

PRRSV, a positive sense and single-stranded RNA virus, is a member of family *Arteriviridae* [1]. Since it was emerged in the United States in 1987 and in Europe in 1990, PRRSV has rapidly spread in the swine producing regions and became one of the most important devastating diseases of swine worldwide. It can cause severe reproductive failure in sows and respiratory distress in young growing pigs [2]. Infection with PRRSV also made pigs easy to secondary infection by other pathogens [3]. Up to date, since there is no efficient method or drugs against PRRSV, it is very important and urgent to develop the effective therapeutic strategies to control PRRS.

The PRRSV genome has nine open reading frames (ORFs) composed of ORF1a, ORF1b, ORF2a, ORF2b, and ORF3-7. ORF1a and ORF1b could produce 16 nonstructural proteins (nsp1α, nsp1β, nsp2 etc.) [4–7]. Previous studies have shown that the nsp11 of equine arteritis virus(EAV), which is another member of family *Arteriviridae*, may play a key role in viral RNA synthesis and additional functions in the viral life cycle [8]. Other and our previous work also demonstrated that PRRSV nsp11 inhibited the host innate immune responses such as the transcription of type I interferon [7], the RNAi innate immune response [9] and the NLR family pyrin domain-containing 3 (NLRP3)-mediated production of IL-1β [10], which indicated that PRRSV nsp11 may play an important role in PRRSV infection. So the purpose of present study is to explore the effect of over-expression of nsp11 on PRRSV infection and whether the siRNAs targeting the PRRSV nsp11 could influence PRRSV infection.

* Correspondence: shixibao@aliyun.com; zhanggaiping2003@163.com
†Equal contributors
[1]College of Life Sciences, Henan Normal University, Xinxiang 453007, China
[2]Key Laboratory of Animal Immunology of the Ministry of Agriculture, Henan Provincial Key Laboratory of Animal Immunology, Henan Academy of Agricultural Sciences, Zhengzhou, Henan 450002, China
Full list of author information is available at the end of the article

Methods

Cell and virus

MARC-145 cells, derived from a monkey fibroblast cell line MA-104 [11], and 293T cells were grown in Dulbecco's Modified Eagle medium (Gibco) plus 10 % heat-inactivated fetal bovine serum (Hyclone). PRRSV strain BJ-4 was a kind gift from Prof. Hanchun Yang (China Agricultural University).

Plasmids

The pcDNA3.1-GFP-nsp11 plasmid was constructed by sub-cloning from the plasmid pcDNA3.1-FLAG-nsp11 [7] to pcDNA3.1-GFP [12] using the restriction endonuclease Hind III and EcoRI. The plasmids pcDNA3.1-FLAG, pcDNA3.1-FLAG-nsp11 and pcDNA3.1-FLAG-nsp11 H129A have been described in our previous work [7].

Transfection of plasmids and viral infection

All newly-prepared plasmids were confirmed correctly by DNA sequencing. Transient transfection was carried out by using Lipofectamine 2000 (Invitrogen). Cells cultured in 24-well plates were transfected with the indicated expression plasmid or control vector (800 ng/well) in triplicate. And 6 h (h) later, the cells were infected with PRRSV at a multiplicity of infection (MOI) of 0.1, and then the cells were lysed by freezing and thawing three times after 24 h infection. The supernatants were collected by centrifugation, and the viral titers were detected by 50 % tissue culture infected dose (TCID50) assay using the method of Reed–Muench in the MARC-145 cells.

Western blots

The 293 T cells were cultured in 24-well plates and transfected with pcDNA3.1-GFP-nsp11 or pcDNA3.1-GFP and nsp11 siRNA (100 nM) or control siRNA (100 nM) in triplicate, and 36 h later, the cells were lysed with lysing buffer (1 % Nonidet P-40, 0.1 % sodium deoxycholate, 0.1 % SDS, 50 mM Tris-HCl (pH 7.4), 150 mM NaCl, 2 mM EDTA, 2 mM Na3VO4, 2 mM NaF and a protease inhibitor cocktail). The detailed procedure for immunoblots has been described in our previous work [13]. Briefly, the proteins were separated by 10 % SDS-PAGE and transferred to polyvinylidene difluoride (PVDF) membranes (Millipore Company, Boston, Massachusetts, USA), and then the PVDF membranes were incubated with anti-GFP (Clontech) or anti-actin (Cell Signaling Technology) antibodies. Subsequently the membranes were incubated with appropriate secondary antibodies and were tested by an ECL detection system (Cell Signaling Technology, Boston, USA).

Transfection of siRNA and viral infection

Cells cultured in 24-well plates were transfected with the indicated siRNA in triplicate (100 nM/well) (chemical synthesis by Bioneer) (Table. 1). And 6 h later, the cells were infected with PRRSV strain BJ-4 at an MOI of 0.1, and 24 h later, the cells were lysed by freezing and thawing three times. The supernatants were collected by centrifugation, and the viral titers were detected by TCID50 using the method of Reed–Muench in the MARC-145 cells.

Real time (RT)-PCR

MARC-145 cells cultured in 24-well plates were transfected in triplicate with nsp11-siRNAs/control siRNA and plasmid pcDNA3.1-GFP-nsp11 (800 ng/well) or pcDNA 3.1-GFP (800 ng/well). And 24 h later, the cells were infected with PRRSV at an MOI of 0.1, and 48 h later, the cellular RNA was extracted with TRIzol (Invitrogen). M-MLV reverse transcriptase was used for the PrimeScript™ RT Reagent Kit with gDNA Eraser (Takara, Dalian, China). Quantitative real-time RT-PCR (qRT-PCR) was performed using SYBR® Premix Ex Taq™ (Takara, Dalian, China) and was tested by the 7500 First real-time PCR system (Applied Biosystems, Foster City, CA, USA). Glyceraldehyde-3-phosphate dehydrogenase (GAPDH) was used as an internal control. The $2^{-\Delta\Delta Ct}$-method was used to analyze the relative amount of target gene expression.

Table 1 Synthesized siRNAs sequences targeting the regions of PRRSV nsp11 and the primers of ORF-7 for RT-PCR

Gene of target	Name of siRNA	Location(bp)	Sequence (5'-3')
nsp 11	nsp11 siRNA 1	244-262	CGTGTCATACTATCTCACA
nsp 11	nsp11 siRNA 2	526-544	CACACTGACAGATGTGTAC
nsp 11	nsp11 siRNA 3	409-427	CACTACCGTTGGAGGATGT
Negative control	Control siRNA		CCTACGCCACCAATTTCGT
ORF-7 For			AAACCAGTCCAGAGGCAAGG
ORF-7 Rev			GCAAACTAAACTCCACAGTGTAA
GAPDH For			TGACAACAGCCTCAAGATCG
GAPDH Rev			GTCTTCTGGGTGGCAGTGAT

Viral titers

MARC-145 cells transfected with the nsp11 siRNA (100 nM) or control siRNA (100 nM) and the plasmid pcDNA3.1-GFP-nsp11 (800 ng/well) or pcDNA3.1-GFP (800 ng/well), and 24 h later, the cells were infected by PRRSV at an MOI of 0.1, the cells were frozen and thawed in three cycles and collected after 48 h infection. Then the viral titers were determined by TCID50 assay using the method of Reed–Muench in the cells of MARC-145.

Statistical analysis

Statistical analyses were performed by Student's t-test, and the differences were considered as statistical significance when $P <0.05$.

Results

Over-expression of nsp11 enhanced PRRSV titers in MARC-145 cells

Firstly, we successfully constructed the expression plasmid of pcDNA 3.1-GFP-nsp11 (GFP-nsp11) and the western blot in Fig. 1a confirmed the successful expression of GFP-nsp11 (Fig. 1). Secondly, the MARC-145 cells were cultured in 24-well plates

overnight, and then the cells were transfected with the expression plasmid pcDNA3.1-GFP-nsp11 or the control plasmid pcDNA3.1-GFP. The results in Fig. 1b showed that the PRRSV titers from the MARC-145 cells transfected with pcDNA3.1-GFP-nsp11 were about one point six times to that from MARC-145 cells transfected with control plasmid, while the RNA levels of PRRSV from the MARC-145 cells transfected with pcDNA3.1-GFP-nsp11 were three times to that from MARC-145 cells transfected with control plasmid (Fig. 1c).

siRNAs targeting nsp11 reduced PRRSV titers in MARC-145 cells

Now that over-expression of nsp11 could enhance PRRSV titers, it was reasonable to design siRNAs targeting nsp11 to investigate whether the siRNAs could reduce PRRSV titers. The sequences of the siRNA special for nsp11 and the control siRNA were listed in Table 1. 293T cells (Fig. 2a) or MARC-145 cells (Fig. 2b) grown in 24-well plates were co-transfected with the nsp11 siRNA (100 nM/well) or control siRNA and the plasmid GFP-nsp11 (800 ng/well).

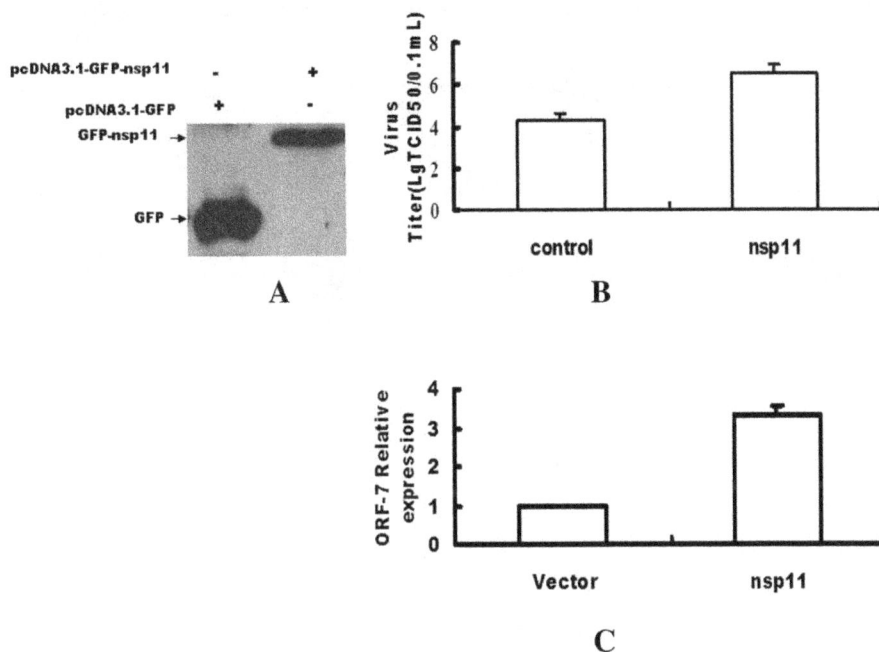

Fig. 1 Over-expression of nsp11enhenced the titers of PRRSV. **a** 293T cells were transfected with pcDNA3.1-GFP (GFP) or pcDNA3.1-GFP-nsp11 (GFP-nsp11), and 48 h later, the cells were analyzed by western blots. **b** MARC-145 cells cultured in 24-well plate were transfected with pcDNA3.1-GFP-nsp11 (nsp11)(800 ng/well) or pcDNA 3.1-GFP (Con). And 6 h later, the cells were infected with PRRSV at an MOI of 0.1 or mock infected, and 24 h later, the cells were lysed by freezing and thawing three times in three cycles, then the viral titers were measured by TCID50. **c** MARC-145 cells cultured in 24-well plate were transfected with pcDNA 3.1-GFP-nsp11 (nsp11) (800 ng/well) or pcDNA3.1-GFP (Con). And 6 h later, the cells were infected with PRRSV at an MOI of 0.1 or mock infected, and 24 h later, the cells were collected and the viral RNA levels were measured by RT-PCR. Data represented means of three replicates, and experiments were repeated three times. Error bars represented the standard deviations. *: $P <0.05$ compared with the results in control. MOI: multiplicity of infection

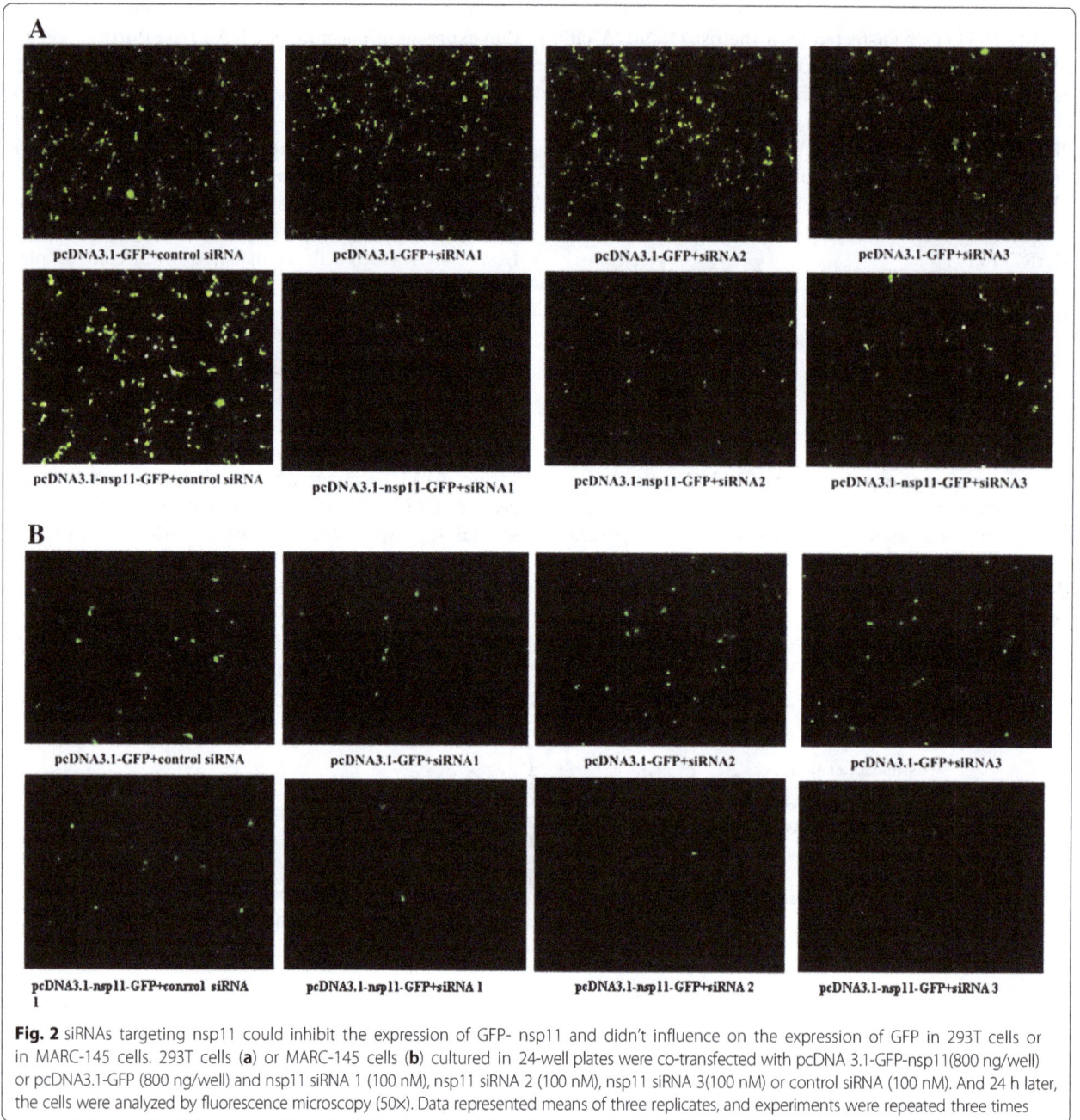

Fig. 2 siRNAs targeting nsp11 could inhibit the expression of GFP- nsp11 and didn't influence on the expression of GFP in 293T cells or in MARC-145 cells. 293T cells (**a**) or MARC-145 cells (**b**) cultured in 24-well plates were co-transfected with pcDNA 3.1-GFP-nsp11(800 ng/well) or pcDNA3.1-GFP (800 ng/well) and nsp11 siRNA 1 (100 nM), nsp11 siRNA 2 (100 nM), nsp11 siRNA 3(100 nM) or control siRNA (100 nM). And 24 h later, the cells were analyzed by fluorescence microscopy (50×). Data represented means of three replicates, and experiments were repeated three times

And 24 h later, the cells in five random fields were analyzed by fluorescence microscopy (50×) and only one of them was shown in Fig. 2. The results in Fig. 2 showed that all of the three siRNAs targeting nsp11 could inhibit the expression of GFP-nsp11 and didn't influence on the expression of GFP.

Finally, we selected two siRNAs (siRNA1 and siRNA2) targeting nsp11 to determine whether siRNAs targeting nsp11 could reduce PRRSV titers in MARC-145 cell. The western blots results in Fig. 3a confirmed that siRNA1 and siRNA2 could efficiently

reduce the nsp11 expression in 293T cells (Fig. 3a). The results in Fig. 3b and c showed that siRNAs targeting nucleic acid sequence of nsp11 could significantly reduce viral titers of PRRSV (Fig. 3b) and reduce RNA levels of PRRSV (Fig. 3c).

The endoribonuclease activity of nsp11 was important for nsp11 to enhance the PRRSV titers

Our previous work has shown that the endoribonuclease activity of nsp11 was important for nsp11 to inhibit the transcription of IFN-β [7] and the secretion of IL-1β

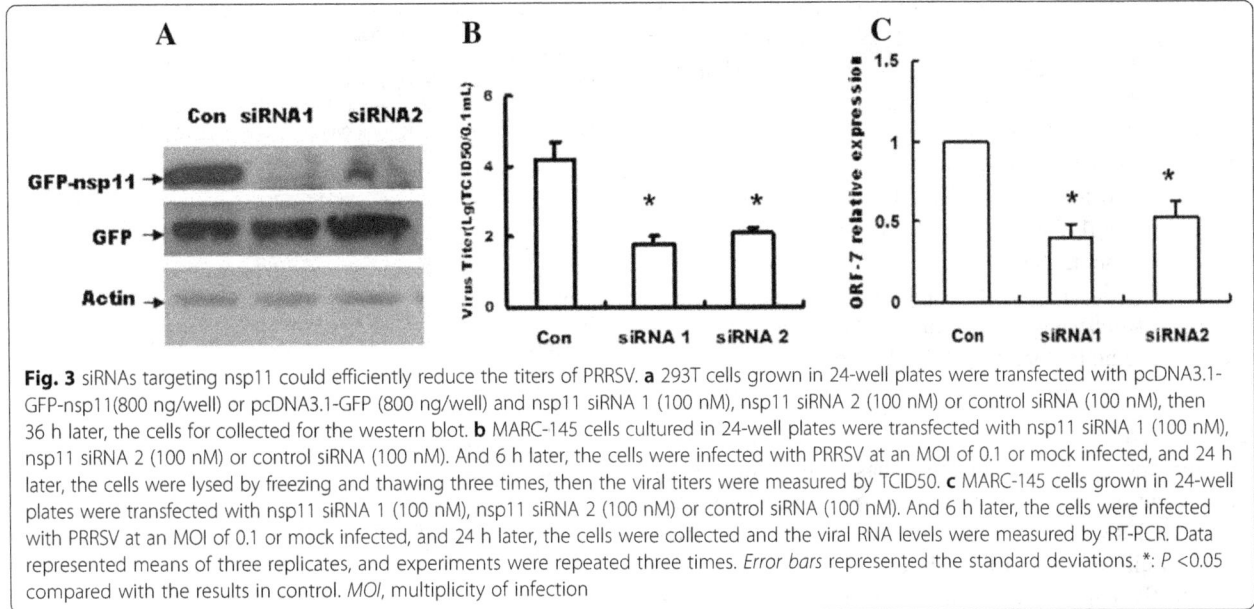

Fig. 3 siRNAs targeting nsp11 could efficiently reduce the titers of PRRSV. **a** 293T cells grown in 24-well plates were transfected with pcDNA3.1-GFP-nsp11(800 ng/well) or pcDNA3.1-GFP (800 ng/well) and nsp11 siRNA 1 (100 nM), nsp11 siRNA 2 (100 nM) or control siRNA (100 nM), then 36 h later, the cells for collected for the western blot. **b** MARC-145 cells cultured in 24-well plates were transfected with nsp11 siRNA 1 (100 nM), nsp11 siRNA 2 (100 nM) or control siRNA (100 nM). And 6 h later, the cells were infected with PRRSV at an MOI of 0.1 or mock infected, and 24 h later, the cells were lysed by freezing and thawing three times, then the viral titers were measured by TCID50. **c** MARC-145 cells grown in 24-well plates were transfected with nsp11 siRNA 1 (100 nM), nsp11 siRNA 2 (100 nM) or control siRNA (100 nM). And 6 h later, the cells were infected with PRRSV at an MOI of 0.1 or mock infected, and 24 h later, the cells were collected and the viral RNA levels were measured by RT-PCR. Data represented means of three replicates, and experiments were repeated three times. *Error bars* represented the standard deviations. *: $P < 0.05$ compared with the results in control. *MOI*, multiplicity of infection

[10]. So next we investigated whether the endoribonuclease activity of nsp11 was important for nsp11 to promote the PRRSV infection. Nedialkova et al. showed that His-129, His-144, and Lys-173 were the catalytic centers, and mutating one of the three amino acids could abolish its enzyme activity. The results in Fig. 4 showed that inactivating the endoribonuclease activity made nsp11 not promote the PRRSV infection.

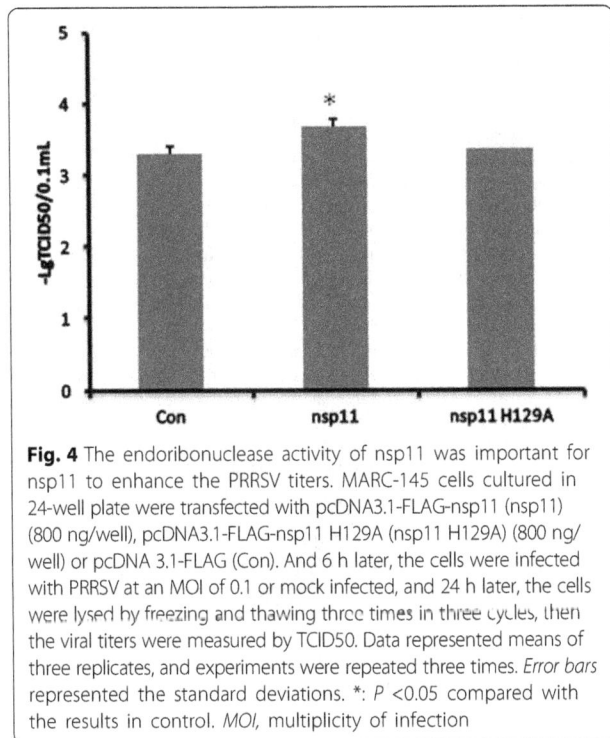

Fig. 4 The endoribonuclease activity of nsp11 was important for nsp11 to enhance the PRRSV titers. MARC-145 cells cultured in 24-well plate were transfected with pcDNA3.1-FLAG-nsp11 (nsp11) (800 ng/well), pcDNA3.1-FLAG-nsp11 H129A (nsp11 H129A) (800 ng/well) or pcDNA 3.1-FLAG (Con). And 6 h later, the cells were infected with PRRSV at an MOI of 0.1 or mock infected, and 24 h later, the cells were lysed by freezing and thawing three times in three cycles, then the viral titers were measured by TCID50. Data represented means of three replicates, and experiments were repeated three times. *Error bars* represented the standard deviations. *: $P < 0.05$ compared with the results in control. *MOI*, multiplicity of infection

Discussion

nsp11 was a multi-functional protein. Both our and other previous studies have shown that PRRSV nsp11 is an interferon antagonist [7, 14] and that nsp11 plays an important role in viral RNA synthesis and in the viral life cycle [8]. Our recent work also demonstrated that PRRSV nsp11 inhibited the RNAi innate immune response [9] and NLRP3-mediated production of IL-1β [10]. While our present work showed that over expression of PRRSV nsp11 could enhance the titers of PRRSV (Fig. 1), so our present work gave a directly evidence that nsp11 was an important viral component for up-regulating the PRRSV titers.

Identification of and targeting viral important components is useful for developing viral vaccine and controlling the virus. For example, both the nonstructural protein 1 of influenza virus and the nonstructural protein 1 of mouse hepatitis virus (MHV) were important for the viral virulence respectively, and both the modified live-viral vaccines that deletion of nonstructural protein 1 resulted in complete protection against challenge with influenza virus infection and MHV infection respectively [15–17].

RNA interference (RNAi) is an exciting method to silence viral genes. Inhibition of specific genes by siRNA has proven to be a potential therapeutic strategy against viral infection [18], especially for the positive single stranded RNA viruses since their genomes function as both the mRNA and the replication template [19, 20]. Up to date, RNAi has been used against several viruses such as hepatitis B virus, foot-and-mouth disease virus, dengue virus and so on [20–22]. In this work, we also explore whether the siRNA, which targeted the nucleic acid sequence of nsp11, influenced the titers of PRRSV, and the results showed that siRNA targeting nsp11 significantly reduced the titers of PRRSV (Fig. 3). A recent improved live PRRSV vaccine has indicated that the ORF1a and ORF1b were the virulence determinants of PRRSV [23]. In addition, our recent work also show that RNAi innate immune response was an antiviral response to PRRSV and PRRSV inhibited this response by PRRSV nsp1α and nsp11, which indicated that targeting nsp11 would be useful for RNAi innate immunity against PRRSV [9]. So it is reasonable to propose that our present results gave a new clue for generating the new PRRSV vaccine by targeting PRRSV nsp11.

Conclusion

In conclusion, our present study has shown that nsp11 was an important viral component for up-regulating the titers of PRRSV and that siRNAs directly targeting nsp11 could inhibit PRRSV infection, which indicated that PRRSV nsp11 may be an interesting target for controlling PRRSV in future.

Abbreviations
EAV, equine arteritis virus; GAPDH, Glyceraldehyde-3-phosphate dehydrogenase; GFP-nsp11, pcDNA 3.1-GFP-nsp11; MHV, mouse hepatitis virus; MOI, multiplicity of infection; NLRP3, NLR family pyrin domain-containing 3; nsp11, nonstructural protein 11; ORF, open reading frame; PRRSV, porcine reproductive and respiratory syndrome virus ; PVDF, polyvinylidene difluoride; qRT-PCR, quantitative real-time RT-PCR; RNAi, RNA interference; siRNA, small interfering RNA; TCID50, 50 % tissue culture infected dose

Acknowledgements
We would like to thank Prof. Hanchun Yang for providing PRRSV strain BJ-4.

Funding
This work was supported by the National Natural Science Foundation of China (grant no. 31302073), the key project of National Natural Science Fund (No.31490600), a grant from the Major State Basic Research Development Program of China (973 Program) (No.2014CB542700), the key scientific research projects of Henan provincial institution of higher education(16A180008), the Doctoral Starting up Foundation of Henan Normal University (5101049170153) and another National Foundation of China (grant no. 31472177), Natural Science Foundation Project of CQ CSTC(cstc2012jjA10108).

Authors' contributions
SX, ZX, ZG designed the study, SX, ZX and CY performed the experiments and wrote the paper, and WA, JB and DR performed statistical analysis. All authors read and approved the final manuscript.

Competing interests
The authors declare that they have no competing interests.

Consent for publication
Our present work does not contain any individual persons' data, so it is not applicable.

Author details
[1]College of Life Sciences, Henan Normal University, Xinxiang 453007, China. [2]Key Laboratory of Animal Immunology of the Ministry of Agriculture, Henan Provincial Key Laboratory of Animal Immunology, Henan Academy of Agricultural Sciences, Zhengzhou, Henan 450002, China. [3]College of Veterinary Medicine and Animal Science, Henan Agricultural University, Zhengzhou, Henan 450002, China. [4]Department of Bioengineering, Zhengzhou University, Zhengzhou, Henan 450000, 450002, China. [5]Office of Science & Technology, Chongqing Police College, Chongqing 401331, China.

References
1. Cavanagh D. Nidovirales: a new order comprising Coronaviridae and Arteriviridae. Arch Virol. 1997;142(3):629–33.
2. Zimmerman JJ, Yoon KJ, Wills RW, Swenson SL. General overview of PRRSV: a perspective from the United States. Vet Microbiol. 1997;55(1–4):187–96.
3. Mateu E, Diaz I. The challenge of PRRS immunology. Vet J. 2008;177(3):345–51.
4. den Boon JA, Snijder EJ, Chirnside ED, de Vries AA, Horzinek MC, Spaan WJ. Equine arteritis virus is not a togavirus but belongs to the coronaviruslike superfamily. J Virol. 1991;65(6):2910–20.
5. Chen Z, Lawson S, Sun Z, Zhou X, Guan X, Christopher-Hennings J, Nelson EA, Fang Y. Identification of two auto-cleavage products of nonstructural protein 1 (nsp1) in porcine reproductive and respiratory syndrome virus infected cells: nsp1 function as interferon antagonist. Virology. 2010;398(1): 87–97.
6. den Boon JA, Faaberg KS, Meulenberg JJ, Wassenaar AL, Plagemann PG, Gorbalenya AE, Snijder EJ. Processing and evolution of the N-terminal region of the arterivirus replicase ORF1a protein: identification of two papainlike cysteine proteases. J Virol. 1995;69(7):4500–5.
7. Shi X, Wang L, Li X, Zhang G, Guo J, Zhao D, Chai S, Deng R. Endoribonuclease activities of porcine reproductive and respiratory syndrome virus nsp11 was essential for nsp11 to inhibit IFN-beta induction. Mol Immunol. 2011;48(12–13):1568–72.
8. Posthuma CC, Nedialkova DD, Zevenhoven-Dobbe JC, Blokhuis JH, Gorbalenya AE, Snijder EJ. Site-directed mutagenesis of the Nidovirus replicative endoribonuclease NendoU exerts pleiotropic effects on the arterivirus life cycle. J Virol. 2006;80(4):1653–61.
9. Chen J, Shi X, Zhang X, Wang L, Luo J, Xing G, Deng R, Yang H, Li J, Wang A, et al. Porcine Reproductive and Respiratory Syndrome Virus (PRRSV) inhibits RNA-mediated gene silencing by targeting ago-2. Viruses. 2015; 7(10):5539–52.
10. Wang C, Shi X, Zhang X, Wang A, Wang L, Chen J, Deng R, Zhang G. The endoribonuclease activity essential for the nonstructural protein 11 of porcine reproductive and respiratory syndrome virus to inhibit NLRP3 inflammasome-mediated IL-1beta induction. DNA Cell Biol. 2015;34(12):728–35.
11. Kim HS, Kwang J, Yoon IJ, Joo HS, Frey ML. Enhanced replication of porcine reproductive and respiratory syndrome (PRRS) virus in a homogeneous subpopulation of MA-104 cell line. Arch Virol. 1993;133(3–4):477–83.
12. Shi X, Zhang X, Wang F, Wang L, Qiao S, Guo J, Luo C, Wan B, Deng R, Zhang G. The zinc-finger domain was essential for porcine reproductive and respiratory syndrome virus nonstructural protein-1alpha to inhibit the production of interferon-beta. J Interferon Cytokine Res. 2013;33(6):328–34.
13. Shi X, Qin L, Liu G, Zhao S, Peng N, Chen X. Dynamic balance of pSTAT1 and pSTAT3 in C57BL/6 mice infected with lethal or nonlethal Plasmodium yoelii. Cell Mol Immunol. 2008;5(5):341–8.
14. Beura LK, Sarkar SN, Kwon B, Subramaniam S, Jones C, Pattnaik AK, Osorio FA. Porcine reproductive and respiratory syndrome virus nonstructural protein 1beta modulates host innate immune response by antagonizing IRF3 activation. J Virol. 2010;84(3):1574–84.
15. Richt JA, Lekcharoensuk P, Lager KM, Vincent AL, Loiacono CM, Janke BH, Wu WH, Yoon KJ, Webby RJ, Solorzano A, et al. Vaccination of pigs against swine influenza viruses by using an NS1-truncated modified live-virus vaccine. J Virol. 2006;80(22):11009–18.
16. Wacheck V, Egorov A, Groiss F, Pfeiffer A, Fuereder T, Hoeflmayer D, Kundi M, Popow-Kraupp T, Redlberger-Fritz M, Mueller CA, et al. A novel type of influenza vaccine: safety and immunogenicity of replication-deficient influenza virus created by deletion of the interferon antagonist NS1. J Infect Dis. 2010;201(3):354–62.
17. Zust R, Cervantes-Barragan L, Kuri T, Blakqori G, Weber F, Ludewig B, Thiel V. Coronavirus non-structural protein 1 is a major pathogenicity factor: implications for the rational design of coronavirus vaccines. PLoS Pathog. 2007;3(8):e109.

18. Mahmood ur R, Ali I, Husnain T, Riazuddin S. RNA interference: the story of gene silencing in plants and humans. Biotechnol Adv. 2008;26(3):202–9.

19. Parashar D, Paingankar MS, Kumar S, Gokhale MD, Sudeep AB, Shinde SB, Arankalle VA. Administration of E2 and NS1 siRNAs inhibit chikungunya virus replication in vitro and protects mice infected with the virus. PLoS Negl Trop Dis. 2013;7(9), e2405.

20. Idrees S, Ashfaq UA, Khaliq S. RNAi: antiviral therapy against dengue virus. Asian Pac J Trop Biomed. 2013;3(3):232–6.

21. Huang W, Li X, Yi M, Zhu S, Chen W. Targeted delivery of siRNA against hepatitis B virus by preS1 peptide molecular ligand. Hepatol Res. 2011;44(8): 897–906.

22. Kahana R, Kuznetzova L, Rogel A, Shemesh M, Hai D, Yadin H, Stram Y. Inhibition of foot-and-mouth disease virus replication by small interfering RNA. J Gen Virol. 2004;85(Pt 11):3213–7.

23. Wang Y, Liang Y, Han J, Burkhart KM, Vaughn EM, Roof MB, Faaberg KS. Attenuation of porcine reproductive and respiratory syndrome virus strain MN184 using chimeric construction with vaccine sequence. Virology. 2008;371(2):418–29.

12

Virulent duck enteritis virus infected DEF cells generate a unique pattern of viral microRNAs and a novel set of host microRNAs

Xianglong Wu[1,3], Renyong Jia[1,2,3*] (iD), Jiakun Zhou[1,3], Mingshu Wang[1,2,3], Shun Chen[1,2,3], Mafeng Liu[1,2,3], Dekang Zhu[1,2,3], Xinxin Zhao[1,2,3], Kunfeng Sun[1,2,3], Qiao Yang[1,2,3], Ying Wu[1,2,3], Zhongqiong Yin[2], Xiaoyue Chen[1,2], Jue Wang[4] and Anchun Cheng[1,2,3*]

Abstract

Background: Duck enteritis virus (DEV) belongs to the family *Herpesviridae* and is an important epornitic agent that causes economic losses in the waterfowl industry. The Chinese virulent (CHv) and attenuate vaccines (VAC) are two different pathogenic DEV strains. MicroRNAs (miRNAs) are a class of non-coding RNAs that regulate gene expression in viral infection. Nonetheless, there is little information on virulent duck enteritis virus (DEV)-encoded miRNAs.

Results: Using high-throughput sequencing, we identified 39 mature viral miRNAs from CHv-infected duck embryo fibroblasts cells. Compared with the reported 33 VAC-encoded miRNAs, only 13 miRNA sequences and 22 "seed sequences" of miRNA were identical, and 8 novel viral miRNAs were detected and confirmed by stem-loop RT-qPCR in this study. Using RNAhybrid and PITA software, 38 CHv-encoded miRNAs were predicted to target 41 viral genes and formed a complex regulatory network. Dual luciferase reporter assay (DLRA) confirmed that viral dev-miR-D8-3p can directly target the 3′-UTR of CHv US1 gene ($p < 0.05$). Gene Ontology analysis on host target genes of viral miRNAs were mainly involved in biological regulation, cellular and metabolic processes. In addition, 598 novel duck-encoded miRNAs were detected in this study. Thirty-eight host miRNAs showed significant differential expression after CHv infection: 13 miRNAs were up-regulated, and 25 miRNAs were down-regulated, which may affect viral replication in the host cell.

Conclusions: These data suggested that CHv encoded a different set of microRNAs and formed a unique regulatory network compared with VAC. This is the first report of DEF miRNAs expression profile and an analysis of these miRNAs regulatory mechanisms during DEV infection. These data provide a basis for further exploring miRNA regulatory roles in the pathogenesis of DEV infection and contribute to the understanding of the CHv-host interaction at the miRNA level.

Keywords: Duck enteritis virus, MicroRNAs, Conservation, Pathogenesis, High-throughput sequencing

* Correspondence: jiary@sicau.edu.cn; chenganchun@vip.163.com
[1]Research Center of Avian Disease, College of Veterinary, Medicine of Sichuan Agricultural University, Wenjiang District, Chengdu 611130, Sichuan Province, China
Full list of author information is available at the end of the article

Background

Duck viral enteritis, also called as duck plague, is an acute, contagious and fatal disease of duck and geese, resulting in considerable economic losses in the waterfowl breeding industry [1–4]. The causative agent of this disease is duck enteritis virus (DEV) which belongs to the species Anatid *herpesvirus* I, genus *Mardivirus*, subfamily *Alphaherpesvirinae*, family *Herpesviridae* [5]. Many countries, such as China, Britain, the United States, Germany, and Netherlands have reported the prevalence of this virus [6–8]. The genome of DEV is a linear double-stranded DNA molecule composed of a unique long region (UL) and a unique short region (US) flanked by an internal repeat sequence (IRS) and a terminal repeat sequence (TRS). Its genomic arrangement pattern (UL-IRS-US-TRS) is consistent with the members of Marek's disease virus 1 and 2 (MDV-1 and MDV-2), herpes simplex virus types 1 and 2 (HSV-1 and HSV-2) and Pseudorabies virus (PRV) [7, 8].

MicroRNAs (miRNAs) are small (18–24 nt), endogenous non-coding RNAs that widely found in plant, animal and viral genomes and are now increasingly recognized as important regulators of gene expression through post-transcriptional mechanisms, leading to mRNA degradation or translational inhibition by binding to fully or partially complementary 3′ untranslated regions (3′UTR) [9]. These small miRNAs participate in a variety of biological processes, including cellular proliferation, differentiation, apoptosis, signal transduction and the process of virus-host interactions [10–14].

Over 300 virus-encoded miRNAs have been identified (miRBase 22.0). They were encoded by multiple virus families [15, 16], such as herpesviruses adenoviruses, polyomaviruses and retroviruses [17–19]. approximately 95% of viral miRNAs were encoded by herpesvirus families [20]. This phenomenon suggested the importance of miRNA-mediated gene regulation in the biology of herpesvirus infections. Some functions of viral miRNAs were validated by experiments in the pathogenesis of herpesvirus infection [21, 22].

As with many other miRNA-encoding α-herpesviruses [23–28], DEV-encoded miRNAs were identified from VAC-infected chicken embryo fibroblast (CEF) by deep sequencing technology [29]. This research identified 24 pre-miRNAs in VAC genome producing 33 mature miRNAs. The VAC strain was attenuated and was widely used against duck viral enteritis [7], while the CHv strain (Chinese virulent DEV strain) can cause epidemical and fatal disease in waterfowl [30]. CHv and VAC are two different pathogenic DEV strains [31]. The mechanism of the two viruses causing different pathogenesis is not well understood. Our aim was to confirm whether the CHv encoded the same miRNAs as VAC and explore those miRNAs regulatory roles in CHv infection.

Moreover, recent studies have demonstrated that host miRNAs play crucial roles in viral infection [20, 21], but DEF-encoded miRNAs have not been reported until now. For the above purposes, we constructed and analysed the miRNA expression profile from CHv-infected and uninfected DEF cells using high-throughput sequencing. The potential targets of viral and host miRNAs were predicted by RNAhybrid and PITA software. These data may contribute to the understanding of CHv pathogenesis and the CHv virus-host interaction at the overall miRNA level.

Methods

Virus and cells

CHv (GenBank accession No. JQ647509), a classic Chinese virulent strain, was isolated from an infected duck farm and kept in our laboratory. Primary duck embryo fibroblast (DEF) cells were made using 10-day-old embryonated duck eggs (Chengdu Egg & Poultry Co. China) for virus propagation. The use of duck embryos in this study was approved by the Animal Ethics Committee of Sichuan Agricultural University (approval No. XF2014–18). Cell monolayers were cultured in Dulbecco's Modified Eagle's Medium (DMEM, Gibco, Grand Island, NY USA) supplemented with 8% foetal bovine serum (FBS, Gibco, USA) and 1% penicillin-streptomycin (Gibco, USA) at 37 °C in a 5% CO_2 humidified incubator.

Isolation and sequence of RNA

Duck embryo fibroblasts (DEF) cells (80% confluency per dish) were infected with CHv at a multiplicity of infection (MOI) of 1.0, with mock-infected DEF as a control. Cells were harvested at 2, 4, 6, 8, 12, 18, 24 and 30 h post-infection (hpi) and resuspended in TRIzol (TIANGEN, Beijing, China). Total RNAs from DEV-infected and uninfected DEF cells at the above time points were extracted according to the manufacturer's directions (TIANGEN, Beijing, China) and quantified using a NanoDrop 2000 Spectrophotometer (Thermo, Carlsbad, CA, USA). The RNA (0.125 μg) extracted from the eight time points was mixed as a group. Our experiments were performed in triplicate and all the infected and control samples were subjected to Huada (Guangdong, China) for high-throughput sequencing of small RNAs (sRNAs). The same mixed RNA samples were used in the subsequent stem-loop RT-qPCR experiments.

Data sources

The CHv genome has been sequenced and the total size is 162,175 bp. The annotated VAC-encoded miRNAs were from miRBase 22.0 (http://www.mirbase.org/). Duck genomic sequences and the 3′UTR of duck genes were downloaded from the Ensembl database (http://

www.ensembl.org). The annotated chicken and Zebra Finch mature miRNAs were from miRBase 22.0 (http://www.mirbase.org/).

Analysis of viral small RNAs

The total raw small RNA (sRNA) reads were detected by an Illumina Genome Analyser. The cleaned sequence reads were obtained after the filtering procedure as previously described [32, 33]. Using the Bowtie algorithm [34], the filtered sRNA reads were aligned to the known DEV pre-miRNA sequences in miRBase 22.0 with no mismatch and then aligned to the corresponding mature miRNA with at least 16 nts overlap allowing offsets. The known CHv-encoded miRNAs including the pre-miRNA sequences, length and count of reads would be obtained. The remaining sRNA reads mapped to genome were subjected for novel miRNA prediction. Mireap software (http://sourceforge.net/projects/mireap/) was used to predict novel miRNA by exploring the secondary structure. Dicer cleavage sites and predicted minimum free energies of unannotated sRNA reads.

Analysis of host small RNAs

There are not any *Anas platyrhynchos* miRNAs annotated in the miRBase 22.0. All host small RNA sequences were aligned with known mature miRNAs of two reference species (*Gallus* and *Taeniopygia guttata*) and *Anas platyrhynchos* genome by the Bowtie algorithm [34]. Different miRNA expression levels were normalized to get the number of transcripts per million (TPM) in two samples (CHv-infected and uninfected). Normalization formula: Normalized expression = Actual miRNA count/Total count of clean reads*1000000. A change of at least 2-fold between libraries was considered significant. Fold-change formula: Fold-change = log2 (treatment/control). *P*-value was set as the reported formula [35]. P-value < 0.05 indicated significance differentially expressed miRNA.

Target prediction and GO analysis of viral and host miRNAs

Target genes of viral and host miRNAs were predicted using RNAhybrid and PITA software, and the parameters were strictly set as a previously reported program in the seed sequence [36]. The potential host target genes were analysed using the Gene Ontology (GO) program (http://www.geneontology.org). Gene Ontology enrichment analysis of the target genes was performed using Goseq [37] to detect the significantly enriched GO terms of the host target. The GO terms with $p < 0.05$ were considered significant. The WEGO software (http://wego.genomics.org.cn) was used to produce histograms of the GO annotations, including three fields: cellular component, biological process and molecular function.

Stem-loop RT-qPCR

The stem-loop RT-qPCR was conducted as previously described [36, 38]. Briefly, 1000 ng of RNA mixture were reverse-transcribed to cDNA and then 2 μL cDNA was used for Real-time PCR amplification according to the company kit instructions (Thermo, Carlsbad, CA, USA). All primers used are listed in (Additional file 1: Table S1). The reaction conditions were as follows: reverse transcription was incubated at 50 °C for 45 min and kept at 85 °C for 5 min. Next, real-time PCR was 95 °C for 5 min, 39 cycles of denaturing at 95 °C for 15 s, annealing and extending 60 °C for 60 s, and the cellular miRNA U6 was used as an internal control. The relative expression values were calculated using the comparative $2^{-\Delta\Delta Ct}$ method [38].

Vector constructs and luciferase assay

The dev-miR-D8-3p mimic and negative control mimic (miR-NC) were synthesized by Ribobio (Guangzhou, China). The CHv US1 gene 3'UTR (nt 136,085–136,248) including the predicted dev-miR-D8-3p binding sites were synthesised and cloned into a pmirGLO vector (Promega, Madison, WI, USA) with SacI and XhoI sites and named pmirGLO-WT-US1, Accordingly, the mutant 3'UTR of the US1 vector was constructed and named pmirGLO-MU-US1. For luciferase assay, COS7 cells were seeded in 96-well plates and co-transfected with dev-miR-D8-3p mimic, miR-NC, pmirGLO-WT-US1 and pmirGLO-MU-US1 with Lipofectamine 3000 (Invitrogen, Carlsbad, CA, USA). We performed site-directed dual luciferase reporter assay (DLRA), and luciferase activity was measured at 36 h post-transfection according to the manufacturer's protocol (Promega, Madison, WI, USA).

Statistical analysis

Each experiment was performed in triplicate and the data were presented as the means (M) ± standard deviations (SD) by the software GraphPad Prism (version7.0). The significance of the variability between different treatment groups was determined by one-way analysis of variance (ANOVA) tests of variance using the GraphPad Prism software (version 7.0). *P*-values < 0.05 was considered statistically significant.

Results

Analysis of sRNA libraries by deep sequencing

In this study, we obtained 12,088,641 and 12,263,713 sRNA reads of 18–30 nucleotides from CHv-infected and uninfected DEF cells. After filtering adapter sequences and low-quality sequences. 11,462,557 (94.82%)

and 11,836,099 (96.51%) high quality reads from infected and uninfected sample were obtained, respectively. Among each sample, approximately 89.36% and 92.85% sRNAs ranged from 20 to 24 nt respectively, and most of the sRNA reads were 22 nt in length (Fig. 1a). In addition to miRNAs, other noncoding sRNAs were also detected and categorized by following the priority rule: microRNA (miRNA) > repeat > rRNA > tRNA > snoRNA > snRNA (Additional file 2: Table S2). Ultimately, 7,446,931 (64.97%) and 7,995,424 (67.55%) miRNA reads from CHv-infected and uninfected libraries respectively were matched to the annotated miRNAs of VAC and the two reference species (*Gallus gallus* and *Taeniopygia guttata*), and remaining 3,158,331 (27.55%) and 3,085,287 (26.07%) unannotated sRNA reads from two libraries were matched to CHv and the duck genome for predicting novel miRNAs (Fig. 1b, c).

Conservation analysis of miRNAs in CHv and VAC

In our study, we obtained 29 pre-miRNAs (Additional file 3) and 39 mature miRNAs from the CHv strain by deep sequencing. The names, sequences, length and location of 39 mature miRNAs are listed in Table 1. Compared with previously reported 33 mature VAC-encoded miRNAs [29], 31 of 33 reported miRNAs were detected and were shown in Table 2. The remaining two miRNAs, dev-miR-D2–3p and dev-miR-D10-3p were not detected in our study. Among

31 detected miRNAs, only 13 miRNA sequences were identical, and 18 were different in contrast to VAC-encoded miRNAs (Table 2). Twenty-two miRNAs were identical in the "seed sequence" and the other 9 were not identical. The difference of the "seed sequence" mostly occurs in 2–8 nucleotides at the 5′ end of miRNAs. For example, dev-miR-D19-5p and dev-miR-D21-5p had one deleted base, dev-miR-D7-5p, dev-miR-D11-3p, dev-miR-D13-5p, dev-miR-D14-3p and dev-miR-D23-3p had two deleted bases, the dev-miR-D4-3p had four deleted bases, and the dev-miR-D17-3p had three inserted bases. In addition, 8 novel CHv-encoded miRNAs were identified and were named from dev-miR-D25-5p to dev-miR-D31-3p (Table 1). The pre-miRNA hairpin structures and isoform expression profile of these novel miRNAs are shown in Additional file 3. Thirty-nine CHv-encoded miRNAs were distributed mostly the unique long region (UL) and the repeat region (IRS and TRS) of the genome (Fig. 2). This result was consistent with the previous report about distribution of VAC-encoded miRNAs [29]. We found that 7 miRNAs were present in two copies, which were located in two loci in the CHv genome. Including dev-miR-D20 to dev-miR-D24 (Table 1). Those miRNAs mapped in the internal repeat sequence (IRS) were marked as 'a' and the homologous miRNAs in terminal repeat sequences (TRS) were marked as 'b' (Fig. 2). including

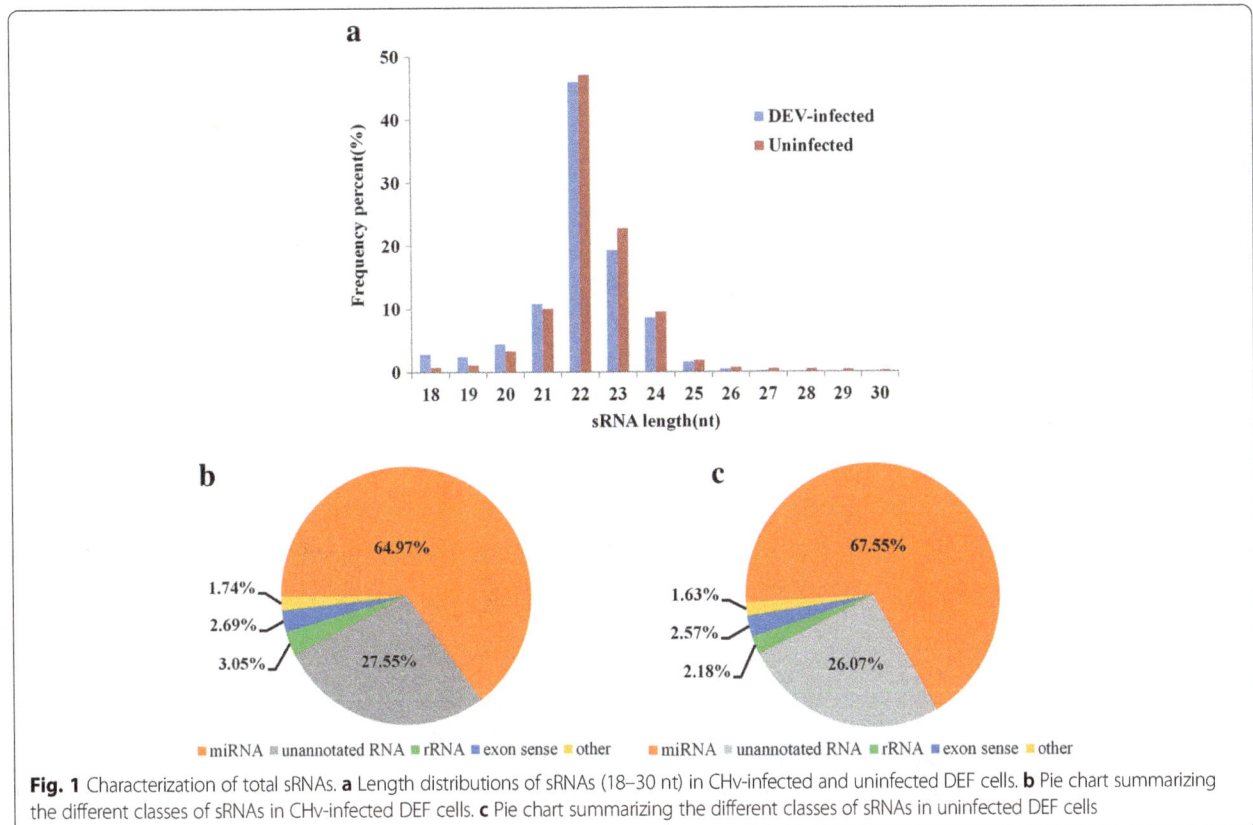

Fig. 1 Characterization of total sRNAs. **a** Length distributions of sRNAs (18–30 nt) in CHv-infected and uninfected DEF cells. **b** Pie chart summarizing the different classes of sRNAs in CHv-infected DEF cells. **c** Pie chart summarizing the different classes of sRNAs in uninfected DEF cells

Table 1 Summary of sequence and genomic position of CHv-encoded miRNAs

Name	Sequence(5'-3')	Length	Reads	Position and Strand
dev-miR-D1-5p	UUGGGAAUGGCGGAAGAGCAGACU	24	628	1328:1351 (−)
dev-miR-D1-3p	UCCUCUUGCGCGAUCCCCACGU	22	479	1294:1315 (−)
dev-miR-D3-3p	AUUGUUGCGUUUGGUGGUUUGUG	23	63	17,761:17783 (+)
dev-miR-D4-3p	UUGUCGGAUUGGUAUGCUUU	20	4	25,758:25777 (−)
dev-miR-D5-5p	UGUCAUCUGCGACGUCCUGCUCG	23	4157	52,654:52676 (−)
dev-miR-D6-5p	UGACACACCACCAUUCUGGCCG	22	904	53,728:53749 (−)
dev-miR-D6-3p	GUCAGAGUGUCGGUGAGUCGA	21	1018	53,695:53715 (−)
dev-miR-D7-5p	CGUAGCGGCGUAUAAUGGUUU	21	20	68,655:68675 (+)
dev-miR-D8-5p	UGCCUCCCGAUUAAACUAUACG	22	12	72,347:72368 (−)
dev-miR-D8-3p	UACAGUUUCGUUGGGCGGUUU	21	18,987	72,309:72329 (−)
dev-miR-D9-5p	CGUUUGAACGUUCUGUACUGCC	22	12,713	72,498:72519 (−)
dev-miR-D9-3p	CAGUCCAGAAUGUUCAAAC	19	1680	72,458:72476 (−)
dev-miR-D11-3p	AAAAGGGCAGCCUGGGCU	18	1	75,095:75112 (+)
dev-miR-D12–5p	UACCUGGGACAGAACCGCGGCCG	23	15,960	79,299:79321 (−)
dev-miR-D12–3p	CUCCGCGGUGAGGUCCCAGAA	21	870	79,263:79283 (−)
dev-miR-D13-5p	CGUGGGGUAGAACGCAUG	18	14	105,693:105710 (−)
dev-miR-D14-3p	GUUAUGUCUGGUUAUUAUGUUUU	23	1	107,259:107281 (−)
dev-miR-D15-3p	CGAGCGUGGGCAAGGUACC	19	700	112,570:112588 (−)
dev-miR-D16-3p	CUAAACACCAACGGAUGAACGU	22	14,930	112,727:112748 (−)
dev-miR-D17-5p	UGCAACGAAGGCGAACGGUUGA	22	5191	117,132:117153 (−)
dev-miR-D17-3p	UCCGACCGCUCGCCUUCGAGGC	22	3	117,098:117119 (−)
dev-miR-D18-5p	GGGAUCGGUGAGGGGGGAUUGUG	23	2676	119,157:119179 (−)
dev-miR-D18-3p	CCAUCCCCUCCGCUGGCCCCAA	22	1819	119,119:119140 (−)
dev-miR-D19-5p	AUGAAAGAGCGGUGCCUUU	19	771	119,180:119198 (−)
dev-miR-D20-5p	AAUGUCGGCCAGCCUCUCCGCUU	23	11,422	125,008:125030 (+)/160,535:160557(−)
dev-miR-D21-5p	GGUUUGGAGACAGCUGCGGUGG	22	651	125,178:125199 (+)/160,366:160387(−)
dev-miR-D21-3p	AUCCAUGCAAUCUCCAAACAAC	22	347	125,218:125239 (+)/160,326:160347(−)
dev-miR-D22-5p	UUACCCGCCCAUGCGUGACUGCC	23	2201	126,494:126516 (+)/159,049:159071(−)
dev-miR-D22–3p	GUCACACAAGGCGGCUAGCAGG	22	11	126,532:126553 (+)/159,012:159033(−)
dev-miR-D23-3p	CGAACCGUCACAGUCUGCAGA	21	3322	128,060:128080 (+)/157,485:157505(−)
dev-miR-D24-3p	AUUGGCUUCAGAGUGCGAACGC	22	21	134,514:134535 (+)/151,030:151051(−)
dev-miR-D25-5p	UGUGGGGACCGUGUAUGAGAUGU	23	145	696:718 (−)
dev-miR-D26-5p	AUCGAAGCGAGGCGAGAUAACCU	23	12	26,368:26390 (−)
dev-miR-D26-3p	GUUCUCCCUUGCUUUGACAU	20	12	26,329:26348 (−)
dev-miR-D27-5P	AUCCUGGACCGAUAUAUGGACA	22	197	73,467:73488 (−)
dev-miR-D28-3P	CUGGUGGGAAGAAUUUUCGC	20	149	77,133:77152 (−)
dev-miR-D29-5p	AACAUAUCUCUUGACCUCUGGCGU	24	2323	87,039:87062 (−)
dev-miR-D30-3P	ACUGGCUGGGGUGCAACUAAGU	22	9	103,962:103983 (−)
dev-miR-D31-3p	AUCACGGGGUGUUAGAUGAACC	22	13,664	123,167:123188 (+)

dev-miR-D20a/b-5p, dev-miR-D21a/b-3P, dev-miR-D21a/b-5p dev-miR-D22a/b-3p, dev-miR-D22a/b-5P, dev-miR-D23a/b-3p and dev-miR-D24a/b-3p. This "two-copy" phenomenon seems to be a common feature in α-herpesviruses.

Self-regulation analysis of viral miRNAs

Prediction results showed that 41 viral genes were targeted by 38 viral miRNAs. Some novel viral miRNAs (like dev-miR-D27-5p and dev-miR-D28-3p) could target multiple CHv genes, and the some CHv genes (like

Table 2 The differences (D) or similarities (S) between the known viral miRNAs from CHv and VAc strain (miRBase)

Name	CHv-Seq(5'-3')	Vac-Seq(5'-3')	Seq(S/D)	Seed Seq(S/D)[a]
dev-miR-D1-3p	UCCUCUUGCGCGAUCCCCACGU	UCCUCUUGCGCGAUCCCCACGU	S	S
dev-miR-D1-5p	UUGGGAAUGGCGGAAGAGCAGACU	UUGGGAAUGGCGGAAGAGCAGACU	S	S
dev-miR-D3-3p	AUUGUUGCGUUUGGUGGUUUGUG	AUUGUUGCGUUUGGUGGUUUGUG	S	S
dev-miR-D4-3p	UUGUCGGAUUGGUAUGCUUU	UUAAUUGUCGGAUUGGUAUGCUUUUU	D	D
dev-miR-D5-5p	UGUCAUCUGCGACGUCCUGCUCG	UGUCAUCUGCGACGUCCUGCUCG	S	S
dev-miR-D6-3p	GUCAGAGUGUCGGUGAGUCGA	GUCAGAGUGUCGGUGAGUCGACG	D	S
dev-miR-D6-5p	UGACACACCACCAUUCUGGCCG	UGACACACCACCAUUCUGGCCG	S	S
dev-miR-D7-5p	CGUAGCGGCGUAUAAUGGUUU	UUCGUAGCGGCGUAUAAUGGUUU	D	D
dev-miR-D8-3p	UACAGUUUCGUUGGGCGGUUU	UACAGUUUCGUUGGGCGGUUUC	D	S
dev-miR-D8-5p	UGCCUCCCGAUUAAACUAUACG	UGCCUCCCGAUUAAACUAUACGC	D	S
dev-miR-D9-3p	CAGUCCAGAAUGUUCAAAC	CAGUCCAGAAUGUUCAAACG	D	S
dev-miR-D9-5p	CGUUUGAACGUUCUGUACUGCC	CGUUUGAACGUUCUGUACUGCCC	D	S
dev-miR-D11-3p	AAAAGGGCAGCCUGGGCU	GCAAAAGGGCAGCCUGGGCUCUAU	D	D
dev-miR-D12-3p	CUCCGCGGUGAGGUCCCAGAA	CUCCGCGGUGAGGUCCCAGAAA	D	S
dev-miR-D12-5p	UACCUGGGACAGAACCGCGGCCG	UACCUGGGACAGAACCGCGGCCG	S	S
dev-miR-D13-5p	CGUGGGGUAGAACGCAUG	CCCGUGGGGUAGAACGCAU	D	D
dev-miR-D14-3p	GUUAUGUCUGGUUAUUAUGUUUU	GCGUUAUGUCUGGUUAUUAUGUUUUU	D	D
dev-miR-D15-3p	CGAGCGUGGGCAAGGUACC	CGAGCGUGGGCAAGGUACCAG	D	S
dev-miR-D16-3p	CUAAACACCAACGGAUGAACGU	CUAAACACCAACGGAUGAACGU	S	S
dev-miR-D17-3p	UCCGACCGCUCGCCUUCGAGGC	GACCGCUCGCCUUCGAGGCCACC	D	D
dev-miR-D17-5p	UGCAACGAAGGCGAACGGUUGA	UGCAACGAAGGCGAACGGUUG	D	S
dev-miR-D18-3p	CCAUCCCCUCCGCUGGCCCCAA	CCAUCCCCUCCGCUGGCCCCAA	S	S
dev-miR-D18-5p	GGGAUCGGUGAGGGGGGAUUGUG	GGGAUCGGUGAGGGGGGAUUGUG	S	S
dev-miR-D19-5p	AUGAAAGAGCGGUGCCUUU	GAUGAAAGAGCGGUGCCUUU	D	D
dev-miR-D20-5p	AAUGUCGGCCAGCCUCUCCGCUU	AAUGUCGGCCAGCCUCUCCGCUU	S	S
dev-miR-D21-3p	AUCCAUGCAAUCUCCAAACAAC	AUCCAUGCAAUCUCCAAACAACC	D	S
dev-miR-D21-5p	GGUUUGGAGACAGCUGCGGUGG	UGGUUUGGAGACAGCUGCGGUGGU	D	D
dev-miR-D22-3p	GUCACACAAGGCGGCUAGCAGG	GUCACACAAGGCGGCUAGCAGG	S	S
dev-miR-D22-5p	UUACCCGCCCAUGCGUGACUGCC	UUACCCGCCCAUGCGUGACUGCC	S	S
dev-miR-D23-3p	CGAACCGUCACAGUCUGCAGA	CGCGAACCGUCACAGUCUGCAG	D	D
dev-miR-D24-3p	AUUGGCUUCAGAGUGCGAACGC	AUUGGCUUCAGAGUGCGAACGC	S	S
dev-miR-D2-3p[b]		AUAAGGCGAUCCGUGGUUU		
dev-miR-D10-3p[b]		CUUUGAGUUCUAGCCCGUCUAUC		

[a]Seed sequence of miRNAs were present in italic font
[b]The dev-miR-D2-3p and dev-miR-D10-3p were not detected in CHv-infected DEF cells

UL24, UL28 and UL52) could be targeted by multiple viral miRNAs. A complex regulatory network was formed according to the regulation interaction between viral miRNAs and target genes (Additional file 4: Figure S2).

Regulatory analysis of viral miRNAs on host genes
Analysis results showed that the 3'UTRs of 4703 host genes were targeted by 39 viral miRNAs using the intersection of the two software programs (Additional file 5: Table S3). Gene Ontology (GO) annotation was performed to analyze biological function of the host target genes. The results reflected that these host target genes were mainly concentrated in the cellular process, metabolic process, signal-organism process, biological regulation process and others (Additional file 6: Table S4). Among of these host target genes, GO enrichment analysis showed that 236 genes were related to signaling processes ($p < 0.05$) and 66 genes were related to immune system processes ($p < 0.05$) (Fig. 3, Additional file 6: Table S4), which implied that viral miRNAs may play important regulatory function during viral infection and immune evasion.

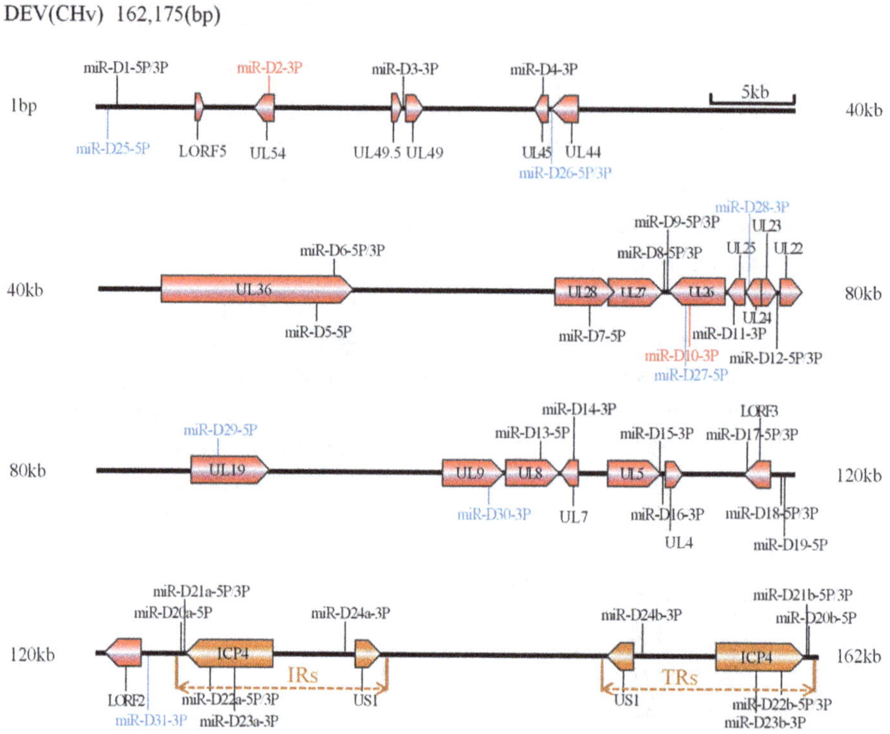

Fig. 2 Location of virus-encoded mature miRNAs in the CHv genome. The relative positions of the known and predicted novel miRNAs in the CHv genome are shown. The linear form indicated DEV CHv genome. The orientations of each of the ORFs in relation to the miRNA location were indicated with red or orange arrows. The internal repeat sequences (IRs) and terminal repeat sequences (TRs) of DEV CHv genome were indicated with orange. The undetected miRNAs were indicated with red font. The known miRNAs were indicated with black font and the novel miRNAs were indicated with blue font

Expression and differential analysis for host miRNAs

Alignment results showed that 598 mature host miRNAs were detected in this study (Fig. 4a). Among these, 386 (64.5%) miRNAs (264 aligned and 122 novel) were co-expressed in both libraries (Additional file 7: Table S5), 108 (18.1%) miRNAs were unique to the DEV-infected group and 104 (17.4%) miRNAs were unique to the un-infected group (Additional file 8: Table S6). Among the co-expressed host miRNAs, 38 miRNAs were differentially expressed between the CHv-infected sample and uninfected sample (Additional file 9: Table S7). Thirteen were significantly up-regulated and 25 were significantly down-regulated after CHv infection (Fig. 4b). Thirty-eight differentially expressed host miRNAs were predicted to target viral genes using the RNAhybrid and PITA software, and the results showed that the 3'UTRs of 40 viral genes were targeted by 36 host miRNAs by the intersection of two software (Additional file 10: Figure S3).

Stem-loop RT-qPCR for miRNAs confirmation

To further validate deep sequencing results, 8 novel viral miRNAs and 10 randomly differentially expressed host miRNAs were confirmed using stem-loop RT-qPCR. The results obtained by RT-qPCR were highly consistent with the deep sequencing data (Fig. 4c, d).

Dev-miR-D8-3p target the 3'UTR of US1 gene

Dual luciferase reporter assay (DLRA) showed that the luciferase level of the pmirGLO-WT-US1 was significantly repressed by dev-miR-D8-3p compared to the negative control miR-NC ($p < 0.05$) (Fig. 5a, b). To further ascertain that the down-regulation of targets by dev-miR-D8-3p is binding sites dependent, the binding sites of US1 were mutated and constructed as pmirGLO-MU-US1 vector (Fig. 5a). As expected, the dev-miR-D8-3p lost its repression effect on the mutant vector of pmirGLO-MU-US1. These results indicated that the dev-miR-D8-3p can directly target the CHv US1 gene by 7 nucleotide complementary seed sequence.

Discussion

Previous research has reported that the VAC encoded 33 mature miRNAs in the viral genome [29]. We obtained 39 mature viral miRNAs from CHv-infected DEF cells, 22 of 39 CHv-encoded miRNAs share identical "seed sequence" with VAC-encoded miRNAs. Another 17 miRNAs (9 different "seed sequence" miRNAs and 8 novel miRNAs) were different in the "seed sequence". As we know, target-gene recognition of viral miRNA is strictly dependent on the full base complementarity of the "seed sequence", which covered 2 to 8 nucleotides from the 5'

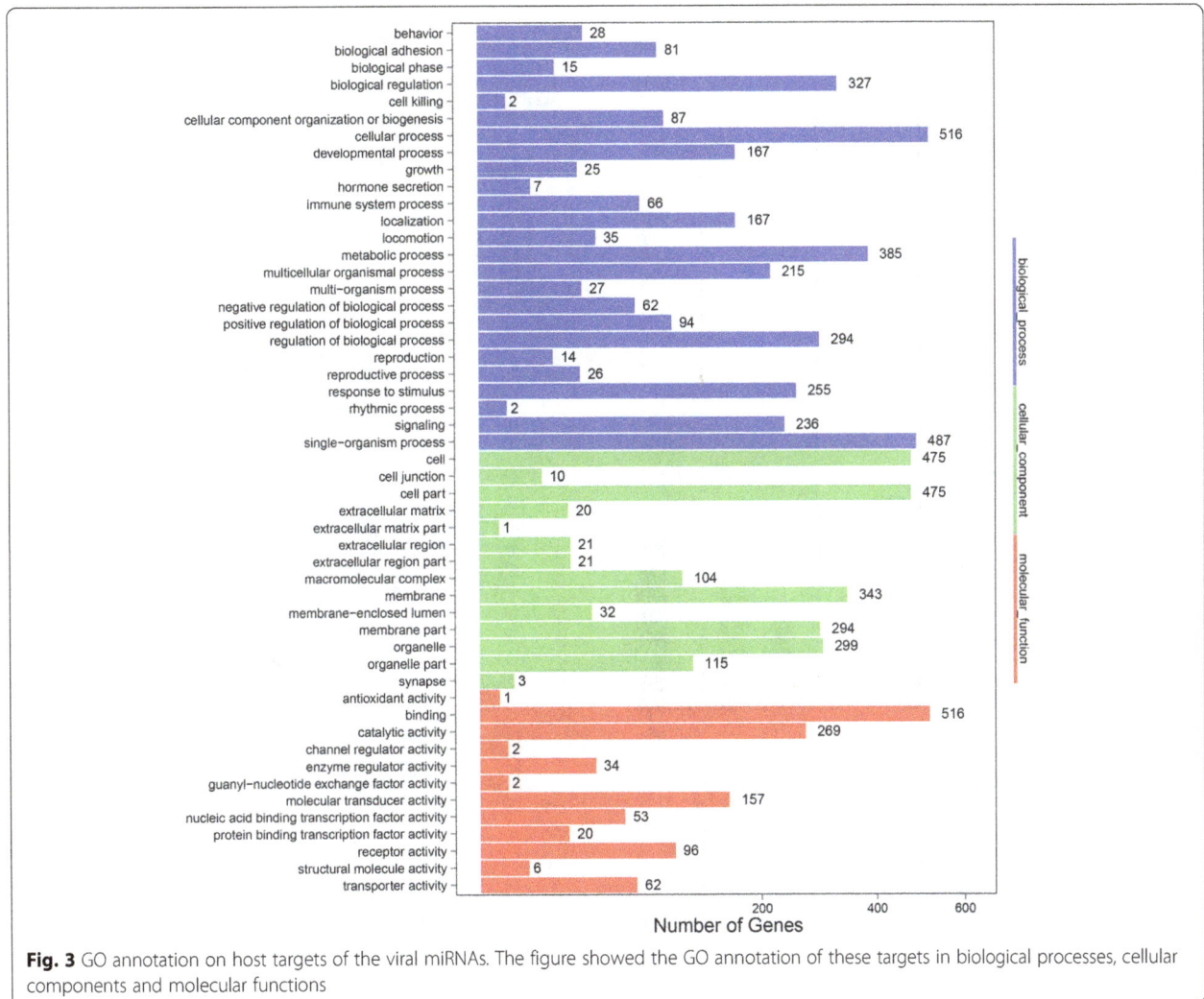

Fig. 3 GO annotation on host targets of the viral miRNAs. The figure showed the GO annotation of these targets in biological processes, cellular components and molecular functions

end of the miRNA [39]. Thus, the stability of the "seed sequence" of viral miRNA is crucial for target-gene discrimination. We speculate that the 22 miRNAs of the identical "seed sequence" play same regulatory roles in DEV-infection. Overall, the data analysis showed that CHv encoded a different pattern of miRNA Compared with VAC, which might form a complex regulatory network between viral miRNAs and their target genes. The differences of miRNAs regulatory network might lead to the differential pathogenesis of these two viruses.

Studies have revealed that viral and host miRNAs play important roles in host-virus interactions [20, 22, 40]. CHv is a virulent herpesvirus that can mainly cause contagious lethal disease in ducks [30, 31] and the VAC is an avirulent virus and has been reported to encode 33 mature miRNAs from VAC-infected CEF cells [29]. However, a precise regulatory network analysis of DEV miRNAs is unlikely to be achieved without the discovery of the virulent DEV miRNAome. In this research, we collected sRNA samples from CHv-infected DEF cells at

eight time points to detect as many viral miRNAs as possible. Using High-throughput sequencing technology, we obtained 29 pre-miRNA sequences with 39 mature miRNAs from CHv-infected DEF cells. Eight novel viral miRNAs were predicted and were confirmed by stem-loop RT-qPCR (Fig. 4c). In addition, we also first made a repertoire of DEF cells miRNAs transcriptome in CHv-infected and uninfected cells and have performed a preliminary analysis of the functions of these miRNAs. These data provide a foundation for further investigations on host-herpesvirus interactions.

Among the 8 novel viral miRNAs, the dev-miR-D27-5p, dev-miR-D28-3p, dev-miR-D29-5p and dev-miR-D30-3p were located in coding region of UL26, UL24, UL19 and UL9 gene, respectively. The remaining four miRNAs were encoded in the in noncoding regions of CHv genome. Several reports revealed that most α-herpesvirus-encoded miRNAs were found clustered in the repeat or other adjacent regions of the viral genome [20, 23–27, 41]. However, the 39 CHv-encoded miRNAs

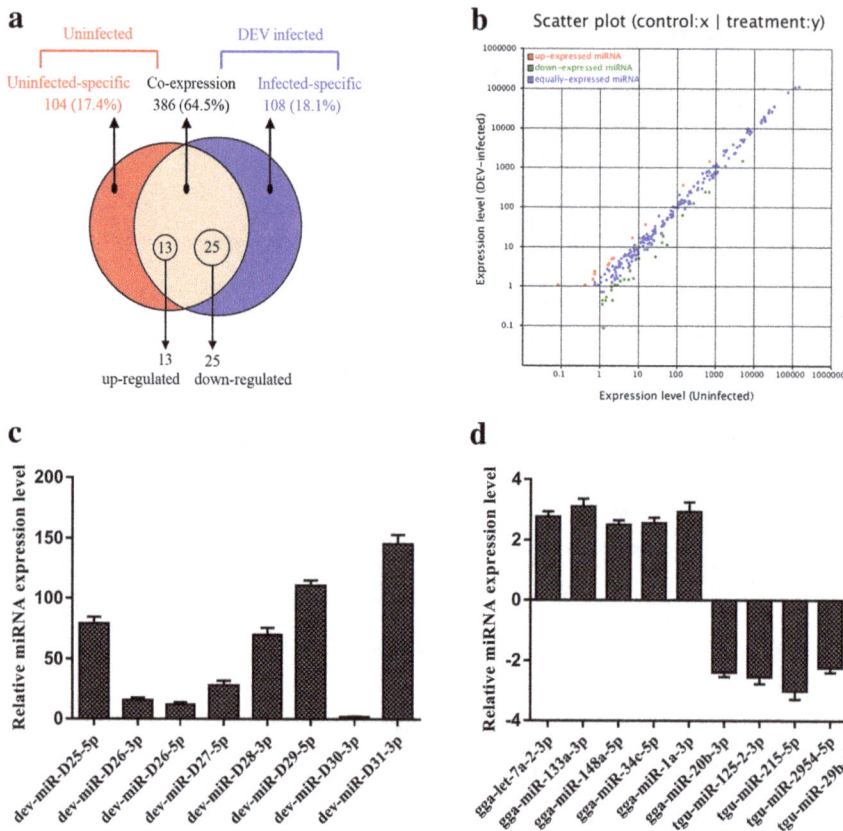

Fig. 4 Characteristics of viral and host miRNAs. **a** The Venn diagram shows the distribution of 598 unique miRNAs between uninfected (left, red circle) and DEV-infected sample (right, blue circle) libraries. **b** Differential expression of host miRNAs as a function of DEV CHv infection. Red, miRNAs with ratio > 2 (infected/uninfected in expression); blue, miRNAs with $1/2 \leq$ ratio ≤ 2; green, ratio < 1/2. **c** Expression levels detection of 8 virus-encoded novel miRNAs using stem-loop RT-qPCR. **d** Confirmation of 10 differentially expressed host miRNAs using stem-loop RT-qPCR

were distributed mostly in the unique long region (UL) and the repeat region (IRS and TRS) of the genome (Fig. 2). This result was consistent with the previous report about the distribution of VAC-encoded miRNAs [29]. Moreover, of the seven miRNAs detected with two copies, miR-D22b-3p, miR-D22b-5p and miR-D23b-3p were located in the coding region of ICP4 in an anti-sense orientation, which could theoretically lead to the cleavage of the transcript and negative regulation of the gene like siRNAs [42–46].

Several studies have confirmed that herpesvirus-encoded miRNAs can target viral immediate-early (IE) genes to regulate viral latent and lytic infection [20–22]. The hsv1-miR-H2–3p and hsv1-miR-H6 target the ICP0 and ICP4 genes of HSV-1 respectively [47], the hsv2-miR-H2–3 target the ICP0 gene of HSV-2 [46, 48], the mdv1-miR-M7-5P target the ICP4 and ICP27 genes of MDV-1 [49] and the litv-miR-I5 target the ICP4 gene of LITV [50]. The above target genes acted as viral IE genes which upregulate early and late genes of herpes-virus subfamilies and downregulate latency-associated transcript (LAT), inducing the virus towards lytic

infection [20, 21]. The targeting of IE genes by viral miRNAs was thought to inhibit entry into viral replica-tion and maintain the latent infection state [22]. In our study, dev-miR-D4-3p, dev-miR-D11-3p, and dev-miR-D20-5p were predicted to target the 3'UTR region of the CHv ICP4 gene. Dev-miR-D1-5p, dev-miR-D8-3p, dev-miR-D12–5p, dev-miR-D17-3p, dev-miR-D26-3p, dev-miR-D28-3p and dev-miR-D30-3p were predicted to target the 3'UTR region of the CHv US1 gene. Our results confirmed that dev-miR-D8-3p could directly target the 3'-UTR of the US1 gene. Dev-miR-D13-5p and dev-miR-D14-3p are predicted to target the 3'UTR region of the CHv UL54 gene (Additional file 4: Fig-ure S2b). The ICP4, US1 and UL54 of CHv were consid-ered the functional equivalent of the immediate-early (IE) genes ICP4, ICP22 and ICP27 of HSV-1 [7, 30, 51, 52]. Thus, we speculate that these viral miRNAs may play key roles in the regulation of the CHv lytic and latent infection.

Some virus-encoded miRNAs could regulate the cellu-lar signal pathway to evade the immune response. For example, the hcmv-miR-UL112–3p was reported to

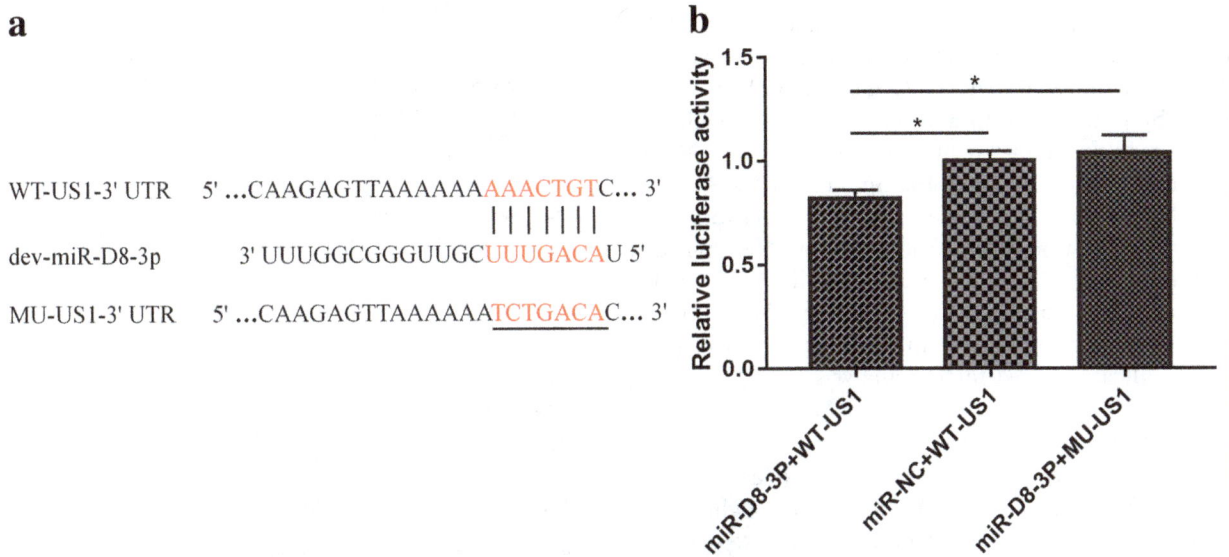

Fig. 5 Luciferase reporter assay for the interaction between dev-miR-D8-3p and US1 gene. **a** The seed sequence of dev-miR-D8-3p and its target site in 3'UTR of the US1 mRNA are shown in red, seven nucleotides were mutated in 3'UTR of the US1 mRNA (underlined). **b** Activity of the luciferase gene linked to the 3'UTR of the US1 mRNA. The wild-type pmirGLO-WT-US1 (WT-US1) and mutant pmirGLO-MU-US1 (MU-US1) were respectively transfected into COS7 cells with the dev-miR-D8-3p (miR-D8-3p) mimic or the negative control (miR-NC). Luciferase activities were measured after 36 h. The data were presented as the means and the standard deviations (SDs) of separate transfections ($n = 3$). Statistical significance was analyzed by one-way analysis of variance (ANOVA). The significant differences ($p < 0.05$) are indicated as single star

target toll-like receptors 2 (TLR2), inhibiting IRAK1/NFκB signaling and avoiding the related inflammatory response [53]. The mdv1-miR-M4-5P regulated the endogenous TLR3 gene that repressed IFN-β production expression and facilitates virus replication [54]. Three virus-encoded miRNAs (e.g., hcmv-miR-UL112–1, ebv-miR-BART2-5p and kshv-miR-K12–7) could repress identical target gene MICB and lead to a similar outcome, evading NK cell recognition and immune response [55]. The kshv-miR-K12–9 and kshv-miR-K12–5 could target IRAK1 and MYD88, respectively, which repressed TLR/IL-1R signaling, resulting in reduced inflammation [56]. The kshv-miR-K12–11 could target IκB kinase epsilon (IKKε), inhibiting type I interferon signal pathway [57]. Moreover, viral miRNAs could regulate cell growth and survival to favour viral replication. For example, kshv-miR-K12–10 could inhibit TWEAK-induced apoptosis by targeting the cellular TWEAKR [4], which contributed to cell survival. In addition, mdv1-miR-M4-5p could target LTBP1, which suppressed the TGF-β signaling [58]. Kshv-miR-K12–11 targeted SMAD5 which interfered with the TGF-β pathway [59]. The suppression of TGF-β signaling ultimately result in increased cell survival and virally induced oncogenesis [58, 59]. In our study, GO analysis on the cellular targets of viral miRNAs showed that these targets were involved in complex cellular processes, including signal-organism processes, the metabolic pathway, biological regulation, immune response and signaling process.

The virus could alter host miRNA expression profiles to favour viral replication. In our study, 38 cellular miRNAs were expressed differentially in both the CHv-infected library and mock library. These dysregulated host miRNAs were identified to play crucial roles in other viral infections. For example, miR-let-7a was downregulated in NPC cells after EBV-infection, which in turn promoted viral replication by targeting the dicer gene [60]. The gga-miR-26a was downregulated in MDV-infected spleens at cytolytic infection, latency and tumour transformation phases. Decreasing the expression of gga-miR-26a had been shown to contribute to MDV-induced lymphomagenesis upon regulation of NEK6 proteins [61]. The previous research showed that the differential expression of gga-miR-181a contributed to MDV-induced lymphomagenesis by targeting IGF2BP3/MYBL1 genes [62]. Cellular microRNA miR-181b inhibited replication of mink enteritis virus (MEV) by repression of non-structural protein 1(NS1) translation [63]. The gga-miR-15b was downregulated in splenic tumours after MDV infection and had a negative effect on the expression of ATF2, facilitating viral replication by increasing the expression of the ATF2 [64]. Expression of miR-146 was upregulated after EBV infection, which could downregulate levels of IRAK1 and TRAF6 proteins, reducing the activity of host immune and inflammatory response [65]. Recently, miR-148 was reported as a novel biomarker in non-small-cell lung cancer screening [66]. In our prediction results, miR-148a-5p could target UL1, UL2 UL3, UL24 and UL25 genes of CHv. MiR-181a-3p could target UL24, UL54, US3, US5 and US8 genes of CHv (Additional file 10: Figure S3a). The ICP4 gene of CHv was targeted by miR-135a-1-3p and miR-135a-2-3p, while the UL54 gene was targeted by miR-124a-3p, miR-135a-1-3p, miR-135a-2-3p, miR-15b-3p, miR-181a-3p and miR-181b-1-3p. A complex regulatory network was formed between 36 differentially expressed host miRNAs and their 40 viral target genes (Additional file 10: Figure S3b). However, the regulatory functions of these dysregulated cellular miRNAs in the process of CHv replication need further analysis.

Conclusion

In this study, we obtained 39 DEV-encoded miRNAs from CHv-infected DEF cells by high-throughput sequencing. Of these, 8 novel viral miRNAs were detected and confirmed through stem-loop RT-qPCR. Conservative analysis showed that CHv encoded a different set of miRNAs and formed a unique regulatory network compared with VAC. In addition, a total of 598 novel duck-encoded miRNAs were detected by aligning with known mature miRNAs of *Gallus gallus* and *Taeniopygia guttata*. This is the first report of a DEF miRNA expression profile and an analysis of these miRNAs regulatory mechanisms during DEV infection.

Additional files

Additional file 1: Table S1. Primers used to amplify virus and host miRNAs by stem-loop RT-qPCR. Stem-loop RT-qPCR was conducted using miRNA specific stem-loop RT primers together with corresponding miRNA specific forward (F) primers and universal reverse (UR) primer. (DOCX 15 kb)

Additional file 2: Table S2. Distribution of sRNAs in DEV-infected and uninfected samples. (DOCX 15 kb)

Additional File 3: The expression profiling of CHv miRNAs and pre-miRNA secondary structures. MiRNA sequences and their coresponding reads mapped on the precursors of CHv miRNA genes. Opening parentheses indicate pairing nucleotides. Inside the closed parentheses indicate the minimum free energy for the secondary structure of the miRNA. The number of reads mapped to the miRNA precursors is indicated in the right side. Mature miRNAs are denoted in red. The hairpin structures of pre-miRNA is shown at the back. (PDF 649 kb)

Additional File 4: Figure S2. Regulatory network of CHv-encoded miRNAs and CHv genes. a Gene regulatory network formed by CHv-encoded miRNAs (red ellipses) and their target genes (yellow ectangles). **b** Gene regulatory network of CHv-encoded miRNAs (red ellipses) and their target immediate-early(IE) genes (yellow rectangles) (PDF 401 kb)

Additional file 5: Table S3. Predicted host target genes of DEV miRNAs. Host target genes were predicted using PITA and RNAhybrid software. (XLSX 832 kb)

Additional file 6: Table S4. Gene ontology analysis on host targets of the viral miRNAs (XLSX 23 kb)

Additional file 7: Table S5. Novel duck miRNA expressed in DEF cells. The 386 mature sequence and sequence read count of novel duck miRNAs predicted with miRDeep in each sequenced sample of DEF cells. (XLSX 26 kb)

Additional file 8: Table S6. Novel duck miRNA expressed in DEF cells. The 212 mature sequence and sequence read count of novel duck miRNAs predicted with miRDeep in each sequenced sample of DEF cells. (XLSX 15 kb)

Additional file 9: Table S7. Differentially expressed host alignment miRNA in CHv-infected DEF cells. Fold-change = log2 (infected/mock in expression) > 1 or < − 1, and p-value < 0.05 indicated significance differentially expressed miRNA. Fold-change < − 1 indicates down-regulated, Fold-change > 1 indicates up-regulated. (XLSX 12 kb)

Additional File 10: Figure S3. Regulatory network of DEF miRNAs and CHv genes. a Gene regulatory network formed by differentially-expressed DEF miRNAs (blue ellipses) and their target genes (yellow rectangles). **b** Gene regulatory network of differentially-expressed DEF miRNAs (blue circles) and target immediate-early (IE) genes (yellow rectangles). (PDF 353 kb)

Abbreviations

CHv: Chinese virulent; DEV: Duck enteritis virus; HCMV: Human Cytomegalovirus; HSV-1: Herpes simplex virus 1; HSV-2: Herpes simplex virus 2; IRS: Inverted repeated sequences; IRS: Inverted repeated sequences; MDV-1: Marek's disease virus 1; MDV-2: Marek's disease virus 2; PRV: Pseudorabies virus; RT-qPCR: Real-time quantitative reverse transcriotion PCR; TRS: terminal repeated sequences; UL: Unique long region; UL: Unique long region; US: Unique short; US: Unique short region; VAC: vaccine

Acknowledgements

We would like to thank Jue Wang (BGI Genomics Co, Ltd., China) for technical assistance.

Funding

This work was supported by National Science and Technology Support Program (2015BAD12B05), Integration and Demonstration of Key Technologies for Duck Industrial in Sichuan Province (2014NZ0030), China Agricultural Research System (CARS-43-8), and Sichuan Province Research Programs (2017JY0014/2014–002).

Authors' contributions

Conceived and designed the experiments: RYJ XLW. Performed the experiments: XLW, JKZ. Analyzed the data: RYJ, MSW, SC, MFL, DKZ, XXZ, KFS, ZQY, XYC, JW, ACC. Contributed reagents and materials: QY, YW. Wrote the paper: XLW, RYJ. All authors read and approved the final manuscript.

Competing interests

The authors declare that they have no competing interests.

Author details

[1]Research Center of Avian Disease, College of Veterinary, Medicine of Sichuan Agricultural University, Wenjiang District, Chengdu 611130, Sichuan Province, China. [2]Key Laboratory of Animal Disease and Human Health of Sichuan Province, Wenjiang District, Chengdu 611130, Sichuan Province, China. [3]Institute of Preventive Veterinary Medicine, Sichuan Agricultural University, Wenjiang District, Chengdu 611130, Sichuan Province, China. [4]BGI Genomics Co,shenzhen Ltd, Shenzhen 518083, Guangdong Province, China.

References

1. Metwally SA SD, Glisson JR, McDougald LR, Nolan, LK, Suarez DL, Nair V,. Duck Virus Enteritis (Duck Plague). In: Dis poultry, 13th edition 2013: 431–440.
2. Converse KA, Kidd GA. Duck plague epizootics in the United States, 1967-1995. J Wildl Dis. 2001;37(2):347–57.

3. Kaleta EF, Kuczka A, Kühnhold A, Bunzenthal C, Bönner BM, Hanka K, et al. Outbreak of duck plague (duck herpesvirus enteritis) in numerous species of captive ducks and geese in temporal conjunction with enforced biosecurity (in-house keeping) due to the threat of avian influenza a virus of the subtype Asia H5N1. Dtw. Dtsch Tierarztl Wochenschr. 2007;114(1):3–11.

4. Abend JR, Uldrick T, Ziegelbauer JM. Regulation of tumor necrosis factor-like weak inducer of apoptosis receptor protein (TWEAKR) expression by Kaposi's sarcoma-associated herpesvirus MicroRNA prevents TWEAK-induced apoptosis and inflammatory cytokine expression. J Virol. 2010;84(23):12139–51.

5. King AMQ. Virus taxonomy: classification and nomenclature of viruses: ninth report of the international committee on taxonomy of viruses: Elsevier/ academic press; 2012.

6. Wang G, Qu Y, Wang F, Hu D, Liu L, Li N, et al. The comprehensive diagnosis and prevention of duck plague in Northwest Shandong province of China. Poultry Sci. 2013;92(11):2892–8.

7. Li Y, Huang B, Ma X, Wu J, Li F, Ai W, et al. Molecular characterization of the genome of duck enteritis virus. Virology. 2009;391(2):151–61.

8. Wang J, Höper D, Beer M, Osterrieder N. Complete genome sequence of virulent duck enteritis virus (DEV) strain 2085 and comparison with genome sequences of virulent and attenuated DEV strains. Virus Res. 2011;160(1): 316–25.

9. Bartel DP. MicroRNAs: genomics, biogenesis, mechanism, and function. Cell. 2004;116(2):281–97.

10. Boss IW, Plaisance KB, Renne R. Role of virus-encoded microRNAs in herpesvirus biology. Trends in Microbiol. 2009;17(12):544–53.

11. Alvarezgarcia I, Miska EA. MicroRNA functions in animal development and human disease. Develop. 2005;132(21):4653–62.

12. Cullen BR. Viral and cellular messenger RNA targets of viral microRNAs. Nature. 2009;457(7228):421–5.

13. Chen Z, Zeng H, Guo Y, Liu P, Pan H. Deng a, et al. miRNA-145 inhibits non-small cell lung cancer cell proliferation by targeting c-Myc. J Experi & Clin Cancer Res. 2010;29(1):1–10.

14. Huang S, He X. The role of microRNAs in liver cancer progression. Bri. J Cancer. 2011;104(2):235–40.

15. Kincaid RP, Sullivan CS. Virus-encoded microRNAs: an overview and a look to the future. PLoS Pathog. 2012;8(12):e1003018.

16. Chen CJ, Cox JE, Kincaid RP, Martinez A, Sullivan CS. Divergent MicroRNA Targetomes of closely related circulating strains of a polyomavirus. J Virol. 2013;87(20):11135–47.

17. Yao Y, Smith LP, Nair V, Watson M. An avian retrovirus uses canonical expression and processing mechanisms to generate viral microRNA. J Virol. 2014;88(1):2–9.

18. Kincaid RP, Burke JM, Cox JC, Villiers EMD, Sullivan CS. A human torque Teno virus encodes a MicroRNA that inhibits interferon signaling. PLoS Pathog. 2013;9(12):e1003818.

19. Kozomara A, Griffithsjones S. miRBase: annotating high confidence microRNAs using deep sequencing data. Nucleic Acids Res. 2014;42:68–73.

20. Yao Y, Nair V. Role of virus-encoded microRNAs in avian viral diseases. Viruses. 2014;6(3):1379–94.

21. Piedade D, Azevedo-pereira JM. The role of microRNAs in the pathogenesis of herpesvirus infection. Viruses. 2016;8(6):156.

22. Grey F. Role of microRNAs in herpesvirus latency and persistence. J Gen Virol. 2015;96(4):739–51.

23. Burnside J, Bernberg E, Anderson A, Lu C, Meyers BC, Green PJ, et al. Marek's disease virus encodes MicroRNAs that map to meq and the latency-associated transcript. J Virol. 2006;80(17):8778–86.

24. Yao Y, Zhao Y, Xu H, Smith LP, Lawrie CH, Watson M, et al. MicroRNA profile of Marek's disease virus-transformed T-cell line MSB-1: predominance of virus-encoded microRNAs. J Virol. 2008;82(8):4007–15.

25. Yao Y, Zhao Y, Xu H, Smith LP, Lawrie CH, Sewer A, et al. Marek's disease virus type 2 (MDV-2)-encoded microRNAs show no sequence conservation with those encoded by MDV-1. J Virol. 2007;81(13):7164–70.

26. Waidner LA, Morgan RW, Anderson AS, Bernberg EL, Kamboj S, Garcia M, et al. MicroRNAs of Gallid and Meleagrid herpesviruses show generally conserved genomic locations and are virus-specific. Virology. 2009;388(1): 128–36.

27. Yao YX, Zhao YG, Smith LP, Watson M, Nair V. Novel microRNAs (miRNAs) encoded by herpesvirus of turkeys: evidence of miRNA evolution by duplication. J Virol. 2009;83(13):6969–73.

28. Rachamadugu R, Lee JY, Wooming A, Kong BW. Identification and expression analysis of infectious laryngotracheitis virus encoding microRNAs. Virus Genes. 2009;39(3):301–8.

29. Yao Y, Smith LP, Petherbridge L, Watson M, Nair V. Novel microRNAs encoded by duck enteritis virus. J Gen Virol. 2012;93(7):1530–6.

30. Wu Y, Cheng A, Wang M, Yang Q, Zhu D, Jia R, et al. Complete genomic sequence of Chinese virulent duck enteritis virus. J Virol. 2012;86(10):5965.

31. Wu Y, Cheng A, Wang M, Zhu D, Jia R, Chen S, et al. Comparative genomic analysis of duck enteritis virus strains. J Virol. 2012;86(24):13841–2.

32. Liu X, Liao S, Xu Z, Zhu L, Yang F, Guo W. Identification and analysis of the porcine MicroRNA in porcine cytomegalovirus-infected macrophages using deep sequencing. PLoS One. 2016;11(3):e0150971.

33. Liu X, Zhu L, Liao S, Xu Z, Zhou Y. The porcine microRNA transcriptome response to transmissible gastroenteritis virus infection. PLoS One. 2015; 10(3):e0120377.

34. Langmead B, Trapnell C, Pop M, Salzberg SL. Ultrafast and memory-efficient alignment of short DNA sequences to the human genome. Genome Biol. 2009;10(3):R25.

35. Audic S, Claverie JM. The significance of digital gene expression profiles. Genome Res. 1997;7(10):986–95.

36. Liu F, Zheng H, Tong W, Li GX, Tian Q, Liang C, et al. Identification and analysis of novel viral and host dysregulated MicroRNAs in variant pseudorabies virus-infected PK15 cells. PLoS One. 2016;11(3):e0151546.

37. Young MD, Wakeeld MJ. Smyth GK. Oshlack A. goseq: Gene Ontology testing for RNA-seq datasets; 2012.

38. Varkonyi-Gasic E, Hellens RP. Quantitative stem-loop RT-PCR for detection of microRNAs. Meth mole. Biol. 2011;744(9):145–57.

39. Bartel DP. MicroRNA target recognition and regulatory functions. Cell. 2009; 136(2):215–33.

40. Frappier L. Regulation of herpesvirus reactivation by host microRNAs. J Virol. 2015;89(5):2456.

41. Morgan R, Anderson A, Bernberg E, Kamboj S, Huang E, Lagasse G, et al. Sequence conservation and differential expression of Marek's disease virus microRNAs. J Virol. 2008;82(24):12213–20.

42. Barth S, Pfuhl T, Mamiani A, Ehses C, Roemer K, Kremmer E, et al. Epstein-Barr virus-encoded microRNA miR-BART2 down-regulates the viral DNA polymerase BALF5. Nucleic Acids Res. 2008;36(2):666–75.

43. Seo GJ, Fink LH, O'Hara B, Atwood WJ, Sullivan CS. Evolutionarily conserved function of a viral microRNA. J Virol. 2008;82(20):9823–8.

44. Seo GJ, Chen CJ, Sullivan CS. Merkel cell polyomavirus encodes a microRNA with the ability to autoregulate viral gene expression. Virology. 2009;383(2): 183–7.

45. Tang S, Bertke AS, Patel A, Wang K, Cohen JI, Krause PR. An acutely and latently expressed herpes simplex virus 2 viral microRNA inhibits expression of ICP34.5, a viral neurovirulence factor. Proc Natl Acad Sci U S A. 2008; 105(31):10931–6.

46. Shuang Tang AP, Krause PR. Novel less-abundant viral MicroRNAs encoded by herpes simplex virus 2 latency-associated transcript and their roles in regulating ICP34.5 and ICP0 mRNAs. J Virol. 2009;83(3):1433–42.

47. Umbach JL, Kramer MF, Jurak I, Karnowski HW, Coen DM, Cullen BR. MicroRNAs expressed by herpes simplex virus 1 during latent infection regulate viral mRNAs. Nature. 2008;454(7205):780–3.

48. Tang S, Bertke AS, Patel A, Margolis TP, Krause PR. Herpes simplex virus 2 MicroRNA miR-H6 is a novel latency-associated transcript-associated MicroRNA, but reduction of its expression does not influence the establishment of viral latency or the recurrence phenotype. J Virol. 2011; 85(9):4501–9.

49. Strassheim S, Stik G, Rasschaert D, Laurent S. mdv1-miR-M7-5p, located in the newly identified first intron of the latency-associated transcript of Marek's disease virus, targets the immediate-early genes ICP4 and ICP27. J Gen Virol. 2012;93(8):1731–42.

50. Waidner LA, Burnside J, Anderson AS, Bernberg EL, German MA, Meyers BC, et al. A microRNA of infectious laryngotracheitis virus can downregulate and direct cleavage of ICP4 mRNA. Virology. 2011;411(1):25–31.

51. Brehm M, Samaniego LA, Bonneau RH, Deluca NA, Tevethia SS. Immunogenicity of herpes simplex virus type 1 mutants containing deletions in one or more α-genes: ICP4, ICP27, ICP22, and ICP0. Virology. 1999;256(2):258–69.

52. Liu C, Cheng A, Wang M, Chen S, Jia R, Zhu D, et al. Duck enteritis virus UL54 is an IE protein primarily located in the nucleus. Virol J. 2015;12(1):1–8.

53. Landais I, Pelton C, Streblow D, Defilippis V, Mcweeney S, Nelson JA. Human cytomegalovirus miR-UL112-3p targets TLR2 and modulates the TLR2/ IRAK1/NFκB signaling pathway. PLoS Pathog. 2015;11(5):e1004881.

54. Hu X, Ye J, Qin A, Zou H, Shao H, Qian K. Both MicroRNA-155 and virus-encoded MiR-155 Ortholog regulate TLR3 expression. PLoS One. 2015;10(5): e0126012.

55. Stern-Ginossar N, Elefant N, Zimmermann A, Wolf DG, Saleh N, Biton M, et al. Host immune system gene targeting by a viral miRNA. Science. 2007; 317(5836):376–81.

56. Abend JR, Ramalingam D, Kieffer-Kwon P, Uldrick TS, Yarchoan R, Ziegelbauer JM. Kaposi's sarcoma-associated herpesvirus MicroRNAs target IRAK1 and MYD88, two components of the toll-like receptor/interleukin-1R signaling Cascade, to reduce inflammatory-cytokine expression. J Virol. 2012; 86(21):11663–74.

57. Liang D, Gao Y, Lin X, He Z, Zhao Q, Deng Q, et al. A human herpesvirus miRNA attenuates interferon signaling and contributes to maintenance of viral latency by targeting IKKε. Cell Res. 2011;21(5):793–806.

58. Chi JQ, Teng M, Yu ZH, Xu H, Su JW, Zhao P, et al. Marek's disease virus-encoded analog of microRNA-155 activates the oncogene c-Myc by targeting LTBP1 and suppressing the TGF-β signaling pathway. Virology. 2015;476:72–84.

59. Yunhua L, Rui S, Xianzhi L, Deguang L, Qiang D, Ke L. Kaposi's sarcoma-associated herpesvirus-encoded microRNA miR-K12-11 attenuates transforming growth factor beta signaling through suppression of SMAD5. J Virol. 2012;86(3):1372–81.

60. Mansouri S, Pan Q, Blencowe BJ, Claycomb JM, Frappier L. Epstein-Barr virus EBNA1 protein regulates viral latency through effects on let-7 microRNA and dicer. J Virol. 2014;88(19):11166–77.

61. Li X, Lian L, Zhang D, Qu L, Yang N. Gga-miR-26a targets NEK6 and suppresses Marek's disease lymphoma cell proliferation. Poultry Sci. 2014; 93(5):1097–105.

62. Lian L. Qu L, Chen Y, Lamont SJ, Yang N, Lian L, et al. a systematic analysis of miRNA transcriptome in Marek's disease virus-induced lymphoma reveals novel and differentially expressed miRNAs. PLoS One. 2011;7(11):e51003.

63. Sun JZ, Wang J, Yuan D, Wang S, Li Z, Yi B, et al. Cellular microRNA miR-181b inhibits replication of mink enteritis virus by repression of non-structural protein 1 translation. PLoS One. 2013;8(12):e81515.

64. Tian F, Luo J, Zhang H, Chang S, Song J. MiRNA expression signatures induced by Marek's disease virus infection in chickens. Genomics. 2012; 99(3):152–9.

65. Taganov KD, Boldin MP, Chang KJ, Baltimore D. NF-kappaB-dependent induction of microRNA miR-146, an inhibitor targeted to signaling proteins of innate immune responses. Proc Natl Acad Sci U S A. 2006;103(33):12481–6.

66. Yang JS, Li BJ, Lu HW, Chen Y, Lu C, Zhu RX, et al. Serum miR-152, miR-148a, miR-148b, and miR-21 as novel biomarkers in non-small cell lung cancer screening. Tumour Biology the Journal of the International Society for Oncodevelopmental Biology & Medicine. 2014;36(4):3035–42.

13

Geospatial and temporal associations of Getah virus circulation among pigs and horses around the perimeter of outbreaks in Japanese racehorses in 2014 and 2015

Hiroshi Bannai[1]*[iD], Manabu Nemoto[1], Hidekazu Niwa[1], Satoshi Murakami[2], Koji Tsujimura[1], Takashi Yamanaka[1] and Takashi Kondo[1]

Abstract

Background: We studied a recent epizootic of Getah virus infection among pigs in the southern part of Ibaraki Prefecture and the northern part of Chiba Prefecture, Japan, focusing on its possible association with outbreaks in racehorses in 2014 and 2015. The genomic sequence of a Getah virus strain from an infected pig was analyzed to evaluate the degree of identity with the strains from horses.

Results: Sera were collected from pigs from September to December 2012 to 2015 in south Ibaraki (380 pigs in 29 batches), and from September to December 2010 to 2015 in north Chiba (538 pigs in 104 batches). They were examined by using a virus-neutralizing test for Getah virus. Seropositivity rates in 2012–2013 in south Ibaraki and 2010–2012 in north Chiba ranged from 0% to 1.6%. In south Ibaraki, seropositivity rates in 2014 (28.8%) and 2015 (65.0%) were significantly higher than those in the previous years ($P < 0.01$); 4/5 batches had positive sera in 2014 and 7/7 in 2015. In north Chiba, seropositivity rates in 2013 (14.1%), 2014 (17.8%), and 2015 (48.0%) were significantly higher than those in the previous years ($P < 0.01$); 6/27 batches had positive sera in 2013, 3/9 in 2014, and 5/5 in 2015. Complete genome analysis revealed that the virus isolated from an infected pig had 99.89% to 99.94% nucleotide identity to the strains isolated from horses during the outbreaks in 2014 and 2015.

Conclusions: Serological surveillance of Getah virus in pigs revealed that the virus was circulating in south Ibaraki and north Chiba in 2014 and 2015; this was concomitant with the outbreaks in racehorses. The Getah virus strain isolated from a pig was closely related to the ones from horses during the 2014 and 2015 outbreaks. To our knowledge, this is the first convincing case of simultaneous circulation of Getah virus both among pigs and horses in specific areas.

Keywords: Getah virus, Japan, Horses, Pigs, Virus circulation

Background

Getah virus is classified in the genus *Alphavirus* in the family *Togaviridae* [1]. It is mosquito-borne and is widespread from Eurasia to Australasia. This virus causes fever, skin eruptions, and limb edema in horses [1], and it causes fetal death and reproductive disorders in pigs [2, 3]. Both pigs and horses can play important roles in the amplification and circulation of Getah virus, because maximum virus titers in the blood of infected animals of these species are sufficient for the vector mosquitoes to be infected, although viremia does not last for a long period [1, 4–6]. *Aedes vaxans nipponii* and *Culex tritaeniorhynchus* are considered to be major species that transmit Getah virus in Japan [7].

Outbreaks of Getah virus infection in horses occurred in autumn 2014 and 2015 at the Miho training center of the Japan Racing Association after 31 years of no reported outbreaks in Japan [8, 9]. The Miho training center is located in Ibaraki Prefecture in eastern Japan (Fig. 1). Surrounding the center, in Ibaraki and the

* Correspondence: bannai@equinst.go.jp
[1]Equine Research Institute, Japan Racing Association, 1400-4 Shiba, Shimotsuke, Tochigi 329-0412, Japan
Full list of author information is available at the end of the article

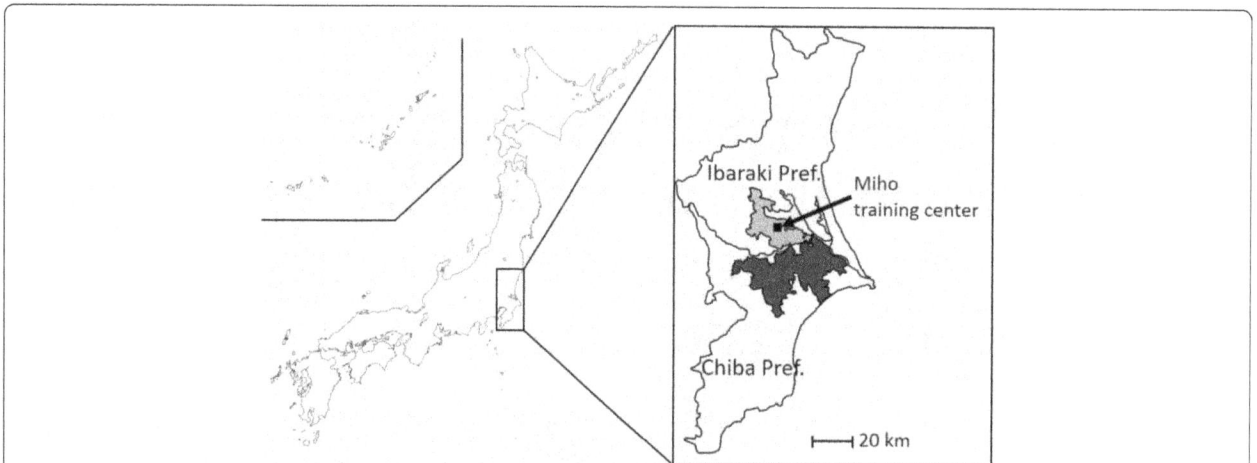

Fig. 1 Locations of the study areas. The area in the southern part of Ibaraki Prefecture (south Ibaraki, gray) includes the village of Miho and the cities of Inashiki, Ushiku, and Tsuchiura and covers an area of 454 km². There are 27 pig farms rearing about 32,000 pigs. The area in the northern part of Chiba Prefecture (north Chiba, black) includes the towns of Tako and Tohnosho and the cities of Inzai, Katori, Sakura, Narita, Shiroi, Yachimata, and Tomisato and covers an area of 988 km². There are 158 pig farms rearing about 230,000 pigs. The location of the Miho training center—the site of the outbreak of Getah virus in racehorses in 2014 and 2015—is indicated by the black square

neighboring Chiba Prefecture, are more than 30 private racehorse farms. Our previous study revealed that the virus was epizootic not only at the training center in 2014 and 2015, but also at the private horse farms surrounding the center [9, 10]. Because these two prefectures are major producers of pigs, the involvement of pigs in Getah virus circulation in the area and their association with outbreaks in horses have been suspected. However, there have been only a few recent reports on the prevalence of Getah virus among pigs. Serological surveillance in 2010–2014 revealed continuous circulation of the virus among pigs in the Kanto region, which includes Ibaraki and Chiba prefectures [*Reproduction Disorders Caused by Getah Virus Infection* (Kyoto Biken Information for Swine Veterinarians, 2013, no. 5) and *Prevention of Pigs from Getah Virus Infection* (Kyoto Biken Information for Swine Veterinarians, 2015, no. 10), in Japanese; http://kyotobiken.sakura.ne.jp/security/index.php]. However, detailed information on sample collection, such as numbers of farms, numbers of pigs, and time periods, were not indicated in these studies, and it is not clear whether the magnitude of the epizootic outbreaks was larger in 2014 and 2015 than in the previous years. Hence, the association of pigs with the Getah virus outbreaks in horses was only speculative.

Here, we investigated Getah virus seropositivity among pigs in the southern part of Ibaraki Prefecture and the northern part of Chiba Prefecture, around the Miho training center, in the period 2010–2015, focusing on a possible association with the outbreaks in racehorses in 2014 and 2015. To further clarify the interactive circulation of the virus between the species, we isolated a

Getah virus strain from a pig and compared its genomic sequence with those of previous equine isolates.

Methods

Study areas and pig sera

The study area in the southern part of Ibaraki Prefecture (south Ibaraki) included the village of Miho and the cities of Inashiki, Ushiku, and Tsuchiura, covering an area of 454 km²; there are 27 pig farms rearing about 32,000 pigs (Fig. 1). The area in the northern part of Chiba Prefecture (north Chiba) included the towns of Tako and Tohnosho and the cities of Inzai, Katori, Sakura, Narita, Shiroi, Yachimata, and Tomisato, covering an area of 988 km²; there are 158 pig farms rearing about 230,000 pigs. Sera were collected from pigs at various times from September to December 2012 to 2015 in south Ibaraki (380 pigs in 29 batches) and from September to December 2010 to 2015 in north Chiba (538 pigs in 104 batches). They were kindly provided by South Ibaraki Livestock Hygiene Service Center and Chiba Central Livestock Hygiene Service Center. Each batch represented 2 to 52 serum samples (mean, 6.5 samples) collected at one farm on one day. Some batches were collected at multiple time points on the same farms, but not in the same years. In this case, the samples collected at different time points were regarded as different batches, because they should reflect different virus-exposure status. The numbers of batches and serum samples collected in each month are shown in Tables 1 and 2. The pigs were 4 to 6 months old and had not been vaccinated with Getah virus vaccine. Also, the maternal antibodies against most pathogens were considered to be undetectable in 4–6 months old piglets.

Table 1 Rates of seropositivity to Getah virus among pigs in south Ibaraki in autumn 2012–2015

Year		Sep.	Oct.	Nov.	Dec.	Total
2012	No. samples	0/17 (0%)[a]	-	1/47 (2.1%)	0/36 (0%)	1/100 (1.0%)
	No. batches	0/2	-	1/4	0/2	1/8
2013	No. samples	0/25 (0%)	0/41 (0%)	0/26 (0%)	0/5 (0%)	0/97 (0%)
	No. batches	0/2	0/4	0/2	0/1	0/9
2014	No. samples	6/30 (20.0%)	12/21 (57.1%)	1/15 (6.7%)	-	19/66 (28.8%)
	No. batches	1/2	2/2	1/1	-	4/5
2015	No. samples	2/6 (33.3%)	53/83 (63.9%)	20/26 (76.9%)	2/2 (100.0%)	77/117 (65.0%)
	No. batches	1/1	2/2	3/3	1/1	7/7

[a]No. of positive samples/No. of tested samples (positivity rate)

Therefore, any antibodies present should be the result of virus exposure within the previous 4 to 6 months.

Cell culture

For virus-neutralization (VN) testing and virus isolation we used Vero cells (Sumitomo Dainippon Pharma, Tokyo, Japan). Cells were cultured in minimum essential medium (MEM, MP Biomedicals, Irvine, CA, USA) containing 10% fetal calf serum (Sigma Aldrich Inc., St. Louis, MO, USA), 100 units/ml penicillin, and 100 µg/ml streptomycin (Sigma Aldrich Inc.). MEM containing 2% fetal calf serum, 100 units/ml penicillin, and 100 µg/ml streptomycin was used as a maintenance medium for VN testing and virus isolation.

VN testing for Getah virus

For VN testing, we used Getah virus strain 14-I-605 isolated from a racehorse during the 2014 Getah virus outbreak [10]. Sera were diluted at 1:4 with the maintenance medium and inactivated for 30 min at 56 °C. They were then transferred on flat-bottomed 96-well plates (Asahi Glass Co. Ltd., Tokyo, Japan) in duplicate and mixed with an equal volume of the prepared virus (one hundred 50%-

tissue-culture infective doses/50 µl/well). After incubation of the wells for 1 h at 37 °C with 5% CO_2, Vero cells were added at a concentration of 2.5×10^4 cells/50 µl/well. After incubation for 3 days at 37 °C with 5% CO_2, the cells were stained with crystal violet – formalin solution. The test was performed with single dilution of 1:4, and the sera were defined as positive for VN antibodies when the cytopathic effect was completely inhibited. Statistical significance for seropositive rates between the years was analyzed by z-test.

Detection of Getah virus RNA in pig sera

To detect Getah virus RNA efficiently, we used seronegative samples collected from farms suspected to be contaminated with Getah virus. Serum batches that were collected in September and October of each year and that included seropositive samples were tested for Getah virus RNA by using RT-PCR. Five sera from each batch were pooled (20 µl for each sample), and each pool included as many seronegative samples as possible. For some batches that included more than five seronegative samples, multiple pools were tested. If the number of seronegative samples was not enough to make a pool,

Table 2 Rates of seropositivity to Getah virus among pigs in north Chiba in autumn 2010–2015

Year		Sep.	Oct.	Nov.	Dec.	Total
2010	No. samples	2/68 (2.9%)[a]	0/35 (0%)	0/15 (0%)	0/5 (0%)	2/123 (1.6%)
	No. batches	2/14	0/7	0/3	0/1	2/25
2011	No. samples	0/45 (0%)	0/27 (0%)	0/5 (0%)	0/34 (0%)	0/111 (0%)
	No. batches	0/9	0/6	0/1	0/7	0/23
2012	No. samples	0/10 (0%)	0/5 (0%)	0/29 (0%)	0/30 (0%)	0/74 (0%)
	No. batches	0/2	0/1	0/6	0/6	0/15
2013	No. samples	0/60 (0%)	6/40 (15.0%)	9/15 (60.0%)	4/20 (20.0%)	19/135 (14.1%)
	No. batches	0/12	3/8	2/3	1/4	6/27
2014	No. samples	0/10 (0%)	8/35 (22.9%)	-	-	8/45 (17.8%)
	No. batches	0/2	3/7	-	-	3/9
2015	No. samples	-	24/50 (48.0%)	-	-	24/50 (48.0%)
	No. batches	-	5/5	-	-	5/5

[a]No. of positive samples / No. of tested samples (positivity rate)

seropositive samples were selected randomly from the batch. In total, 24 pools were tested. RNA was extracted from pooled sera by using a nucleic acid isolation kit (MagNA Pure LC Total Nucleic Acid Isolation Kit, Roche Diagnostics, Mannheim, Germany), and viral gene detection was performed by using an RT-PCR for the Getah virus *non-structural protein 1 (nsP1)* gene using primer sets M_2W-S and M_3W-S [11].

Virus isolation

In the case of those pooled sera that were positive for Getah virus RNA, 25 µl of original sera that had been used to make the corresponding pool was inoculated simultaneously with Vero cells (5×10^5 cells/2 ml/well) on each well of a six-well plate (Asahi Glass Co. Ltd.). The next day, the cells were washed three times with phosphate-buffered saline and cultured in maintenance medium. To identify Getah virus–specific nucleotide sequences, the supernatants of inoculated cells that showed cytopathic effects were tested by using RT-PCR as described above and sequenced by Filgen Inc. (Nagoya, Japan).

Complete-genome sequencing of Getah viruses, and phylogenetic analysis

Strain 15-I-752, which was isolated from a horse in 2015 [9], and strain 15-I-1105, which was isolated from a pig in this study, were sequenced. Viral RNA was extracted by using a MagNA Pure LC Total Nucleic Acid Isolation Kit (Roche Diagnostics, Mannheim, Germany). RT-PCR was performed with three primer sets (Getah F1–21 and Getah R5366–5345; Getah F5009–5028 and Getah R8649–8630; and Getah F8401–8424 and Getah R111305–11286), as described previously [12], and was conducted with a PrimeScript II High Fidelity RT-PCR Kit (Takara Bio Inc., Shiga, Japan). PCR amplicons were sequenced by using Ion Torrent technology (Thermo Fisher Scientific, MA, USA). Libraries were constructed by using an Ion Xpress Plus Fragment Library Kit and an Ion Xpress Barcode Adapters Kit. Emulsion PCR, enrichment and loading onto an Ion 316 Chip were performed automatically with Ion Chef, and sequencing was conducted with the Ion PGM system according to the manufacturer's instructions. The average depth was 6402.5 times. The raw signal data were analyzed by using Torrent Suite version 5.0.5, and 14-I-605-C1 (LC079088) was used as a reference sequence. Torrent Variant Caller version 5.0.4.0 was used to call the mutated sites, and the most frequently observed alleles were selected in each of the called positions. Bases within the reference sequence were substituted with the observed alleles if frequencies exceeded 50%. The 5′- and 3′-end sequences were determined by Rapid Amplification of cDNA Ends as described previously [12]. Sequences

were analyzed and assembled by using Geneious software (Biomatters, Auckland, New Zealand). Nucleotide sequences for 15-I-752 (LC212972) and 15-I-1105 (LC212973) were deposited in the GenBank/EMBL/DDBJ databases.

Phylogenetic analysis of nucleic acid sequences was performed with MEGA 7 software [13]. A phylogenetic tree based on complete genome sequences was constructed by using the maximum-likelihood method, and statistical analysis was performed by using bootstrap tests (1000 replicates). Nucleotide sequence accession numbers for the Getah virus strains used in the phylogenetic analysis were as follows: MI-110-C1 (LC079086), MI-110-C2 (LC079087), 14-I-605-C1 (LC079088), and 14-I-605-C2 (LC079089) were isolated from horses; HB0234 (EU015062), YN0540 (EU015063), SC1210 (LC107870), LEIV 17741 MPR (EF631999), M1 (EU015061), LEIV 16275 Mag (EF631998), and Sagiyama (AB032553) were isolated from mosquitoes; and South Korea (AY702913) and Kochi/01/2005 (AB859822) were isolated from pigs (Fig. 2).

Results

Rates of seropositivity to Getah virus among pigs in south Ibaraki and north Chiba

The rates of seropositivity to Getah virus in autumn 2012–2015 in south Ibaraki and 2010–2015 in north Chiba are summarized in Tables 1 and 2. For those serum batches that contained at least one seropositive serum, the exact sampling date, the location, and the numbers of positive/tested samples are summarized in Tables 3 and 4. Seropositivity rates in 2012–2013 in south Ibaraki and 2010–2012 in north Chiba ranged from 0% to 1.6% (Tables 1 and 2). In these years, each positive serum sample was the only one in each batch, and no batches had multiple seropositive samples (Tables 3 and 4). In contrast, seropositivity rates in 2014 and 2015 in south Ibaraki were significantly higher than those in the previous years: 28.8% in 2014 ($P < 0.01$) and 65.0% in 2015 ($P < 0.01$). The proportions of batches that had positive sera were also high (4/5 in 2014 and 7/7 in 2015; Table 1). Among the 11 batches that were positive in 2014–2015 in south Ibaraki, nine had multiple seropositive samples with VN antibodies (Table 3). Seropositivity rates in north Chiba from 2013 to 2015 were significantly higher than those in the previous years: 14.1% in 2013 ($P < 0.01$), 17.8% in 2014 ($P < 0.01$), and 48.0% in 2015 ($P < 0.01$). In 2013, 6/27 batches had positive sera; this value was 3/9 in 2014 and 5/5 in 2015 (Table 2). Eleven out of 14 seropositive batches in 2013–2015 in north Chiba included multiple seropositive samples (Table 4).

Detection of viral RNA in pig sera and virus isolation

For efficient isolation of Getah virus, we first performed RT-PCR to detect Getah virus RNA in pooled sera.

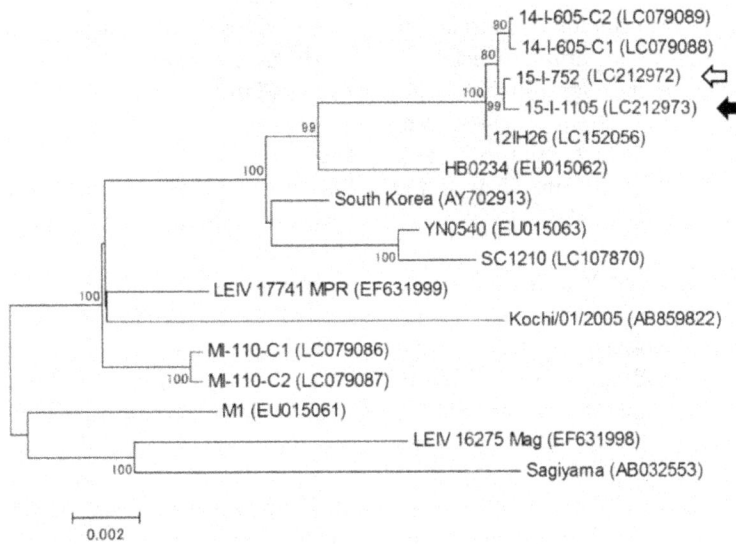

Fig. 2 Phylogenetic analysis of complete Getah virus genome sequences. The Getah virus strain isolated from a pig (15-I-1105) is indicated by the black arrow. The strain isolated from a horse during the 2015 outbreak (15-I-752) is indicated by the white arrow. The percent bootstrap support is indicated by the values at each node; values less than 70% were omitted. MI-110-C1, MI-110-C2, 14-I-605-C1, and 14-I-605-C2 are Getah virus isolates from horses; 12IH26, HB0234, YN0540, SC1210, LEIV17741 MPR, M1, LEIV16275 Mag, and Sagiyama are isolates from mosquitoes; South Korea and Kochi/01/2005 are isolates from pigs

Because the epidemic season of Getah virus is generally from September to October, we set pools from the batches collected in these two months of each year and that included seropositive samples. Five sera from each batch were pooled ($n = 24$), and each pool included as many seronegative samples as possible. This was because virus recovery might not have been achieved from antibody-positive sera: the presence of antibodies should indicate that the viremic phase has finished. Getah virus RNA was detected in four out of 24 pools: two pools collected on 8 September 2014 in Inashiki city (Ibaraki Pref.), one pool collected on 28 September 2015 in Tsuchiura city (Ibaraki Pref.), and one pool collected on 23 October 2015 in Tomisato town (Chiba Pref.) (Tables 3 and 4).

Table 3 Serum batches that were collected in south Ibaraki in autumn 2012–2015 and that included VN antibody-positive pigs

Year	Date	City	Virus-neutralization test (positive/tested samples)	RT-PCR (positive/ tested pools)
2012	Nov 8	Tsuchiura	1/17	ND
2014	Sep 8	Inashiki	6/20	2/2
	Oct 20	Inashiki	2/6	0/1
	Oct 31	Inashiki	10/15	0/1
	Nov 25	Ushiku	1/15	ND
2015	Sep 28	Tsuchiura	2/6	1/1[a]
	Oct 1	Tsuchiura	46/52	0/1
	Oct 1	Tsuchiura	7/31	0/4
	Nov 10	Inashiki	14/14	ND
	Nov 17	Tsuchiura	5/5	ND
	Nov 27	Tsuchiura	1/7	ND
	Dec 15	Tsuchiura	2/2	ND

ND, not done

[a]Getah virus was isolated

Table 4 Serum batches that were collected in north Chiba in autumn 2010–2015 and that included VN antibody-positive pigs

Year	Date	City/Town	Virus-neutralization test (positive/tested samples)	RT-PCR (positive/ tested pools)
2010	Sep 1	Katori	1/3	ND
	Sep 14	Katori	1/4	ND
2013	Oct 18	Tako	1/5	0/1
	Oct 24	Katori	3/5	0/1
	Oct 31	Katori	2/5	0/1
	Nov 19	Tohnosho	5/5	ND
	Nov 21	Katori	4/5	ND
	Dec 13	Katori	4/5	ND
2014	Oct 1	Tomisato	3/5	0/1
	Oct 1	Tomisato	2/5	0/1
	Oct 1	Narita	3/5	0/1
2015	Oct 23	Tomisato	1/10	1/2
	Oct 27	Narita	1/10	0/2
	Oct 29	Katori	9/10	0/1
	Oct 29	Sosa	9/10	0/1
	Oct 29	Sosa	4/10	0/2

ND, not done

The sera included in the Getah virus RNA-positive pools were inoculated with Vero cells for virus isolation. Out of 20 sera from four pools, one serum collected on 28 September 2015 in Tsuchiura city (Ibaraki Pref.) showed cytopathic effects; the strain isolated was confirmed as Getah virus by RT-PCR for the *nsP1* gene with the predicted length of 434 base pairs (Fig. 3) and sequencing; it was designated strain 15-I-1105.

Complete genome sequencing of Getah virus strains isolated from pig and horses

To clarify the interactions between the Getah virus strains isolated from the pig and the horses, we compared the complete genome sequence of strain 15-I-1105 with those of other strains, including the ones isolated during the outbreaks in horses. Phylogenetic analysis showed that strain 15-I-1105 was located in the same cluster as the strains isolated from horses during the 2014 outbreak (14-I-605-C1, 14-I-605-C2), one strain isolated from a horse during the 2015 outbreak (15-I-752), and one strain isolated from mosquitoes in Nagasaki Prefecture in 2012 (12IH26, [14]) (Fig. 2). Strain 15-I-1105 had extremely high nucleotide sequence identity to each strain in the same cluster, with 99.90% to 14-I-605-C1, 99.89% to 14-I-605-C2, 99.94% to 15-I-752, and 99.91% to 12IH26. Small numbers of amino acid differences between 15-I-1105 strain and the strains from horses were found mostly in the nsP1–2–3–4

polyprotein (Table 5). 15-I-1105 strain had different amino acids in five sites against 15-I-752 strain, and seven sites against both 14-I-605-C1 and 14-I-605-C2 strains (Table 5).

Discussion

Pigs are regarded as natural hosts and amplifiers of Getah virus because of the high virus titers found in the viremic phase after experimental infection and the high seroprevalence among pigs in the field [4, 15, 16]. As described above, Ibaraki and Chiba prefectures, where our survey areas were located, are major producers of pigs in Japan. Unlike in Japan's other pig production areas, in south Ibaraki and north Chiba many racehorse farms and piggeries are located close together. The numbers of horse farms in which racehorses stayed before introduction to the Miho training center in 2015 were 20 in south Ibaraki and 11 in north Chiba. The existence of mosquito vectors of Getah virus has been confirmed in these areas [7]. Therefore, it is likely that transmission of Getah virus between pigs and horses can occur easily there.

As we showed here, Getah virus was epizootic among pigs in south Ibaraki and north Chiba in autumn 2014 and 2015; this was consistent with the epizootic periods among racehorses [8, 9]. The number of affected horses at the Miho training center was 33 in 2014 and 30 in 2015, and the maximum seropositivity rates among horses that were introduced from farms in Ibaraki and Chiba to the center were 42.9% in 2014 and 34.9% in 2015 [8–10]. Although we tested a portion of the pig farms in these areas, seropositive batches were found at multiple sites during these epizootic seasons. In most of the seropositive batches, multiple seropositive samples

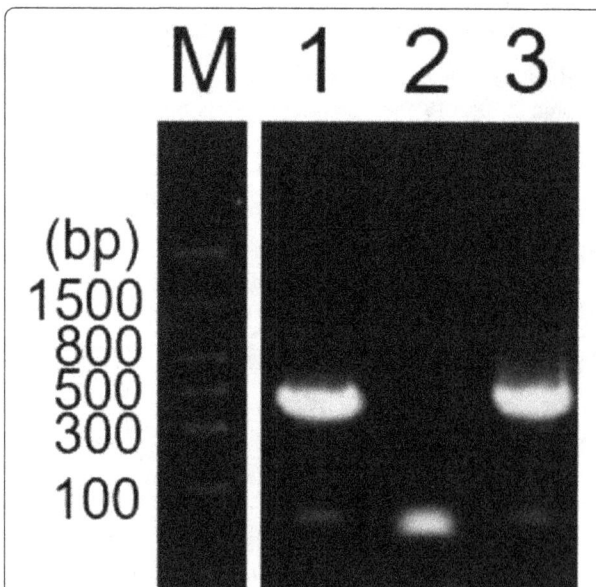

Fig. 3 Detection of viral RNA by RT-PCR for Getah virus from the supernatant of Vero cells inoculated with a pig serum. After virus isolation in Vero cells, one serum collected on 28 September 2015 in Tsuchiura city (Ibaraki Pref.) showed cytopathic effects; the nucleic acid was extracted from the culture supernatant, and RT-PCR for the *nsP1* gene was performed. M, marker; lane 1, supernatant of the cytopathic effects-positive culture; lane 2, negative control; lane 3, positive control

Table 5 Amino acid differences between the strains from the pig and horses

Protein	Position	Pig strain	Horse strains		
		15-I-1105	14-I-605-C1	14-I-605-C2	15-I-752
Non-structural polyprotein[a]					
nsP1	58	I	V	V	V
	354	A	A	V	A
	385	V	L	V	V
	427	I	T	T	I
	475	I	M	M	M
nsP3	1694	A	A	A	V
	1814	V	E	E	E
nsP4	2114	T	A	A	T
Structural polyprotein[b]					
E1	1113	S	N	N	N

[a]Non-structural polyprotein (nsP1–2–3-4) has 2467 amino acids
[b]Structural polyprotein (C-E3-E2-6 K-E1) has 1253 amino acids

were found at each site. We therefore considered that individual seropositive pigs were not infected sporadically, but that the virus was circulating in the pig population in these areas. However, to our knowledge, there have been no published reports of an increased incidence of disease potentially caused by Getah virus in local pigs. Despite the viral spread, most of the infections in pigs seemed to be sub-clinical or not reported.

In north Chiba, a high seropositivity rate was found in pigs not only in 2014 and 2015 but also in 2013, when there was no detectable outbreak in horses. In our previous study we investigated seropositivity rates among racehorses introduced from Ibaraki and Chiba prefectures to the training center; the maximum seropositivity rate from June to October 2013 was 2.0% [10]. In that study, of 250 horse sera examined over a total of five months, 62 were from horses introduced from Chiba, and all of them were seronegative for Getah virus. Therefore, it seemed that virus circulation in north Chiba in 2013 occurred only among pigs, and not among horses in the area. The absence of a detectable outbreak in horses in 2013 might be attributable to the fact that virus circulation among pigs did not occur in south Ibaraki, which contains the Miho training center and greater numbers of horse farms than are found in north Chiba.

Genomic comparison revealed that the strain isolated from a pig in 2015 (15-I-1105) was closely related to the strains isolated from horses in 2014 (14-I-605-C1 and 14-I-605-C2) and 2015 (15-I-752), with extremely high nucleotide sequence identities. This finding, together with the serological data, indicates this was the first convincing case that Getah virus circulation among horses and pigs occurred interactively in a specific area. On the basis of the theory that pigs are natural hosts of Getah virus, the outbreak among racehorses in 2014 and 2015 was likely preceded by viral amplification in pigs. However, the virus might not be transferred in a one-way manner from pigs to horses; instead, it might circulate interactively between the two species, because the high-level viremia required for vector transmission has also been observed in infected horses [5]. Therefore, we could not conclude that the epizootics in 2014 and 2015 were initiated from infected pigs, and it was also possible that the virus was derived from infected horses that were transferred from other epizootic areas. Because horses are moved from one farm to the other farms frequently, such movement in the area should be responsible for virus transfer among horse population. However, animal and human movements between horse farms and pig farms are considered to be rarely occurred, and may not be the major route of viral transfer between the two species. As we described above, the pig farms and horse farms are closely located in the area, and some of them were located within several hundred meters distance. In this regard, the virus should be transferred by flight activities of infected mosquitoes.

Our phylogenetic analysis of various Getah virus strains revealed that the strain from a pig (15-I-1105) and strains from recent outbreaks in horses (14-I-605-C1, 14-I-605-C2, and 15-I-752) formed a cluster with a strain isolated from mosquitoes in Nagasaki Prefecture in 2012 (12IH26, [14]) (Fig. 2). Phylogenetic analyses by Kobayashi et al. using the sequences for the *E2*, *capsid*, and *nsP1* genes suggested that this cluster containing 12IH26, 14-I-605-C1, and 14-I-605-C2 was closely related to the recent Chinese and South Korean strains but only distantly related to Japanese strains from the 1970s and 1980s [14]. Although it is still unclear how and when the virus was brought into Japan from overseas and spread into Ibaraki and Chiba prefectures, this cluster of viruses seems to have been established in these areas and was circulating among both pigs and horses in recent years.

Conclusions

Serological investigation of Getah virus in pigs revealed that the virus was circulating in south Ibaraki and north Chiba in 2014 and 2015, during the same period as outbreaks in racehorses. The Getah virus strain isolated from an infected pig was closely related to those from horses during that period. This was the first convincing case of interactive circulation of Getah virus among both pigs and horses in a specific area. Consistent with the findings of previous studies, our current data suggest that pigs are important natural hosts and amplifiers. Periodic surveillance of pigs over a larger area would be useful for estimating the risk of epizootics in horses.

Abbreviations
MEM: minimum essential medium; nsP1: non-structural protein 1; VN: virus-neutralization

Acknowledgments
We thank Mr. Takao Saitou of South Ibaraki Livestock Hygiene Service Center and Mr. Takehiko Ohtsubo of Chiba Central Livestock Hygiene Service Center for providing the pig sera, and investigated the relevant information of the study areas. We thank Mr. Akira Kokubun, Ms. Akiko Suganuma, Ms. Akiko Kasagawa, Ms. Kazue Arakawa, and Ms. Kaoru Makabe of the Equine Research Institute for their technical assistance. We thank Dr. Charles Mark El-Hage of the University of Melbourne for discussing the manuscript.

Funding
This study was supported by the Japan Racing Association.

Authors' contributions
HB performed VN testing and virus isolation from pig sera and wrote the draft of the manuscript. MN, HN, and SM performed complete genome sequencing and phylogenetic analysis. KT, TY, and TK participated in the design of the experiment and discussed the manuscript. We all read and approved the final manuscript.

Competing interests
The authors declare that they have no competing interests.

Consent for publication
Not applicable.

Ethics approval
Ethics regarding this study were approved by The Promotion and Ethics Committee for Research of the Equine Research Institute under the identification number H25–4, and also authorized ex post facto by the Director of Equine Research Institute under the identification number 17–1. The prefectural Livestock Hygiene Service Centers are the government authorities for disease prevention in domestic animal, and the veterinary officers of South Ibaraki Livestock Hygiene Service Center and Chiba Central Livestock Hygiene Service Center collected clinical samples and provided them for this study as a part of regular activities for disease prevention under the Act on Domestic Animal Infectious Disease Control (Act 166, May 31, 1951; Amendment Act No. 68, 2004).

Author details
[1]Equine Research Institute, Japan Racing Association, 1400-4 Shiba, Shimotsuke, Tochigi 329-0412, Japan. [2]Thermo Fisher Scientific, Life Technologies Japan Ltd, 4-2-8 Shibaura, Minato-ku, Tokyo, Japan.

References
1. Fukunaga Y, Kumanomido T, Kamada M. Getah virus as an equine pathogen. Vet Clin North Am Pract. 2000;16:605–17.
2. Yago K, Hagiwara S, Kawamura H, Narita M. A fatal case in newborn piglets with Getah virus infection: isolation of the virus. Jpn J Vet Sci. 1987;49:989–94.
3. Izumida A, Takuma H, Inagaki S, Kubota M, Hirahara T, Kodama K, et al. Experimental infection of Getah virus in swine. Jpn J Vet Sci. 1988;50:679–84.
4. Kumanomido T, Wada R, Kanemaru T, Kamada M, Hirasawa K, Akiyama Y. Clinical and virological observations on swine experimentally infected with Getah virus. Vet Microbiol. 1988;16:295–301.
5. Kamada M, Wada R, Kumanomido T, Imagawa H, Sugiura T, Fukunaga Y. Effect of viral inoculum size on appearance of clinical signs in equine Getah virus infection. J Vet Med Sci. 1991;53:803–6.
6. Sentsui H, Kono Y. An epidemic of Getah virus infection among racehorses: isolation of the virus. Res Vet Sci. 1980;29:157–61.
7. Kumanomido T, Fukunaga Y, Ando Y, Kamada M, Imagawa H, Wada R, et al. Getah virus isolations from mosquitoes in an enzootic area in Japan. Jpn J Vet Sci. 1986;48:1135–40.
8. Nemoto M, Bannai H, Tsujimura K, Kobayashi M, Kikuchi T, Yamanaka T, et al. Outbreak of Getah virus infection among racehorses in Japan in 2014. Emerg Infect Dis. 2015;21:883–5.
9. Bannai H, Ochi A, Nemoto M, Tsujimura K, Yamanaka T, Kondo T. A 2015 outbreak of Getah virus infection occurring among Japanese racehorses sequentially to an outbreak in 2014 at the same site. BMC Vet Res. 2016;12:98.
10. Bannai H, Nemoto M, Ochi A, Kikuchi T, Kobayashi M, Tsujimura K, et al. Epizootiological investigation of Getah virus infection among racehorses in Japan in 2014. J Clin Microbiol. 2015;53:2286–91.
11. Wekesa SN, Inoshima Y, Murakami K, Sentsui H. Genomic analysis of some Japanese isolates of Getah virus. Vet Microbiol. 2001;83:137–46.
12. Nemoto M, Bannai H, Tsujimura K, Yamanaka T, Kondo T. Genomic, pathogenic, and antigenic comparisons of Getah virus strains isolated in 1978 and 2014 in Japan. Arch Virol. 2016;161:1691–5.
13. Kumar S, Stecher G, Tamura K. MEGA7: molecular evolutionary genetics analysis version 7.0 for bigger datasets. Mol Biol Evol. 2016;33:1870–4.
14. Kobayashi D, Isawa H, Ejiri H, Sasaki T, Sunahara T, Futami K, et al. Complete genome sequencing and phylogenetic analysis of a Getah virus strain (genus *Alphavirus*, family *Togaviridae*) isolated from *Culex tritaeniorhynchus* mosquitoes in Nagasaki, Japan in 2012. Vector Borne Zoonotic Dis. 2016;16:769–76.
15. Kumanomido T, Fukunaga Y, Ando Y, Kamada M, Imagawa H, Wada R, et al. Ecological survey on Getah virus among swine in Japan. Bull Equine Res Inst. 1982;19:89–92.

14

Monitoring for bovine arboviruses in the most southwestern islands in Japan between 1994 and 2014

Tomoko Kato[1], Tohru Yanase[1*], Moemi Suzuki[2], Yoshito Katagiri[2], Kazufumi Ikemiyagi[3], Katsunori Takayoshi[2], Hiroaki Shirafuji[1], Seiichi Ohashi[4], Kazuo Yoshida[5], Makoto Yamakawa[5] and Tomoyuki Tsuda[6]

Abstract

Background: In Japan, epizootic arboviral infections have severely impacted the livestock industry for a long period. Akabane, Aino, Chuzan, bovine ephemeral fever and Ibaraki viruses have repeatedly caused epizootic abnormal births and febrile illness in the cattle population. In addition, Peaton, Sathuperi, Shamonda and D'Aguilar viruses and epizootic hemorrhagic virus serotype 7 have recently emerged in Japan and are also considered to be involved in abnormal births in cattle. The above-mentioned viruses are hypothesized to circulate in tropical and subtropical Asia year round and to be introduced to temperate East Asia by long-distance aerial dispersal of infected vectors. To watch for arbovirus incursion and assess the possibility of its early warning, monitoring for arboviruses was conducted in the Yaeyama Islands, located at the most southwestern area of Japan, between 1994 and 2014.

Results: Blood sampling was conducted once a year, in the autumn, in 40 to 60 healthy cattle from the Yaeyama Islands. Blood samples were tested for arboviruses. A total of 33 arboviruses including Akabane, Peaton, Chuzan, D' Aguilar, Bunyip Creek, Batai and epizootic hemorrhagic viruses were isolated from bovine blood samples. Serological surveillance for the bovine arboviruses associated with cattle diseases in young cattle (ages 6–12 months: had only been alive for one summer) clearly showed their frequent incursion into the Yaeyama Islands. In some cases, the arbovirus incursions could be detected in the Yaeyama Islands prior to their spread to mainland Japan.

Conclusions: We showed that long-term surveillance in the Yaeyama Islands could estimate the activity of bovine arboviruses in neighboring regions and may provide a useful early warning for likely arbovirus infections in Japan. The findings in this study could contribute to the planning of prevention and control for bovine arbovirus infections in Japan and cooperative efforts among neighboring countries in East Asia.

Keywords: Arbovirus, Bovine ephemeral fever, Cattle, *Culicoides* biting midges, Congenital abnormality, Epizootic hemorrhagic disease, Orbivirus, Orthobunyavirus, Rhabdovirus, Serosurveillance

* Correspondence: tyanase@affrc.go.jp
[1]Kyushu Research Station, National Institute of Animal Health, NARO, 2702 Chuzan, Kagoshima 891-0105, Japan
Full list of author information is available at the end of the article

Background

Arthropod-borne viruses (arboviruses) are transmitted by hematophagous arthropod vectors, such as mosquitoes, ticks and *Culicoides* biting midges. Nearly 500 arboviruses have been recorded so far and some of them can seriously harm animal health [1]. Arbovirus infections often affect wide regions in a short period of time. The infected vectors are easily disseminated by air streams [2, 3], and their long-distance migrations (hundreds of kilometers) have been successfully estimated [4–7]. The recent emergence of bluetongue and Schmallenberg virus (SBV) infection in northern Europe demonstrated such rapid and wide expansion and caused huge economic damage to the livestock industry of countries that had been previously free from these arboviral infections [8, 9].

In Japan, epizootic abortion, stillbirth, premature birth and congenital malformations in cattle caused by arboviruses have severely impacted the livestock industry for a long period [10]. Akabane virus (AKAV) and Aino virus (AINOV) of the genus *Orthobunyavirus* in the family *Bunyaviridae* is principally associated with recurring epizootics of abnormal births [10–12]. It was estimated that approximately 42,000 abnormal calves caused by AKAV were born during the largest outbreak between 1972 and 1975. Chuzan virus (CHUV) of the genus *Orbivirus* in the family *Reoviridae* is also known to be an etiological agent of congenital abnormalities in cattle [13, 14]. In recent years, bovine encephalomyelitis caused by postnatal AKAV infection reemerged in Japan after a 21-year absence [15, 16]. Furthermore, incursions of Peaton virus (PEAV), Sathuperi virus (SATV) and Shamonda virus (SHAV) of the genus *Orthobunyavirus* were confirmed in Japan in the past 16 years [17–19]. Although these viruses potentially have teratogenicity in ruminants, little is known about their pathogenicity [20, 21]. As well as CHUV, D'Aguilar virus (DAGV) of the genus *Orbivirus* is a member of Palyam virus (PALV) group and has repeatedly been isolated in Japan since 1987 [22, 23]. Its etiological role in epizootic congenital abnormalities in cattle between 2001 and 2002 was serologically confirmed with colostrum-free calves. Bovine ephemeral fever virus (BEFV) of the genus *Ephemerovirus* in the family *Rhabdoviridae* causes a febrile illness in cattle and water buffalo, and is associated with reduction of milk production in dairy cattle and loss of condition in beef cattle [24]. The last occurrence of bovine ephemeral fever in mainland Japan was reported in 1991, but its periodic epizootics continue in the southwestern islands [25, 26]. Ibaraki virus (IBAV) is a strain of epizootic hemorrhagic disease virus (EHDV) serotype 2 of the genus *Orbivirus*, and difficulty swallowing is its main manifestation [27]. The contribution of EHDV serotype 7 (EHDV-7) to epizootic abortion and stillbirth in pregnant cows was also reported in the southern part of Japan in

1997 [28–30]. The above-mentioned arboviruses are principally transmitted by *Culicoides* biting midges [23, 31] while mosquitoes are strong candidates for BEFV vectors [24]. Because the vector activity ceases in winter, the overwintering of arboviruses is seemingly unrealistic in Japan. In fact, appearances of the same strain/genotype of arboviruses in consecutive years rarely happen in Japan [10, 32, 33]. Development of efficient commercial vaccines to AKAV, AINOV, CHUV, IBAV and BEFV has contributed to reducing diseases in cattle, but arbovirus infections still break out in unvaccinated cattle, and no effective prevention measures are prepared for newly emerged arboviruses.

The ideal climate for vector insects in the tropical zone permits arbovirus circulation through most of the year. Therefore, the lower-latitude regions are considered a potential source of arboviruses in the temperate zone, where vector insects are usually absent during the colder months. *Culicoides*- and mosquito-borne arboviruses are considered to be introduced from lower latitudes in Asia to Japan with the infected vectors carried by airstreams [10, 34]. Although former studies revealed the prevalence of several ruminant arboviruses in Southeast Asian countries [35–37], details of arbovirus activity remain unknown. The Yaeyama Islands in Okinawa Prefecture are located in the subtropical zone and are 100–200 km, 350–450 km, 700 km and 1000 km away from Taiwan, Southern China, Philippines and mainland Japan, respectively. Due to their geographic location, the islands may be well suited for playing a role as an early warning system for other parts of Japan. Indeed, it was suggested that previous epizootics of bovine ephemeral fever in the Yaeyama Islands were epidemiologically linked to those in Taiwan [25, 26, 38]. The enhanced surveillance of ruminant arboviruses in the bordering area could support the planning of preventive and control strategies for arbovirus infections in the remaining part of Japan. To monitor the annual incursion of arboviruses into the Yaeyama Islands, we conducted virus isolation from sentinel cattle and serosurveillance for the above-mentioned arboviruses in the cattle population for 20 years. This long-term monitoring indicates a high risk of incursion by multiple arboviruses from overseas to the Yaeyama Islands. We also discuss whether the monitoring in the islands could contribute to earlier detection of arbovirus activities prior to their spread in mainland Japan.

Methods
Study area
The Yaeyama Islands comprise an archipelago located in the most southwestern part of Japan (24.00-24.67° N, 122.75-124.50°E) from Ishigaki Island to Yonaguni Island (Fig. 1). Approximately 36,000 beef cattle are reared

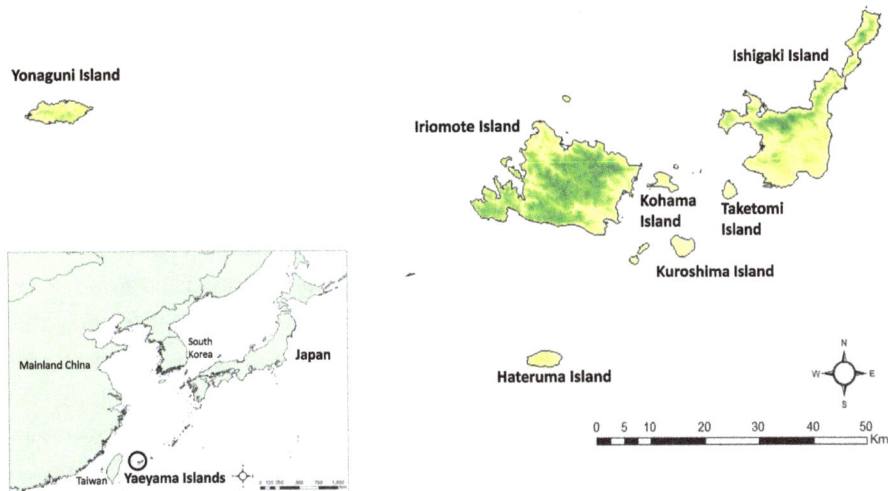

Fig. 1 Geographical location of the study area. The maps were generated using Arc GIS 10 (Esri, Redlands, CA) based on data retrieved from DIVA-GIS and the National Land Numerical Information download service provided by Ministry of Land, Infrastructure, Transport and Tourism, Japan

through the islands, and over 70 % of this cattle population exists on Ishigaki Island, which is the second largest island in the archipelago. The climate of the Yaeyama Islands is subtropical; the annual mean temperature was 24 °C between 1994 and 2014; mean temperatures of the hottest and coldest months are about 29 °C and 19 °C, respectively. Annual precipitation was about 2200 mm and humidity ranged from 70 to 80 % during the study period. Blood sampling was continuously conducted in Ishigaki, Iriomote and Yonaguni Islands, but was optional in Taketomi, Kohama, Kuroshima and Hateruma Islands (Table 1).

Blood collection

Blood sampling was conducted once a year between the middle of October and beginning of December in 1994–2003, 2005–2009 and 2011–2014 (Table 1). Heparinized blood and serum samples were obtained from 40–60 healthy cattle. The blood samples were separated into plasma and blood cells by centrifugation and the blood cells were washed three times with phosphate buffered saline to eliminate the antibodies. The plasmas and blood cells were stored at −80 °C until virus isolation. The serum samples were kept at −20 °C until they were used for virus neutralization tests (VNTs). A total of 977 blood samples were subjected to arbovirus screening. To detect antibodies against arboviruses, VNTs were conducted for 813 serum samples harvested after 1995. The tested sera were selectively collected from cattle that were 6–12 months old and had only been alive for one summer. The blood samples were voluntarily collected by local veterinary officers for monitoring of arbovirus infections and were permitted to be used in this study from the animal health authority of Okinawa Prefecture.

Virus isolation

The processed plasmas and blood cells were inoculated into monolayer cultures of baby hamster kidney (BHK-21) and hamster lung (HmLu-1) cells as described previously [23]. Briefly, cells were incubated into test tubes with Eagle's minimum essential medium (MEM) (Nissui, Tokyo, Japan) supplemented with 0.295 % tryptose phosphate broth (Becton Dickinson and Company, Franklin Lakes, NJ, USA), 0.015 % sodium bicarbonate and 10 % of bovine serum overnight at 37 °C and were washed three times with Earl's solution before inoculation. The cell cultures inoculated with blood samples were maintained in serum-free medium by rotation at 37 °C for 7 days and were collected if a cytopathic effect (CPE) was observed. Two further blind passages were conducted in the same manner if CPE was not observed.

Dot immunobinding assay

Cytopathic agents in the cultured cells were characterized by a dot immunobinding assay (DIA) as described previously [23, 39]. Briefly, the supernatants of cell cultures showing CPE were blotted onto the Immobilon PVDF transfer membrane (Millipore, Billerica, MA, USA) using a slot blotting apparatus. The membrane was immersed in blocking buffer [8 % skim milk in Tris-buffered saline (TBS; 20 mM Tris–HCl pH 7.5, 0.15 M NaCl)] for more than 2 h and then reacted with monoclonal antibodies to the nucleocapsid protein and viral surface glycoprotein (Gc) of AKAV, the Gc protein of AINOV and PEAV, and mouse antisera to CHUV, IBAV and bluetongue virus for 1 h. The membrane was reacted with horseradish peroxidase-conjugated goat antibody to mouse IgG in blocking buffer for 30 min, and the immune complex was detected by color development with 0.027 % 3,3-diaminobenzidine tetrahydrochloride and 0.016 % H_2O_2 in TBS.

Table 1 Blood samples used for virus isolation and neutralization tests

Year	Collection date	No. of blood samples Virus isolation/VNTs[b]	Blood collection sites (No. of blood samples by collection site: virus isolation/VNTs)
1994	21.October – 7.November	58/0	Ishigaki (10/0), Taketomi (10/0), Kohama (10/0), Hateruma (10/0), Kuroshima (8/0), Yonaguni (10/0)
1995	7 – 17.November	50/49	Ishigaki (40/39), Yonaguni (10/10)
1996	17 – 28.October	50/30	Ishigaki (25/25), Iriomote (15/5), Yonaguni (10/0)
1997	22–31.October	60/49	Ishigaki (35/30), Iriomote (15/9), Yonaguni (10/10)
1998	10–11.November	50/39	Ishigaki (20/20), Iriomote (20/10), Yonaguni (10/9)
1999	16–18.November	50/40	Ishigaki (20/13), Iriomote (10/9), Kuroshima (10/9), Yonaguni (10/9)
2000	15–17.November	50/36	Ishigaki (20/13), Iriomote (10/9), Kuroshima (10/7), Yonaguni (10/7)
2001	12–13.November	40/39	Ishigaki (20/20), Yonaguni (20/19)
2002	25. November	50/48	Ishigaki (20/20), Iriomote (10/8), Kuroshima (10/10), Yonaguni (10/10)
2003	26–28.November	55/53	Ishigaki (20/19), Taketomi (10/10), Iriomote (15/14), Yonaguni (10/10)
2004[a]	-	-	-
2005	7–12. December	55/52	Ishigaki (10/9), Taketomi (10/10), Iriomote (12/12), Kuroshima (10/9), Yonaguni (13/12)
2006	27. November-8.December	50/45	Ishigaki (24/23), Iriomote (11/10), Kuroshima (10/7), Yonaguni (5/5)
2007	27. November-5.December	50/45	Ishigaki (26/23), Taketomi (3/3), Iriomote (17/16), Yonaguni (4/3)
2008	11. November-11.December	51/48	Ishigaki (21/20), Iriomote (8/7) Taketomi (4/4), Kohama (3/3), Hateruma (5/5), Kuroshima (5/4), Yonaguni (5/5)
2009	17–25.November	52/49	Ishigaki (15/15), Taketomi (7/7), Iriomote (10/9), Kohama (5/5), Kuroshima (5/5), Yonaguni (10/8)
2010[a]	-	-	-
2011	29. November-1.December	50/49	Ishigaki (25/24), Iriomote (20/20), Yonaguni (5/5)
2012	3–5.December	51/47	Ishigaki (21/19), Iriomote (21/20), Kohama (9/8)
2013	11.November-4.December	50/50	Ishigaki (11/11), Taketomi (3/3), Iriomote (11/11), Kohama (3/3), Hateruma (5/5), Kuroshima (7/7), Yonaguni (10/10)
2014	27.October-26.November	55/45	Ishigaki (12/12), Taketomi (3/2), Iriomote (11/5), Kohama (6/6), Hateruma (6/6), Kuroshima (10/8), Yonaguni (7/6)
Total		977/813	

[a]Sampling was not done in 2004 and 2010
[b]VNTs: virus neutralization tests

RT-PCR and sequence analysis

Viral RNA was extracted from the supernatant of cell cultures showing CPE with the High Pure Viral RNA Kit (Roche Diagnostics, Mannheim, Germany). Group-specific RT-PCRs targeting the S RNA segment of orthobunyaviruses, segment 3 of PALV group viruses and segment 7 of EHDV were performed using the Titan One tube RT-PCR Kit (Roche Diagnostics) in accordance with previous studies [40, 41]. Specific detections of segment 2 of CHUV and DAGV by RT-PCR were conducted with the primer sets CHUVL2F-2/CHUVL2R-2 and DAGVL2F/DAGVL2R-2, respectively (Additional file 1) [23]. An isolate of EHDV was tested in an RT-PCR assay with the serotype-specific primer sets targeting segment 2 [41]. Sample denaturation at 94 °C for 4 min was performed for the orbivirus detection before cDNA synthesis. The cDNA synthesis was conducted at 50 °C for 30 min followed by 94 °C for 2 min. The PCR profile was 10 cycles of 94 °C for 30 s, 55 °C for

30 s and 68 °C for 45 s, followed by 25 cycles of 94 °C for 30 s, 55 °C for 30 s and 68 °C for 45 s, with the latter time increased by 5 s per cycle.

Genome RNA of an orbivirus that was unidentified by the above-mentioned methods was extracted from the infected cells using the TRIZOL LS Reagent (Life Technologies, Carlsbad, CA, USA). Amplification of the cDNA of segment 2 was performed by Full-Length Amplification of cDNAs (FLAC) as described previously [42]. In brief, double-stranded RNA was ligated to the anchor primer (5-15-1) with T4 RNA ligase (New England Bio Labs, Ipswich, MA, USA) at 4 °C overnight and separated in 1 % agarose gel by electrophoresis. Segment 2 was recovered from the gel and purified using the RNaid Kit with SPIN (Bio 101, Vista, CA, USA). First-strand cDNA synthesis was performed using the Superscript III First-Strand Synthesis System for RT-PCR (Life Technologies) in accordance with the manufacturer's instructions. The cDNA was amplified using KOD-plus-ver.2 (TOYOBO, Osaka, Japan) in a reaction mixture under the following conditions: 30 cycles of 98 °C for 10 s, 62 °C for 30 s and 68 °C for 3.5 min.

The PCR products were purified with the QIAquick PCR Purification Kit (Qiagen, Hilden, Germany) and directly sequenced with the BigDye Terminator Cycle sequencing Kit v3.1 (Life Technologies) on the ABI 3100-Avanti Genetic Analyzer (Life Technologies). The nucleotide (nt) sequences were edited by DNASIS Pro Ver. 3.0 (Hitachi Solutions, Tokyo, Japan), and a sequence similarity search was conducted with the basic local alignment search tool (BLAST). Pairwise nt sequence identities were calculated with GENETYX software ver. 10 (GENETYX, Tokyo, Japan).

Virus neutralization test

Virus neutralization tests with AKAV OBE-1, AINOV JaNAr28, PEAV KSB-1/P/06, SATV KSB-2/C/08, SHAV KSB-6/C/02, BEFV YHL, CHUV C31, DAGV KSB-29/E/01, IBAV No.2 and EHDV-7 KSB-14/E/97 were performed. After heat inactivation at 56 °C for 30 min, bovine sera were serially diluted twofold in serum-free Eagle's MEM containing 10 µg/ml gentamicin sulfate from 1:2 to 1:64 in the 96-well microplates. Fifty microliters of serum dilution was mixed with an equal volume of virus inoculum containing 100 × the 50 % tissue culture infective doses and incubated at 37 °C under 5 % CO_2 for 1 h. Then,

100 µl of the suspension of HmLu-1 cells in GIT medium (Wako Pure Chemical Industries, Ltd., Osaka, Japan) was added into each well and incubated at 37 °C under 5 % CO_2 for 7 days. The antibody titer was calculated as the reciprocal of the highest serum dilution inhibiting the CPE. Samples were deemed positive if they had neutralizing antibodies to the viruses in at least a dilution of 1/8.

Results

Isolation and identification of arboviruses from bovine blood samples

A total of 33 arbovirus isolates were obtained from bovine blood samples during the study period (Table 2). Some of the isolated viruses had been genetically analyzed in previous studies [22, 32, 33]. Nine isolates obtained in 1994, 1998 and 2001 were identified as AKAV by DIA. Four PEAV isolates were also found in the viruses obtained in 2001 and 2009. Eighteen isolates were reacted with the polyclonal antibodies against CHUV in DIAs and the group-specific and strain-specific RT-PCRs sorted themselves into CHUV (6 isolates in 1998, 2002 and 2006), DAGV (5 isolates in 2000, 2006 and 2012) and an unidentified PALV group virus (7 isolates in 2008 and 2009). Segment 2 of ON-1/E/08, which is an isolate of the unidentified PALV group viruses, was amplified by the FLAC method. The whole sequence of segment 2 (GenBank accession no. AB973440) showed the highest identity with

Table 2 Arboviruses isolated in the Yaeyama Islands from 1994 to 2014

Virus	Year	Date	Place	No. of isolates Plasma	No. of isolates Blood cells	Sensitive cell line
Genus *Orthobunyavirus*						
Akabane virus	1994	25. October	Ishigaki	3		BHK-21, HmLu-1
	1998	11. November	Iriomote	2	2	BHK-21, HmLu-1
	2001	12. November	Ishigaki	2		BHK-21, HmLu-1
Peaton virus	2001	13. November	Yonaguni		2	BHK-21
	2009	24. November	Ishigaki	1	1	BHK-21
Batai virus	1994	27. October	Yonaguni		1	BHK-21, HmLu-1
Genus *Orbivirus*						
Chuzan virus	1998	10. November	Ishigaki		1	BHK-21
	2002	25. November	Yonaguni		1	BHK-21
	2006	27. November - 8. December	Iriomote, Ishigaki		4	BHK-21
D' Aguilar virus	2000	15. November	Ishigaki		2	BHK-21, HmLu-1
	2006	8. December	Ishigaki		1	BHK-21
	2012	3, 4. December	Ishigaki		2	BHK 21
Bunyip Creek virus	2008	11. December	Yonaguni	1	2	BHK-21, HmLu-1
	2009	18, 24. November	Ishigaki, Taketomi		4	BHK-21
EHD[a] virus	2003	27. November	Taketomi		1	BHK-21

[a]*EHD*, epizootic hemorrhagic disease

Bunyip Creek virus (BCV) CSIRO58 (90.8 % nt and 94.9 % aa identities). The other 6 isolates of the PALV group viruses tested positive by RT-PCR with the BCV segment 2-specific primer pair. DIA and the group-specific RT-PCRs revealed that a 2003 isolate designated as ON-1/E/03 belongs to EHDV. High levels of similarity (82–84 % nt identity) in segment 7 between ON-1/E/03 (GenBank accession No. LC066302) and other EHDV strains were also revealed. However, ON-1/E/03 tested negative by RT-PCRs with the serotype-specific primer sets. The remaining isolate obtained in 1994 could

not be identified by the serological tests. However, detailed genetic analyses in a previous study determined the isolate as Batai virus (BATV) of the genus *Ortho-buyavirus* [43].

Detection of antibodies against bovine arboviruses

All arboviruses that were investigated in this study were observed as present in cattle blood in the Yaeyama Islands between 1995 and 2014 (Fig. 2). Neutralizing antibodies to the above-mentioned arboviruses were often detected in the 4-fold diluted serum. These results

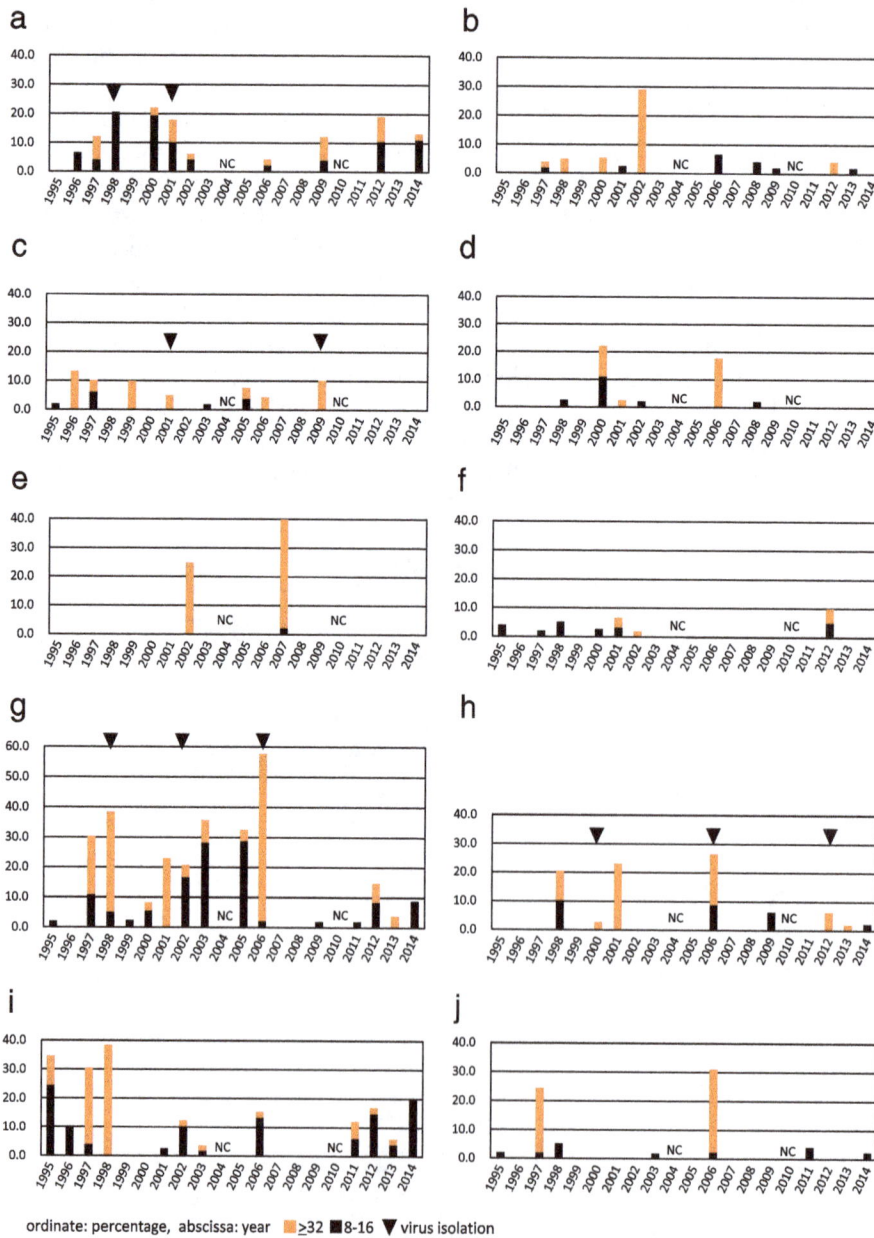

ordinate: percentage, abscissa: year ▮≥32 ▮8-16 ▼virus isolation

Fig. 2 Seroprevalence of bovine arboviruses including AKAV (**a**), AINOV (**b**), PEAV (**c**), SATV (**d**), SHAV (**e**), BEFV (**f**), CHUV (**g**), DAGV (**h**), IBAV (**i**) and EHDV-7 (**j**) in sentinel cattle in the Yaeyama Islands. Down-pointing triangle indicates the year of isolation of each virus. NC: not conducted

may indicate infection by another closely related virus [44–46] or the presence of maternally derived antibodies included in colostrum [47, 48]. Also, non-specific re-activity due to the low dilution of the sera in the VNTs could not be ruled out. Therefore, neutralizing antibody titers < 1:8 were defined as negative. Neutralizing antibodies to AKAV, AINOV, CHUV and IBAV were detected in tested cattle for more than 10 years during the study period. The seroprevalence varied a great deal according to the tested viruses (4.4–22.2 % for AKAV; 2.0–29.2 % for AINOV; 2.0–57.8 % for CHUV and 2.6–38.5 % for IBAV) and the year (Additional file 2). PEAV, BEFV, DAGV and EHDV-7 were each observed in 7–9 years. Neutralizing antibodies against SHAV and SATV were detected in 2–6 years. Serum samples collected in the Yaeyama Islands in 1994 were attempted to detect the antibodies against AKAV and BATV. A certain level of seroprevalence for both viruses (10.3 % for AKAV and 50.0 % for BATV) was observed in examined cattle (data not shown).

Arboviruses were frequently detected in the blood of cattle on Ishigaki and Iriomote Islands, but only occasionally in those of the other islands (Table 3). However, the seroprevalence in Yonaguni Island was sometimes independent of those in Ishigaki and Iriomote Islands (e.g. against PEAV in 2001 and 2006). Synchronization of seroprevalence between Ishigaki Island and surrounding islands, Taketomi, Kohama, Kuroshima and Hateruma, was generally observed during the study period.

Discussion

In Japan, bovine diseases associated with AKAV, AINOV, CHUV, IBAV and BEFV are notifiable diseases under the domestic animal infectious disease control law, and nationwide serosurveillance for these arboviruses with sentinel cattle has been conducted since 1985 [47, 49, 50]. AKAV activity has been observed on mainland Japan almost every year. Also, on the Yaeyama Islands, AKAV was frequently observed and was sometimes linked with occurrences of disease. The isolated viruses in 1994 were genetically undistinguishable from AKAVs isolated on mainland Japan the following year [32, 33]. The finding indicated that the earlier detection of AKAV activity in the Yaeyama Islands could be available in some cases prior to its spread in the mainland. Although the AINOV isolation was not made in this study, its prevalence in the Yaeyama Islands was observed by serological surveillance. The epizootics of abnormal births caused

Table 3 Seroprevalence of arboviruses by each island between 1995 and 2014

Year	AKAV			AINOV			PEAV			SATV			SHAV			BEFV			CHUV			DAGV			IBAV			EHDV-7		
	IS	IR	YO	IS	IR	YO	IS	IR	YO	IS	IR	YO	IS	IR	YO	IS	IR	YO	IS	IR	YO	IS	IR	YO	IS	IR	YO	IS	IR	YO
1995	—			—			●	—		—			—			—			●	●	—	—			●	—	●	●	—	
1996	●	—		—			●	●	—				—			—			●			—			●	●	—			
1997	●	●		●			●									●			●	●					●	●				
1998	●	●				●				●						●			●	●		●	●		●	●		●	●	
1999																					●									
2000				●	●					●	●					●			●	●										
2001	●	—	●	—	●		—	●	●	—			—			●	—		●	—	●	●	—		●	●	—			—
2002	●	●		●	●	●				●			●	●		●			●	●						●	●			
2003												●							●	●								●		
2004	—	—	—	—	—	—	—	—	—	—	—	—	—	—	—	—	—	—	—	—	—	—	—	—	—	—	—	—	—	—
2005										●	●								●	●	●									
2006	●			●	●					●	●	●							●	●	●	●	●		●	●		●	●	●
2007													●	●	●															
2008							●			●																				
2009	●	●					●												●											
2010	—	—	—	—	—	—	—	—	—	—	—	—	—	—	—	—	—	—	—	—	—	—	—	—	—	—	—	—	—	—
2011																			●						●	●		●		
2012	●	●	—	●	—				—			—			—	—	●	●	●	●		●			●	●	—			
2013			●																●	●			●				●			
2014	●	●																	●	●					●	●		●		

IS Ishigaki Island, *IR* Iriomote Island, *YO* Yonaguni Island
●: Seroprevalence was detected. —: Blood sampling was not conducted

by AINOV occurred in mainland Japan, in 1998–1999, 2002–2003 and 2005–2006 seasons. In contrast, the seroprevalence in the Yaeyama Islands was concurrently found in 1998 and 2002. Although AINOV has not been detected in the mainland since 2006, neutralizing antibodies against AINOV were continuously detected in the islands, indicating that AINOV is still active in the lower latitudes in Asia and its reemergence in the mainland is highly possible.

On mainland Japan, the last occurrence of congenital abnormalities by CHUV was reported in 2000 and, since then, CHUV has not been detected. However, the CHUV activity in the neighboring regions could be estimated from the frequent sero-detection and the virus isolations in the Yaeyama Islands. Although no clinical case was reported, the activity of DAGV had been detected several times in the Yaeyama Islands since 1991. It is noteworthy that a DAGV isolate in Ishigaki Island in 2000 shared 99.5 % nt identity in segment 2 with DAGV isolated in mainland Japan during the 2001 outbreak [22]. Furthermore, DAGV returned to the mainland in 2013 after its activity was observed in the Yaeyama Islands in the previous year [23].

Neutralizing antibodies to BEFV were present in the tested cattle in 2001 and 2012 when outbreaks of bovine ephemeral fever occurred. Low seroprevalences (2.0–5.1 %) of BEFV were also detected without clinical cases in 1995, 1997, 1998, 2000 and 2002, suggesting that the incursion and small-scale circulation of BEFV often occurs in the Yaeyama Islands. This also reflects the pattern of frequent epizootics of BEFV in neighboring regions, such as Taiwan and mainland China, in recent years [38, 51–53].

The resurgence of IBAV in the mainland was confirmed in 2013 after a 26-year absence [30]. However, the neutralizing antibodies to IBAV have been frequently detected in the Yaeyama Islands during the study period. Although the clinical evidence and the virus isolation were reported in Taiwan and Korea, respectively [54, 55], there is currently little information on this virus outside of Japan. It should be noted that IBAV caused an extensive outbreak of disease in cattle in Japan in 1959–1960 and resulted in the deaths of over 4000 cattle [27]. The epizootic of EHDV-7 once occurred in the mainland in 1997 [28]. Relatively high seroprevalences (24.5 and 31.1 %) of the virus were detected in the Yaeyama Islands in both 1997 and 2006. Although the serological surveillance suggests the continuous circulation of EHDV-7 in Asian countries, the epizootic abortion and still birth in cattle caused by the virus has not been recorded outside of Japan. However, this virus poses a serious potential threat to the livestock industry because no effective vaccine is currently available. Lower neutralization antibodies (≤1:8) to each of IBAV and

EHDV-7 are often detected, probably due to the cross-reactivities with heterogeneous serotypes of EHDV [46]. The possible incursions of other EHDV serotypes into the cattle population in the Yaeyama Islands should be investigated in the future.

The pathogenicities of PEAV, SATV, and SHAV in cattle remain uncertain. However, colostrum-free calves with congenital abnormalities sometimes carry neutralizing antibodies to these viruses, suggesting their teratogenicity [17–19]. Although the detection of PEAV in Japan was first reported in 1999 [17], retrospective analysis for an unassigned virus from Ishigaki Island revealed that PEAV was already introduced as early as 1987 [56]. Repeated isolation of PEAV in the Yaeyama Islands, other islands in Okinawa Prefecture and mainland Japan also may link to its widespread circulation in Asian countries. Seroprevalences for SATV and SHAV were lower than those of the other arboviruses. However, the surveillance in the Yaeyama Island agreed with the recent incursion of these viruses to mainland Japan [18, 19, 21]. Although suspected clinical cases caused by SATV and SHAV have been very few in Japan, their close relationship with SBV has been clearly indicated by previous studies [21, 57, 58]. It is uncertain whether Japanese strains of SATV and SHAV, as well as SBV, maintain high pathogenicity to ruminants. However, the possible presence of SATV and SHAV should be considered to avoid misdiagnosis of SBV infection in the regions commonly affected by insect-borne arboviruses.

Attempts at virus isolation proved the incursions of BATV, BCV and an EHDV isolate to the Yaeyama Islands. Serological surveillance in 1994 and 1995 also revealed the transient incursion of BATV in 1994 (data not shown). It was reported that BATV occasionally causes a febrile illness in humans and ruminants [59]. It was noteworthy that BATV probably provided a genome segment to Ngari virus, associated with severe hemorrhagic disease in humans [43, 60]. Epizootic activity of BATV is supposed to continue in Asian countries [43, 61] and thus its involvement in human and animal diseases should be monitored hereafter. Although BCV infection in the cattle population was evident in this and previous studies [62, 63], its contribution to cattle illness is not yet clear. The BCV isolation was made in mainland Japan in 2009, one year after the initial isolation in the Yaeyama Islands [23]. The co-circulation of BCV and other PALVs, such as CHUV and DAGV, potentially generate reassortants between these viruses. Because the reassortment may cause changes in viral pathogenicity as well as viral genetic and serological properties, the emergence of reassortants among PALVs should be monitored. The EHDV isolate was obtained from bovine blood in 2003. Unfortunately, serotyping for the isolated EHDV could not be concluded by the current serotype-specific RT-PCRs [41]. The existence of multiple sero-types of EHDV makes the diagnostics of EHDV-related disease more complicated. To our knowledge, at least three serotypes of EHDV have been detected in Japan so far [30]. Further investigation, such as the use of FLAC, should be conducted to investigate the detailed properties of the virus to develop proper diagnostic systems in this region.

Annual seroprevalence patterns differed among the islands. Several factors could contribute to this result. The sampling size and cattle population on Ishigaki Island are significantly larger than those on other islands, probably resulting in the higher frequency of the virus and detection of neutralizing antibodies. Because the distance between Ishigaki Island and surrounding islands including Iriomote, Taketomi, Kohama and Kuroshima Islands is not so far (maximally 15 km apart), wind dispersal of infected vectors likely occurs between the islands. However, Yonaguni Island is approximately 65–110 km from the other islands in the archipelago. This geographical feature might cause independent epizootics there. In addition, Yonaguni Island is the part of Japan closest to Taiwan and mainland China, and thus more sensitive detection of arbovirus activities in adjacent regions could be conducted in some cases. Although the source of arboviruses remains unknown, long-distance migration of the infected insect vectors by air flows is the most probable route by which bovine arboviruses are introduced to the Yaeyama Islands. Atmospheric dispersal modeling would be useful to estimate plausible incursion events and possible source sites of infected vectors.

It is of interest that fewer clinical cases have been reported in the Yaeyama Islands, despite the frequent incursion of ruminant-pathogenic arboviruses. Most cows kept for reproduction in the islands probably gain humoral immunity to the viruses by vaccination or natural infection before their pregnancy. Therefore, the abnormal births might be prevented in most cases if the teratogenic arboviruses were introduced to the islands. Also, constant and high seroprevalence in the cattle population might limit acute and massive epizootics of arboviruses.

Because viremia of bovine arboviruses is generally short, the occasion of isolation is highly limited from sentinel cattle [23]. Therefore, the frequency of blood sampling in this study might not be enough for effective virus isolation. *Culicoides* biting midges are still active in the Yaeyama Islands between October and December [64] and thus the viral transmission possibly occurred after the blood samplings. To obtain the conclusive seroprevalence in each season, additional sampling from the sentinel cattle should have be conducted few months later. It would be necessary to modify the frequency and period of blood sampling for future monitoring in this area.

Too often ruminant arbovirus diseases are ignored and neglected in tropical countries because they have not yet impacted the livestock industry on the surface. However, these diseases might have caused continuous, but unrecognized losses in these countries. Further investigation regarding the ruminant arboviruses will be necessary to assess their impact on the livestock industry in the regions where they persist. Moreover, environmental change through global warming will enhance the virus spread from the tropical to the temperate zone [65]. Monitoring for exotic arboviruses increases in importance in the high-latitude regions where such viruses would do more serious damage to the livestock industry. The Yaeyama Islands were identified as a high-risk zone for arbovirus incursions in this study and thus would be an ideal point for monitoring.

Conclusions

We showed that long-term monitoring in the Yaeyama Islands could estimate the activity of bovine arboviruses in neighboring regions and may provide a useful early warning of risk of arbovirus infections in Japan. The findings in this study could contribute to planning of prevention and control for bovine arbovirus infections in Japan and also to beneficial information-sharing on monitoring efforts among neighboring countries. Enhancement of collaborative work in this region should be essential to reduce the livestock loss caused by arbovirus infections.

Additional files

Additional file 1: Primers used for RT-PCR and sequencing in this study. (XLSX 12 kb)

Additional file 2: Seropositive rate to arboviruses in the Yaeyama Islands from 1995 to 2014. (XLSX 11 kb)

Abbreviations

AINOV, Aino virus; AKAV, Akabane virus; BATV, Batai virus; BCV, Bunyip Creek virus; BEFV, bovine ephemeral fever virus; CHUV, Chuzan virus; CPE, cytopathic effects; DAGV, D'Aguilar virus; DIA, dot immunobinding assay; EHDV, epizootic hemorrhagic disease virus; FLAC, Full-length Amplification of cDNAs; IBAV, Ibaraki virus; MEM, minimum essential medium; PALV, Palyam virus; PEAV, Peaton virus; SATV, Sathuperi virus; SBV, Schmallenberg virus; SHAV, Shamonda virus; VNT, virus neutralization test

Acknowledgements

We thank all staff of Yaeyama Livestock Hygiene Service Center and Okinawa Prefectural Animal Health Institute for blood sample collection and data acquisition. We are also grateful to Dr. Yoko Hayama for providing the supporting illustrating figures in this manuscript. This work was financially supported by the Research Project for Improving Food Safety and Animal Health of the Ministry of Agriculture, Forestry and Fisheries of Japan.

Funding

This work was financially supported by the Research Project for Improving Food Safety and Animal Health of the Ministry of Agriculture, Forestry and Fisheries of Japan.

Authors' contributions

TK performed the virus isolation and serological tests in the study and drafted the manuscript; TY participated in the design of the study and the interpretation, and contribute to the writing of the manuscript; MS, YK, KI and KT participated in the data acquisition and contributed to the improvement of the manuscript; HS, SO, KY and MY participated in the interpretation and contributed to the improvement of the manuscript; TT participate in the design and coordination of the study and contributed to the improvement of the manuscript. All authors read and approved the final manuscript.

Competing interests

The authors declare that they have no competing interests.

Consent to publish

Not applicable.

Author details

[1]Kyushu Research Station, National Institute of Animal Health, NARO, 2702 Chuzan, Kagoshima 891-0105, Japan. [2]Okinawa Prefectural Institute of Animal Health, 1-24-29 Kohagura, Naha, Okinawa 900-0024, Japan. [3]Yaeyama Livestock Hygiene Service Center, 1-2 Miyara, Ishigaki, Okinawa 907-0022, Japan. [4]Viral Disease and Epidemiology Research Division, National Institute of Animal Health, NARO, 3-1-5 Kannondai, Tsukuba, Ibaraki 305-0856, Japan. [5]Exotic Disease Research Station, National Institute of Animal Health, 6-20-1 Josuihoncho, Kodaira, Tokyo 187-0222, Japan. [6]National Institute of Animal Health, NARO, 3-1-5 Kannondai, Tsukuba, Ibaraki 305-0856, Japan.

References

1. Hubálek Z, Rudolf I, Nowotny N. Arboviruses pathogenic for domestic and wild animals. Adv Virus Res. 2014;89:201–15.
2. Hendrickx G, Gilbert M, Staubach C, Elbers A, Mintiens K, Gerbier G, Ducheyne E. A wind density model to quantify the airborne spread of Culicoides species during north-western Europe bluetongue epidemic, 2006. Prev Vet Med. 2008;87:162–81.
3. Sedda L, Rogers DJ. The influence of the wind in the Schmallenberg virus outbreak in Europe. Sci Rep. 2013;3:3361.
4. Ritchie SA, Rochester W. Wind-blown mosquitoes and introduction of Japanese encephalitis into Australia. Emerg Infect Dis. 2001;7:900–3.
5. Gloster J, Burgin L, Witham C, Athanassiadou M, Mellor PS. Bluetongue in the United Kingdom and northern Europe in 2007 and key issues for 2008. Vet Rec. 2007;162:298–302.
6. Aziz-Boaron O, Klausner Z, Hasoksuz M, Shenkar J, Gafni O, Gelman B, David D, Klement E. Circulation of bovine ephemeral fever in the Middle East — strong evidence for transmission by winds and animal transport. Vet Microbiol. 2012;158:300–7.
7. Eagles D, Melville L, Weir R, Davis S, Bellis G, Zalucki MP, Walker PJ, Durr PA. Long-distance aerial dispersal modelling of Culicoides biting midges: case studies of incursions into Australia. BMC Vet Res. 2014;10:135.
8. Wilson AJ, Mellor PS. Bluetongue in Europe: past, present and future. Philos Trans R Soc Lond B Biol Sci. 2009;364:2669–81.
9. Wernike K, Conraths F, Zanella G, Granzow H, Gache K, Schirrmeier H, Valas S, Staubach C, Marianneau P, Kraatz F, Böntgen HD, Reimann I, Zientara S, Beer M. Schmallenberg virus-two years of experiences. Prev Vet Med. 2014;116:423–34.
10. Forman S, Hungerford N, Yamakawa M, Yanase T, Tsai HJ, Joo YS, Yang DK, Nha JJ. Climate change impacts and risks for animal health in Asia. Rev Sci Tech. 2008;27:581–97.
11. Kurogi H, Inaba Y, Goto Y, Miura Y, Takahashi H. Serologic evidence for etiologic role of Akabane virus in epizootic abortion-arthrogryposis-hydranencephaly in cattle in Japan, 1972–1974. Arch Virol. 1975;47:71–83.
12. Tsuda T, Yoshida K, Ohashi S, Yanase T, Sueyoshi M, Kamimura S, Misumi K, Hamana K, Sakamoto H, Yamakawa M. Arthrogryposis, hydranencephaly and cerebellar hypoplasia syndrome in neonatal calves resulting from intrauterine infection with Aino virus. Vet Res. 2004;35:531–8.
13. Goto Y, Miura Y, Kono Y. Serologic evidence for the etiologic role of Chuzan

14. virus in an epizootic of congenital abnormalities with hydranencephaly-cerebellar hypoplasia syndrome of calves in Japan. Am J Vet Res. 1988;49:2026–9.
14. Miura Y, Kubo M, Goto Y, Kono Y. Hydranencephaly-cerebellar hypoplasia in a newborn calf after infection of its dam with Chuzan virus. Jpn J Vet Sci. 1990;52:689–94.
15. Miyazato S, Miura Y, Hase M, Kubo M, Goto Y, Kono Y. Encephalitis of cattle caused by Iriki isolate, a new strain belonging to Akabane virus. Jpn J Vet Sci. 1989;51:128–36.
16. Kono R, Hirata M, Kaji M, Goto Y, Ikeda S, Yanase T, Kato T, Tanaka S, Tsutsui T, Imada T, Yamakawa M. Bovine epizootic encephalomyelitis caused by Akabane virus in southern Japan. Vet Rec. 2008;4:20.
17. Matsumori Y, Inai K, Yanase T, Ohashi S, Kato T, Yoshida K, Tsuda T. Serological and genetical characterization of newly isolated Peaton virus in Japan. Arch Virol. 2002;147:401–10.
18. Yanase T, Fukutomi T, Yosida K, Kato T, Ohashi S, Yamakawa M, Tsuda T. The emergence in Japan of Sathuperi virus, a tropical Simbu serogroup virus of the genus Orthobunyavirus. Arch Virol. 2004;149:1007–13.
19. Yanase T, Maeda K, Kato T, Nyuta S, Kamata H, Yamakawa M, Tsuda T. The resurgence of Shamonda virus of the genus Orthobunyavirus, in Japan. Arch Virol. 2005;150:361–9.
20. Parsonson IM, Della-Porta AJ, McPhee DA. Pathogenesis and virulence studies of Australian simbu serogroup bunyaviruses. In: Mackenzie JS, editor. The viral diseases in Southeast Asia and the western pacific. Sydney: Academic Press; 1982. p. 644–7.
21. Yanase T, Kato T, Aizawa M, Shuto Y, Shirafuji H, Yamakawa M, Tsuda T. Genetic reassortment between Sathuperi and Shamonda viruses of the genus Orthobunyavirus in nature: implications for their genetic relationship to Schumallenberg virus. Arch Virol. 2012;157:1611–6.
22. Ohashi S, Matsumori Y, Yanase T, Yamakawa M, Kato T, Tsuda T. Evidence of an antigenic shift among Palyam serogroup orbiviruses. J Clin Microbiol. 2004;42:4610–4.
23. Kato T, Shirafuji H, Tanaka S, Sato M, Yamakawa M, Tsuda T, Yanase T. Bovine arboviruses in Culicoides biting midges and sentinel cattle in Southern Japan from 2003 to 2013. Transbound Emerg Dis. 2015. doi:10.1111/tbed.12324.
24. Walker PJ. Bovine ephemeral fever in Australia and the world. Curr Top Microbiol Immunol. 2005;292:57–80.
25. Kato T, Aizawa M, Takayoshi K, Kokuba T, Yanase T, Shirafuji H, Tsuda T, Yamakawa M. Phylogenetic relationships of the G gene sequence of bovine ephemeral fever virus isolated in Japan, Taiwan and Australia. Vet Microbiol. 2009;137:217–23.
26. Niwa T, Shirafuji H, Ikemiyagi K, Nitta Y, Suzuki M, Kato T, Yanase T. Occurrence of bovine ephemeral fever in Okinawa Prefecture, Japan, in 2012 and development of a reverse-transcription polymerase chain reaction assay to detect bovine ephemeral fever virus gene. J Vet Med Sci. 2015;77:455–60.
27. Omori T, Inaba Y, Morimoto T, Tanaka Y, Ishitani R. Ibaraki virus, an agent of epizootic disease of cattle resembling bluetongue. I. Epidemiologic, clinical and pathologic observations and experimental transmission to calves. Jpn J Microbiol. 1969;13:139–57.
28. Ohashi S, Yoshida K, Watanabe Y, Tsuda T. Identification and PCR-restriction fragment length polymorphism analysis of a variant of the Ibaraki virus from naturally infected cattle and aborted fetuses in Japan. J Clin Microbiol. 1999;37:3800–3.
29. Ohashi S, Yoshida K, Yanase T, Tsuda T. Analysis of intratypic variation evident in an Ibaraki virus strain and its epizootic hemorrhagic disease virus serogroup. J Clin Microbiol. 2002;40:3684–8.
30. Hirashima Y, Kato T, Yamakawa M, Shirafuji H, Okano R, Yanase T. Reemergence of Ibaraki disease in southern Japan in 2013. J Vet Med Sci. 2015;77:1253–9.
31. Yanase T, Kato T, Kubo T, Yoshida K, Ohashi S, Yamakawa M, Miura Y, Tsuda T. Isolation of bovine arboviruses from Culicoides biting midges (Diptera: Ceratopogonidae) in southern Japan: 1985–2002. J Med Entomol. 2005;42:63–7.
32. Yamakawa M, Yanase T, Kato T, Tsuda T. Chronological and geographical variations in the small RNA segment of the teratogenic Akabane virus. Virus Res. 2006;121:84–92.
33. Kobayashi T, Yanase T, Yamakawa M, Kato T, Yoshida K, Tsuda T. Genetic diversity and reassortments among Akabane virus field isolates. Virus Res. 2007;130:162–71.
34. Morita K. Molecular epidemiology of Japanese encephalitis in East Asia. Vaccine. 2009;27:7131–2.
35. Miura Y, Inaba Y, Tsuda T, Tokuhisa S, Sato K, Akashi H, Matumoto M. A survey of antibodies to arthropod-borne viruses in Indonesian cattle. Jpn J Vet Sci. 1982;44:857–63.

36. Daniels PW, Sendow I, Soleha E, Sukarsih, Hunt NT, Bahri S. Australian-Indonesian collaboration in veterinary arbovirology-a review. Vet Microbiol. 1995;46:151–74.

37. Bryant JE, Crabtree MB, Nam VS, Yen NT, Duc HM, Miller BR. Isolation of arboviruses from mosquitoes collected in northern Vietnam. Am J Trop Med Hyg. 2005;73:470–3.

38. Ting LJ, Lee MS, Lee SH, Tsai HJ, Lee F. Relationships of bovine ephemeral fever epizootics to population immunity and virus variation. Vet Microbiol. 2014;173:241–8.

39. Yoshida K, Tsuda T. Rapid detection of antigenic diversity of Akabane virus isolates by dot immunobinding assay using neutralizing monoclonal antibodies. Clin Diagn Lab Immunol. 1998;5:192–8.

40. Ohashi S, Yoshida K, Yanase T, Kato T, Tsuda T. Simultaneous detection of bovine arboviruses using single-tube multiplex reverse transcription-polymerase chain reaction. J Virol Methods. 2004;120:79–85.

41. Maan NS, Maan S, Nomikou K, Johnson DJ, El Harrak M, Madani H, Yadin H, Incoglu S, Yesilbag K, Allison AB, Stallknecht DE, Batten C, Anthony SJ, Mertens PP. RT-PCR assays for seven serotypes of epizootic haemorrhagic disease virus & their use to type strains from the Mediterranean region and North America. PLoS One. 2010;5:e10323.

42. Maan S, Rao S, Maan NS, Anthony SJ, Attoui H, Samuel AR, Mertens PP. Rapid cDNA synthesis and sequencing techniques for the genetic study of bluetongue and other dsRNA viruses. J Virol Methods. 2007;143:132–9.

43. Yanase T, Kato T, Yamakawa M, Takayoshi K, Nakamura K, Kokuba T, Tsuda T. Genetic characterization of Batai virus indicates a genomic reassortment between orthobunyaviruses in nature. Arch Virol. 2006;151:2253–60.

44. Kinney RM, Calisher CH. Antigenic relationships among Simbu serogroup (Bunyaviridae) viruses. Am J Trop Med Hyg. 1981;30:1307–18.

45. Knudson DL, Tesh RB, Main AJ, St George TD, Digoutte JP. Characterization of the Palyam serogroup viruses (Reoviridae: Orbivirus). Intervirology. 1984;22:41–9.

46. Anthony SJ, Maan S, Maan N, Kgosana L, Bachanek-Bankowska K, Batten C, Darpel KE, Sutton G, Attoui H, Mertens PP. Genetic and phylogenetic analysis of the outer-coat proteins VP2 and VP5 of epizootic haemorrhagic disease virus (EHDV): comparison of genetic and serological data to characterise the EHDV serogroup. Virus Res. 2009;145:200–10.

47. Tsutsui T, Yamamoto T, Hayama Y, Akiba Y, Nishiguchi A, Kobayashi S, Yamakawa M. Duration of maternally derived antibodies against Akabane virus in calves: survival analysis. J Vet Med Sci. 2009;71:913–8.

48. Elbers AR, Stockhofe-Zurwieden N, van der Poel WH. Schmallenberg virus antibody persistence in adult cattle after natural infection and decay of maternal antibodies in calves. BMC Vet Res. 2014;10:103.

49. Ministry of Agriculture, Forestry and Fisheries (MAFF). Statics on animal hygiene 2014. Tokyo: MAFF; 2015.

50. National Institute of Animal Health. Serological surveillance for arbovirus infections with sentinel cattle. 2015. http://www.naro.affrc.go.jp/niah/arbo/index.html. Accessed 22 Jun 2015 (in Japanese).

51. Hsieh YC, Chen SH, Chou CC, Ting LJ, Itakura C, Wang FI. Bovine ephemeral fever in Taiwan (2001–2002). J Vet Med Sci. 2005;67:411–6.

52. Zheng F, Qiu C. Phylogenetic relationships of the glycoprotein gene of bovine ephemeral fever virus isolated from mainland China, Taiwan, Japan, Turkey, Israel and Australia. Virol J. 2012;9:268.

53. Li Z, Zheng F, Gao S, Wang S, Wang J, Liu Z, Du J, Yin H. Large-scale serological survey of bovine ephemeral fever in China. Vet Microbiol. 2015;176:155–60.

54. Bak UB, Cheong CK, Choi HI, Lee CW, Oh HS. An outbreak of Ibaraki disease in Korea. Korean J Vet Res. 1983;23:81–9.

55. Liao YK, Inaba Y, Li NJ, Chain CY, Lee SL, Liou PP. Epidemiology of bovine ephemeral fever virus infection in Taiwan. Microbiol Res. 1998;153:289–95.

56. Yanase T, Aizawa M, Kato T, Yamakawa M, Shirafuji H, Tsuda T. Genetic characterization of Aino and Peaton virus field isolates reveals a genetic reassortment between these viruses in nature. Virus Res. 2010;153:1–7.

57. Hoffmann B, Scheuch M, Höper D, Jungblut R, Holsteg M, Schirrmeier H, Eschbaumer M, Goller KV, Wernike K, Fischer M, Breithaupt A, Mettenleiter TC, Beer M. Novel orthobunyavirus in Cattle, Europe, 2011. Emerg Infect Dis. 2012; 18:469–72.

58. Goller KV, Höper D, Schirrmeier H, Mettenleiter TC, Beer M. Schmallenberg virus as possible ancestor of Shamonda virus. Emerg Infect Dis. 2012;18: 1644–6.

59. Hubálek Z. Mosquito-borne viruses in Europe: vector-borne diseases and climate change. Parasitol Res. 2008;103 Suppl 1:29–43.

60. Briese T, Bird B, Kapoor V, Nichol ST, Lipkin WI. Batai and Ngari viruses: M segment reassortment and association with severe febrile disease outbreaks in East Africa. J Virol. 2006;80:5627–30.

61. Liu H, Shao XQ, Hu B, Zhao JJ, Zhang L, Zhang HL, Bai X, Zhang RX, Niu DY, Sun YG, Yan XJ. Isolation and complete nucleotide sequence of a Batai virus strain in Inner Mongolia, China. Virol J. 2014;11:138.

62. Cybinski DH, St George TD. Preliminary characterization of D'Aguilar virus and three Palyam group viruses new to Australia. Aust J Biol Sci. 1982;35:343–51.

63. Littlejohns IR, Burton RW, Sharp JM. Bluetongue and related viruses in New South Wales: isolations from, and serological tests on samples from sentinel cattle. Aust J Biol Sci. 1988;41:579–87.

64. Hoshino C. Notes on biting midges collected by light traps at a cowshed in Ishigaki-jima, Ryukyu Islands. Med Entomol Zool. 1985;36:55–8 (in Japanese with English summary).

65. Purse BV, Carpenter S, Venter GJ, Bellis G, Mullens BA. Bionomics of temperate and tropical Culicoides midges: knowledge gaps and consequences for transmission of Culicoides-borne viruses. Annu Rev Entomol. 2015;60:373–92.

15

Real-time fluorescence loop-mediated isothermal amplification assay for direct detection of egg drop syndrome virus

Makay Zheney[1,2], Zhambul Kaziyev[2], Gulmira Kassenova[2], Lingna Zhao[1], Wei Liu[1], Lin Liang[1] and Gang Li[1*]

Abstract

Background: Egg drop syndrome (EDS), caused by the adenovirus "egg drop syndrome virus" (EDSV) causes severe economic losses through reduced egg production in breeder and layer flocks. The diagnosis of EDSV has been done by molecular tools since its complete genome sequence was identified. In order to enhance the capabilities of the real-time fluorescence loop-mediated isothermal amplification (RealAmp) assay, we aimed to apply the method for direct detection of the EDSV without viral DNA extraction. In order to detect the presence of the EDSV DNA, three pairs of primers were designed, from the conserved region of fiber gene of the EDSV.

Results: For our assay, test and control samples were directly used in the reaction mixture in 10-fold serial dilution. The target DNA was amplified at 65 °C, which yield positive results in a relatively short period of 40–45 min. The method reported in this study is highly sensitive as compared to polymerase chain reaction (PCR) and showed no sign of cross-reactivity or false positive results. The RealAmp accomplished specific identification of EDSV among a variety of poultry disease viruses.

Conclusions: The direct RealAmp can be used to detect the presence of EDSV. As our result showed, the RealAmp method could be suitable for the direct detection of other DNA viruses.

Keywords: Egg drop syndrome virus, Real-time fluorescence loop-mediated isothermal amplification, Sensitivity, Specificity

Background

Egg drop syndrome is a viral disease, caused by the egg drop syndrome virus (EDSV), officially called duck adenovirus 1 (DAdV-1), belonging to species Duck adenovirus A, genus *Atadenovirus*, family *Adenoviridae*. EDSV was first reported in 1976, it has also been known as adenovirus 127 and egg-drop-syndrome-76 (EDS-76) virus [1]. EDS is characterized by the production of soft-shelled, thin shelled, shell-less, and discolored eggs in otherwise healthy chickens [2]. The natural hosts of the EDSV are ducks and geese, however, the virus can also infect chickens, resulting in major economic losses on egg production [3, 4]. EDSV was involved in severe respiratory disease in 1-day-old goslings where the presence of EDSV DNA was found in different organs of the

naturally and experimentally infected goslings [5]. Severe acute respiratory symptoms with coughing, dyspnea, and gasping were reported in 9-day-old Pekin ducklings in 2013 [6]. For diagnosis of EDSV, five serological methods have been used and tested [7]. In the recent years, several PCR studies have been published, for diagnosis of all avian adenoviruses that are of relevance for poultry production [8–10]. Molecular amplification methods were commonly used to diagnose EDSV infection [11].

Loop-mediated isothermal amplification (LAMP) is a method that can amplify DNA under isothermal conditions. It was first developed by the Japanese researchers, the LAMP employs a DNA polymerase and a set of four specially designed primers that recognize a total of six distinct sequences on the target DNA [12]. Later, LAMP was supplemented by using additional primers, termed loop primers which prime strand displacement DNA synthesis. Moreover, LAMP has some advantages in comparison with PCR methods, including improved sensitivity and specificity, as well as time efficiency [13].

---

* Correspondence: ligang03@caas.cn
[1]State Key Laboratory of Animal Nutrition, Institute of Animal Science, Chinese Academy of Agricultural Sciences, Beijing 100193, People's Republic of China
Full list of author information is available at the end of the article

Since LAMP was published, a range of LAMP methods have been developed. The RealAmp is one of them, which attempted to improve the method for diagnosis by using a simple and portable device capable of performing both the amplification and detection by fluorescence in one platform [14]. Currently, the LAMP assays are utilized to detect bacterial and viral pathogens including *Mycobacterium tuberculosis*, *Acinetobacter baumannii*, avian influenza virus, Middle East respiratory syndrome coronavirus and hemorrhagic enteritis virus [15–19].

Various LAMP procedures have been successfully employed for DNA amplification using DNA templates extracted from the samples. The purpose of this study was to evaluate the usability of the RealAmp method for a rapid detection of the EDSV in a diverse range of samples without a prior need for nucleic acid extraction. Therefore, we infected both duck embryos and duck fibroblast cell culture with EDSV, then the viral samples were collected and employed to the assay directly by serial dilutions.

Methods
Chemicals and reagents
Enzymes including Bst2.0 DNA polymerase (8000 U/ml) and BsrGI-HF (20,000 U/ml) were obtained from New England Bio labs (NEB, USA). Primers for RealAmp and PCR (oligonucleotides) were obtained from Huada (Beijing, China) and suspended in deionized water with appropriate concentrations and stored at -20 °C. The concentrations of each DNA suspension used in this study were measured by NanoVue Plus spectrophotometer (GE Healthcare, USA).

Description of the equipment
The fluorescence reader ESE-Quant Tube Scanner used for this study was developed by a commercial manufacturer (ESE Gmbh, Stockach, Germany). It has an eight tube holder heating block with adjustable temperature settings and spectral devices to detect amplified product using fluorescence spectra. This equipment is easy to handle, and completely portable. The results can be seen in a small monitor or on a computer screen connected to the equipment.

Viruses
Fowl pox virus (FPV isolate FPV4), duck viral enteritis virus (DVEV isolate DPV-F37), Marek's disease virus (MDV isolate CVI 988/Rispens) were obtained from China Institute of Veterinary Drug Control; duck circovirus (DCV isolate DuCV-AH1) was obtained from Beijing Experimental Station of Veterinary Biotechnology and Diagnostic Technology, Ministry of Agriculture,

China. The viruses were kept in tissue culture supernatant in the laboratory at -80 °C.

Inoculation of embryonated duck eggs with the EDSV
For virus propagation 9-day-old embryonated duck eggs were obtained from Beijing Dayinghongguang Duck Farm, and incubated for 2 days in an egg incubator. EDSV (EDS-NE4) used in this experiment was isolated in 1992 [20], and kept in the Animal Disease Control Laboratory of Institute of Animal Science, Chinese Academy of Agricultural Sciences. The virus was diluted with sterile PBS at a ratio of 1/100, followed by inoculation of the allantoic sac of embryonated duck eggs with the viral dilution (200 µL). The eggs were incubated at 37 °C and examined twice each day. After 6 day of incubation the eggs were chilled at 4 °C for 4 h, and then the allantoic fluid was collected from each embryo and stored at -80 °C. The haemagglutination (HA) titer of EDSV in collected allantoic fluid was in average $\log_2 12$.

Cell culture and virus inoculation
Duck fibroblast cell cultures were prepared from 11-day-old duck embryos and cultured in 75 cm^2 flasks on Dulbecco's modified Eagle's medium (DMEM, Gibco, Shanghai, China) supplemented with 10% fetal bovine serum (FBS, Gibco, USA) and 1% gentamycin (Sigma-Aldrich). Cells were inoculated with 1 mL of diluted (1:100) EDSV collected from allantoic fluid (HA $\log_2 12$) and incubated at 37 °C under 5% CO$_2$ for 1 h. The viral inoculum was removed from the cell layer, and replaced by 10 mL of fresh DMEM supplemented with 2% FBS and 1% gentamycin, followed by incubation for 48 h. PBS was used as a negative control, without virus inoculum in the control flask. The cells were examined daily for any cytopathic effect (CPE). Infected supernatant was harvested after 46 h of incubation and then used for RealAmp analysis. The HA titer of EDSV in harvested cell culture supernatant was $\log_2 9$.

Viral DNA extraction
Viral DNA was isolated from the allantoic fluid collected from infected duck embryos and the supernatant of duck embryo fibroblasts cultured cells. Total nucleic acids were extracted using the AxyPrep™ Body Fluid viral DNA/RNA Miniprep Kit (Axygen, USA) according to the manufacturer's instructions and stored at -20 °C.

Design of primers for the RealAmp and PCR
Six specific RealAmp primers (F3, B3, FIP, BIP, LF and LB) were designed using PrimerExplorer V5 software (Eiken Chemical Co. Ltd., Tokyo, Japan) based on the fiber gene sequence of the EDSV (GenBank accession No.Y09598.1). The genome positions of RealAmp primers in EDSV genome are shown in Fig. 1. PCR primers were

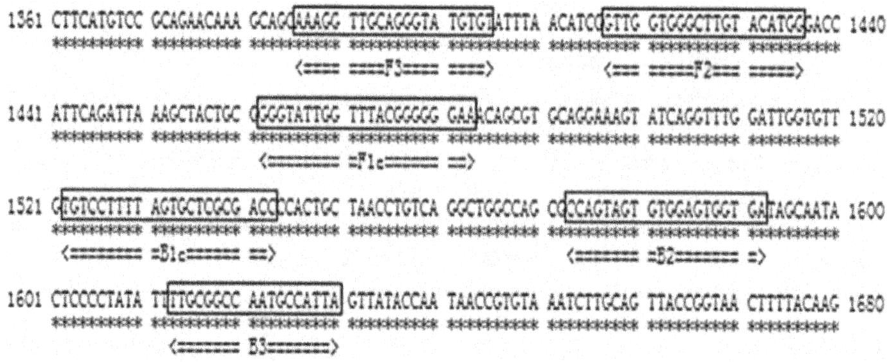

Fig. 1 Position of the RealAmp primers in EDSV fiber gene (1935 bp)

designed using Oligo7 Primer Analysis software (Molecular Biology Insights, Inc. USA). The sequences of the RealAmp and PCR primers are listed in Table 1.

RealAmp assay

RealAmp reactions were performed using EDSV DNA purified from allantoic fluid and cell culture supernatant. The viral DNA was diluted from 10^{-1} to 10^{-6} prior to use. The allantoic fluid and cell culture supernatants were also used directly to the reaction without viral DNA extraction by diluting the samples (10^{-1} to 10^{-5}). The 25 μL reaction includes 1 μL of template (DNA and virus liquid) in 1X isothermal buffer (Bio Labs, USA; 20 mM Tris–HCl, 50 mM KCl, 10 mM $(NH4)_2SO_4$, 2 mM $MgSO_4$, 0.1% Tween 20, pH 8.8) containing 1.2 mM dNTPs, 1 M betaine, 1.6 μM FIP and BIP, 0.2 μM F3 and B3, 0.8 μM LF and LB, and 8 U Bst2.0 DNA polymerase, 9 μL of ddH_2O and 0.5 μL of EvaGreen (10X) (Invitrogen, Carlsbad, CA). Reactions were carried out at 65 °C, and the total run times were 40–45 min for every RealAmp reaction. The graph of fluorescence units and time was plotted using an ESE-Quant Tube Scanner (Qiagen, Germany).

The graph shows the fluorescence in millivolts (mV) on the y-axis and time in minutes on the x-axis. Results can be read in real time using Tube Scanner Studio software.

Specificity and sensitivity of the RealAmp assay

To validate the specificity of RealAmp for EDSV detection, additional DNA viruses (described in methods section) were tested. To check the expected EDSV targeted RealAmp amplicon, 2 μL of the RealAmp product was digested in 25 μL reaction containing 20 units of BsrGI-HF (20,000 U/ml) restriction enzyme, 2.5 μL of Cut-Smart® buffer and 19.5 μL of ddH_2O. The reaction mixture was incubated at 37 °C for 2 h and separated by electrophoresis in a 1.5% agarose gel.

Sensitivity of the RealAmp assay was tested using 10-fold serial dilutions (10^{-1} to 10^{-6}) of constructed pMD19T-fiber plasmid DNA (26 ng /μL) and in parallel by conventional PCR method.

Conventional PCR

Conventional PCR reactions were performed to amplify the fiber gene for construction of pMD19T-fiber plasmid

Table 1 RealAmp and PCR primers

Method	Primer name	Length (bp)	Sequence (5'-3')	Location of the primers[a]
RealAmp	F3	20	AAAGGTTGCAGGGTATGTGT	24,070–24,089
	B3	18	TAATGGCATTGGCCGCAA	24,297–24,314
	FIP (F2)	20	GTTGGTGGGCTTGTACATGG	24,101–24,120
	FIP (F1c)	22	TTCCCCCCGTAAACCAATACCC	24,146–24,167
	BIP (B2)	20	TCACCACTCCACACTACTGG	24,257–24,276
	BIP (B1c)	22	TGTCCTTTTAGTGCTCGCGACC	24,206–24,227
	LF	25	CGCAGTAGCTTTAATCTGAATGGTC	24,121–24,145
	LB	20	CCACTGCTAACCTGTCAGGC	24,228–24,247
PCR	Fiber-F	20	ATGAAGCGACTACGGTTGGA	22,685–22,704
	Fiber-R	26	CTACTGTGCTCCAACATATGTAAAGG	24,594–24,619

F3- forward outer primer, B3- backward outer primer, LF- loop forward primer, LB- loop backward primer, FIP- forward inner primer and BIP- backward inner primer. F denotes forward primer and R denotes reverse primer, [a]locations of primers in EDSV genome

and to compare the established RealAmp sensitivity with conventional PCR. In our study, the size of the target sequence was 245 bp by using the outer primers F3 and B3 and fiber gene 1935 bp by using PCR primers (Table 1). PCR reactions were conducted with a total reaction volume of 25 μL, which contain 12.5 μL of 2X Easy Taq PCR Mix (TransGen; 0.2 mM of each dNTP, 1.5 mM MgCl$_2$), 0.4 μM forward primer, 0.4 μM reverse primer, 9.5 μL of ddH$_2$O and 1 μL of template DNA. The PCR program consisted of an initial denaturation step at 94 °C for 4 min, followed by 30 cycles of denaturation at 94 °C for 30s, primer annealing at 58 °C for 30s, extension at 72 °C for 2 min (fiber gene primers), 30s for (F3 and B3 primers) and a final extension step at 72 °C for 10 min. PCR products were analyzed by electrophoresis and photographed under UV light.

Results

RealAmp of EDSV DNA

In this assay, the EDSV DNA was successfully amplified from diluted viral DNA samples extracted from infected allantoic fluid and cell culture supernatants within 40 min (Fig. 2a and b). No amplification was obtained from uninfected control samples. The RealAmp reactions

were also analysed using agarose gel electrophoresis and as anticipated a ladder-like DNA banding pattern was observed (Fig. 2c and d).

Direct RealAmp assay

By using infected allantoic fluid and cell culture supernatants, as well as undiluted samples directly in the assay, we successfully amplified EDSV DNA. Analysis of each sample was carried out three times independently. The results obtained were similar to those obtained by using EDSV nucleic acids (Fig. 2). A graph of fluorescence units and time was produced for all sample dilutions. For the allantoic fluid samples, all dilutions from 10^{-1} to 10^{-5} showed normal amplification began after 15 min of incubation at 65 °C. However, amplification of the undiluted sample began after 40 min (Fig. 3a). Serial diluted cell culture supernatants (10^{-1} to 10^{-4}) and undiluted sample provided results after 20–30 min of scanning whereas the 10^{-5} dilution and uninfected (as a negative control) samples provided no amplification (Fig. 3b). The RealAmp products from both samples were separated in 1.5% agarose gel (Fig. 3c and d).

Specificity of the RealAmp method

To evaluate the specificity of the utilized RealAmp assay, we used several poultry disease viruses (described in

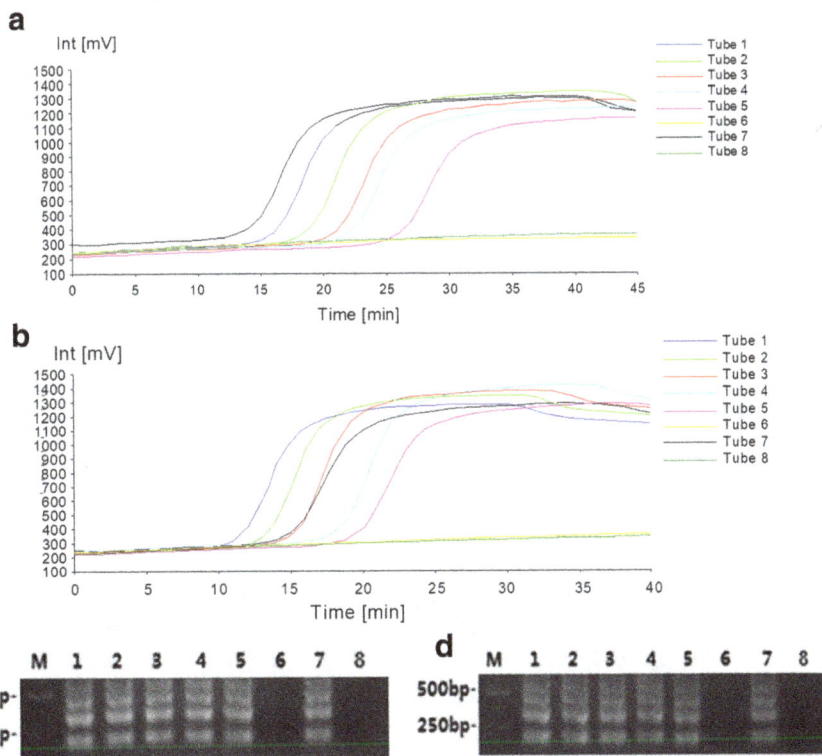

Fig. 2 RealAmp using viral DNA isolated from infected allantoic fluid and cell suspension. **a** RealAmp of viral DNA extracted from allantoic fluid; (**b**) RealAmp of viral DNA from infected duck fibroblast cell culture supernatant used with serial dilutions. For both reactions, tube 1–6 DNA sample dilutions (10^{-1} to 10^{-6}); tube 7 for positive (EDSV DNA) and tube 8 negative (uninfected) controls. **c** and (**d**) 1.5% agarose gel electrophoresis results of (**a**) and (**b**), lane 1–6, sample dilutions (10^{-1} to 10^{-6}); lane 7 positive control; lane 8 negative control; lane M- Trans2K plus DNA marker

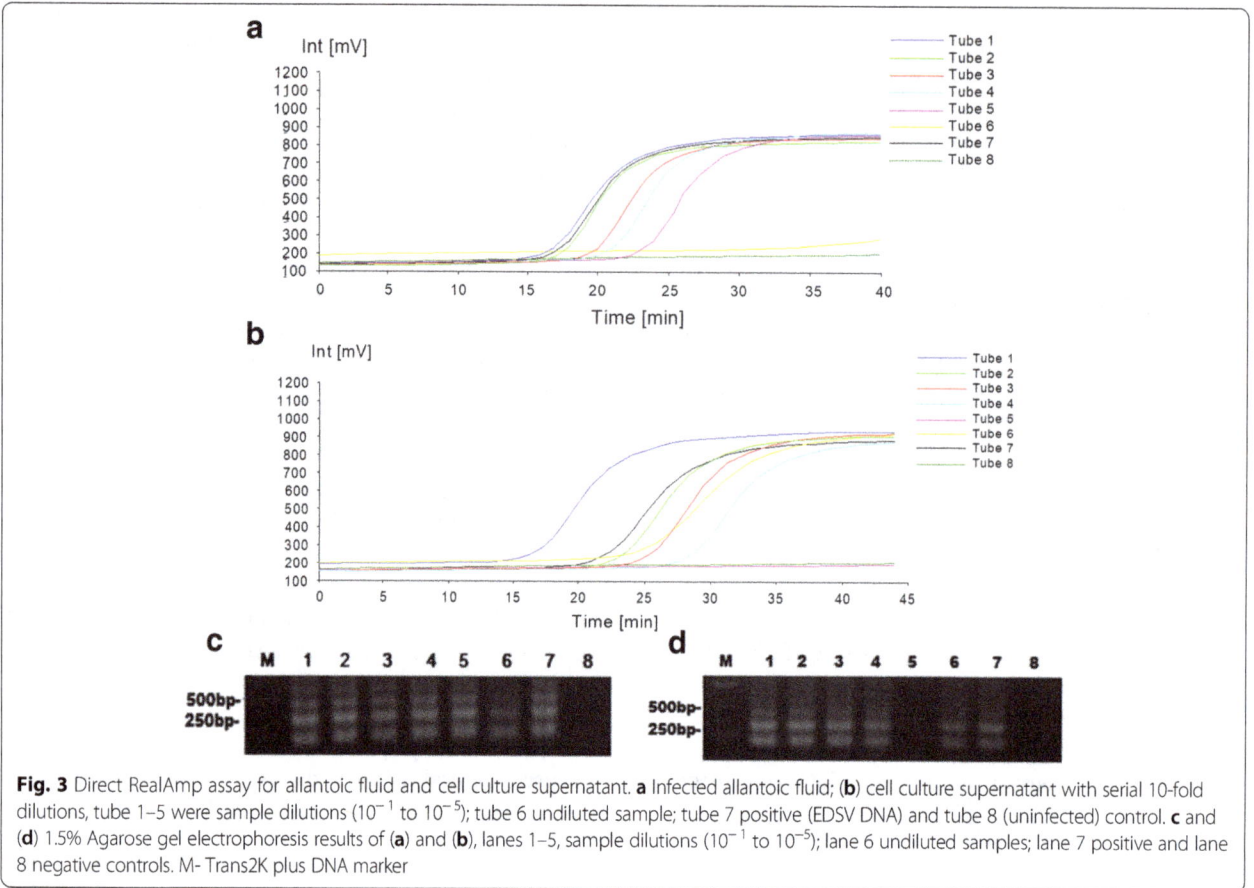

Fig. 3 Direct RealAmp assay for allantoic fluid and cell culture supernatant. **a** Infected allantoic fluid; (**b**) cell culture supernatant with serial 10-fold dilutions, tube 1–5 were sample dilutions (10^{-1} to 10^{-5}); tube 6 undiluted sample; tube 7 positive (EDSV DNA) and tube 8 (uninfected) control. **c** and (**d**) 1.5% Agarose gel electrophoresis results of (**a**) and (**b**), lanes 1–5, sample dilutions (10^{-1} to 10^{-5}); lane 6 undiluted samples; lane 7 positive and lane 8 negative controls. M- Trans2K plus DNA marker

methods section). As expected, the typical amplification curve was only obtained in the test using EDSV sample as template. The results showed that the RealAmp could amplify and differentiate EDSV gene within other DNA viruses (Fig. 4a and b). The amplified product was digested with restriction enzyme BsrGI-HF, which results in the digestion of target region out from a total amplified DNA. The target RealAmp region was cleaved by specific enzyme which is located in the EDSV fiber gene, then separated on a 1.5% agarose gel and shown in Fig. 4c.

Sensitivity of the RealAmp assay

The lower limit of detection of the RealAmp assay determined using plasmid DNA constructed by cloning the PCR amplified fiber gene of EDSV into pMD19T cloning vector (Takara, Japan). The pMD19T-fiber DNA (26 ng) was used to the assay with 10-fold serial dilutions (2.6 ng, 260 pg, 26 pg, 2.6 pg, 260 fg, and 26 fg per microliter). The RealAmp assay demonstrated 100-fold more sensitive than the conventional PCR. The DNA detection limit of the RealAmp was 26 fg and 5.2×10^3 copies/ µL while lower limit of detection of conventional PCR was 2.6 pg and 5.2×10^5 copies/µL. The procedure was monitored using an ESE-Quant Tube Scanner

(Fig. 5a). Both RealAmp and PCR results were assessed by 1.5% agarose gel electrophoresis (Fig. 5b and c).

Discussion

Diagnosis of EDS is being performed using molecular technologies in many virus affected countries. The complete genome sequence of egg drop syndrome virus allowed the development of PCR assays for EDSV detection [21]. Since then the hexon based PCR assay was used to detect and differentiate the EDSV from fowl adenoviruses [22]. The conventional PCR methods have been employed in previous studies to diagnose the EDSV infection as a specific and sensitive method as compared to the serological methods [23]. A quantitative real-time PCR (q-PCR) assay based on hexon gene for the rapid detection of EDSV has also been reported [24]. While in 2014 a novel q-PCR assay was used to detect EDSV DNA in samples of interest [25]. Currently, a 151 bp fragment of the EDSV strain 127 penton base gene amplified by PCR with 100% nucleotide identity and confirmed by q-PCR [26]. In the present study, the ESE-Quant tube scanner was capable of detecting samples in real-time, and it could analyse melting curves using computer software [14].

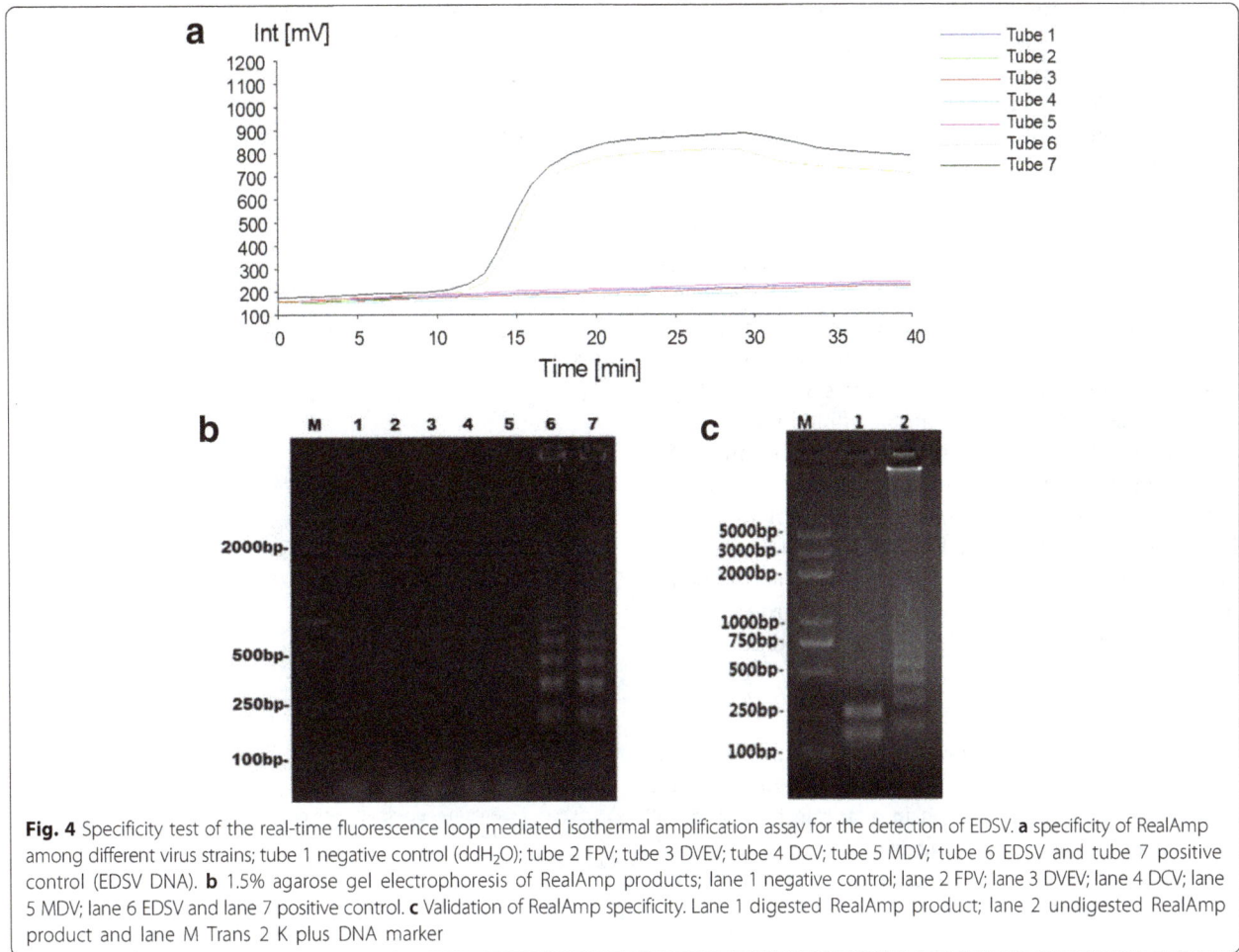

Fig. 4 Specificity test of the real-time fluorescence loop mediated isothermal amplification assay for the detection of EDSV. **a** specificity of RealAmp among different virus strains; tube 1 negative control (ddH$_2$O); tube 2 FPV; tube 3 DVEV; tube 4 DCV; tube 5 MDV; tube 6 EDSV and tube 7 positive control (EDSV DNA). **b** 1.5% agarose gel electrophoresis of RealAmp products; lane 1 negative control; lane 2 FPV; lane 3 DVEV; lane 4 DCV; lane 5 MDV; lane 6 EDSV and lane 7 positive control. **c** Validation of RealAmp specificity. Lane 1 digested RealAmp product; lane 2 undigested RealAmp product and lane M Trans 2 K plus DNA marker

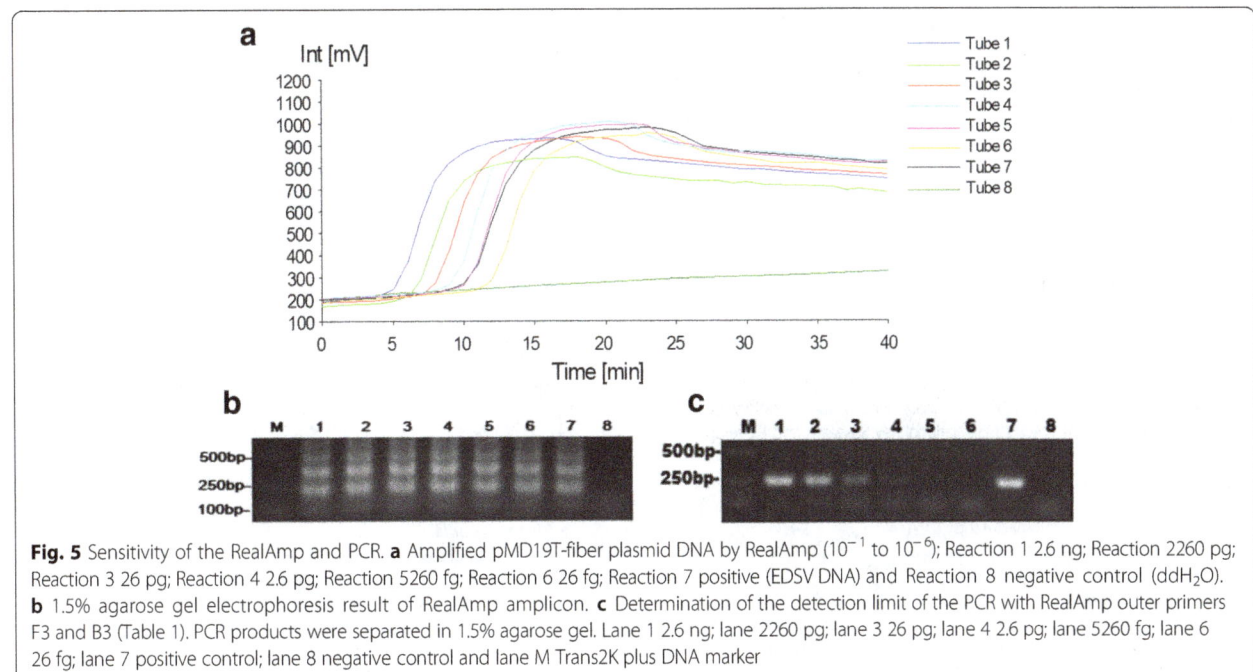

Fig. 5 Sensitivity of the RealAmp and PCR. **a** Amplified pMD19T-fiber plasmid DNA by RealAmp (10^{-1} to 10^{-6}); Reaction 1 2.6 ng; Reaction 2 260 pg; Reaction 3 26 pg; Reaction 4 2.6 pg; Reaction 5 260 fg; Reaction 6 26 fg; Reaction 7 positive (EDSV DNA) and Reaction 8 negative control (ddH$_2$O). **b** 1.5% agarose gel electrophoresis result of RealAmp amplicon. **c** Determination of the detection limit of the PCR with RealAmp outer primers F3 and B3 (Table 1). PCR products were separated in 1.5% agarose gel. Lane 1 2.6 ng; lane 2 260 pg; lane 3 26 pg; lane 4 2.6 pg; lane 5 260 fg; lane 6 26 fg; lane 7 positive control; lane 8 negative control and lane M Trans2K plus DNA marker

To test large number of samples, the real-time PCR methods are too expensive and the q-PCR machines are not always available. Therefore we designed the first EDSV detection by utilizing an assay based on the direct RealAmp. The results of this study clearly indicated the superiority of the RealAmp assay in the detection of EDSV compared to a conventional PCR assay. In order to facilitate improved virus detection, diluted samples of allantoic fluid, and cell culture supernatants were used in the direct RealAmp assay and the results have showed that our method was successfully identified the EDSV from both samples (Fig. 3a and b). The RealAmp cannot give a clear and rapid result, when the sample was highly concentrated (undiluted allantoic fluid sample); the problem lies in the quantity of primer to be much dispersed on different and many DNA pieces. So the self-limiting process could happen (Fig. 3a). In this study, RealAmp has amplified the EDSV fiber gene by using diluted recombinant plasmid DNA as low as 26 fg per microliter in 40 min, while the PCR was around 2.6 pg per microliter in 1 h and 30 min (Fig. 5a and 5c).

Conclusion

To conclude, the rapid, sensitive and specific RealAmp method can be directly employed to detect EDSV within short time span of 40–45 min in allantoic fluid and cell supernatant by using diluting samples up to 1/10000. Further, this cost effective technique is more sensitive than conventional PCR for detection of EDSV.

Abbreviations

CPE: Cytopathic effects; DCV: Duck circovirus; DMEM: Dulbecco's modified Eagle's medium; DVEV: Duck viral enteritis virus; EDS: Egg drop syndrome; EDSV: Egg drop syndrome virus; FBS: Fetal bovine serum; FPV: Fowl pox virus; HA: Haemagglutination; LAMP: Loop mediated isothermal amplification; MDV: Marek's disease virus; PBS: Phosphate buffer saline; PCR: Polymerase chain reaction; qRT-PCR: Quantitative real-time polymerase chain reaction; RealAmp: Real-time fluorescence loop mediated isothermal amplification; RT-PCR: Real-time polymerase chain reaction

Acknowledgments

We would like to thank to Dr. Hermann Unger for his critical revision on this manuscript and Dr. Hamama Islam Butt for her assistance in writing and Dr. Tussipkan Dilnur for her helpful discussions.

Funding

This study was financially supported by the National Natural Science Foundation of China (No. 31472203,31172342); the National Science and Technology Support Program of China (No. 2013BAD12B05); National Key Research and Development Plan (No.2016YFD0501102); Genetically Modified Organisms Breeding Major Projects of P.R. China grant (No. 2014ZX0801203B) and IAEA CRP (No.17453). The funders had no role in the study design, data analysis, and decision to publish, or preparation of the manuscript.

Author's contributions

LG, ZK, and GK designed the experiments and gave suggestions. MZ, WL, LZ and LL performed the experiments and analyzed the results. MZ wrote the paper. All authors have read and approved the manuscript.

Consent for publication
Not applicable

Competing interests
The authors declare that they have no competing interests.

Author details
[1]State Key Laboratory of Animal Nutrition, Institute of Animal Science, Chinese Academy of Agricultural Sciences, Beijing 100193, People's Republic of China. [2]Faculty of Veterinary, Kazakh National Agrarian University, Almaty 050013, Republic of Kazakhstan.

References

1. Harrach B, Benkő M, Both GW, Brown M, Davison AJ, Echavarría M, Hess M, Jones MS, Kajon A, Lehmkuhl HD, Mautner V, Mittal SK, Wadell G. Family _Adenoviridae_. King AMQ. In: Adams MJ, Carstens EB, Lefkowitz EJ, editors. Virus taxonomy: classification and nomenclature of viruses. Ninth report of the international committee on taxonomy of viruses. San Diego: Elsevier; 2011. p. 125–41.
2. Eck JHHV, Davelaar FG, Kol NV, Kouwenhoven B, Guldie FHM. Dropped egg production, soft shelled and shell-less eggs associated with appearance of precipitins to adenovirus in flocks of laying fowls. Avian Pathology. 1976; 5(4):261–72.
3. McFerran JB, McCracken RM, McKillop ER, McNulty MS, Collins DS. Studies on a depressed egg production syndrome in Northern Ireland. Avian Pathology. 1978;7(1):35.
4. Hafez HM. Avian adenoviruses infections with special attention to inclusion body hepatitis/hydropericardium syndrome and egg drop syndrome. Pak Vet J. 2011;31(2):85–92.
5. Ivanics E, Palya V, Glavits R, Dan A, Palfi V, Revesz T, Benko M. The role of egg drop syndrome virus in acute respiratory disease of goslings. Avian Pathology. 2001;30(3):201–8.
6. Cha SY, Kang M, Moon OK, Park CK, Jang HK. Respiratory disease due to current egg drop syndrome virus in Pekin ducks. Vet Microbiol. 2013; 165(3–4):305–11.
7. Adair BM, Todd D, Mcferran JB, Mckillop ER. Comparative serological studies with egg drop syndrome virus. Avian Pathology. 1986;15(4):677–85.
8. Hess M. Detection and differentiation of avian adenoviruses: a review. Avian Pathology. 2000;29(3):195–206.
9. Raj GD, Sivakumar S, Sudharsan S, Mohan AC, Nachimuthu K. Genomic characterization of Indian isolates of egg drop syndrome 1976 virus. Avian Pathology. 2001;30(1):21–6.
10. Ballmann MNZ, Harrach BZ. Detection and partial genetic characterisation of novel avi- and siadenoviruses in racing and fancy pigeons (Columba Livia Domestica). Acta Vet Hung. 2016;64(4):514–28.
11. Begum JA, Chowdhury EH, Parvin R, Matin MA, Giasuddin M, Bari ASM, Islam MR. Detection of Egg Drop Syndrome Virus by Polymerase Chain Reaction. Int J Livest Res. 2013;3(2):112–6.
12. Notomi T, Okayama H, Masubuchi H, Yonekawa T, Watanabe K, Amino N, Hase T. Loop-mediated isothermal amplification of DNA. Nucleic Acids Res. 2000;28(12):E63.
13. Nagamine K, Hase T, Notomi T. Accelerated reaction by loop-mediated isothermal amplification using loop primers. Mol Cell Probes. 2002;16(3): 223–9.
14. Lucchi NW, Demas A, Narayanan J, Sumari D, Kabanywanyi A, Kachur SP, Barnwell JW, Udhayakumar V. Real-time fluorescence loop mediated isothermal amplification for the diagnosis of Malaria. PLoS One. 2010;5(10):73.
15. Imai M, Ai N, Minekawa H, Notomi T, Ishizaki T, Tashiro M, Odagiri T. Development of H5-RT-LAMP (loop-mediated isothermal amplification) system for rapid diagnosis of H5 avian influenza virus infection. Vaccine. 2006;24(44–46):6679–82.
16. Bhadra S, Jiang YS, Kumar MR, Johnson RF, Hensley LE, Ellington AD. Real-time sequence-validated loop-mediated isothermal amplification assays for detection of Middle East respiratory syndrome coronavirus (MERS-CoV). PLoS One. 2015;10(4):e0123126.

17. Liu X, Li Y, Xu C, Qin J, Hao J, Feng M, Tan L, Jia W, Liao M, Cao W. Real-time fluorescence loop-mediated isothermal amplification for the diagnosis of hemorrhagic enteritis virus. Virus Res. 2014;183(7):50–5.

18. Ou X, Wang S, Dong H, Pang Y, Li Q, Xia H, Qu Y, Zhang Z, Li J, Zhang J. Multicenter evaluation of a real-time loop-mediated isothermal amplification (RealAmp) test for rapid diagnosis of mycobacterium tuberculosis. J Microbiol Methods. 2016;129:39–43.

19. Wang Q, Zhou Y, Li S, Zhuo C, Xu S, Huang L, Yang L, Liao K. Real-time fluorescence loop mediated isothermal amplification for the detection of acinetobacter baumannii. PLoS One. 2013;8(7):e66406.

20. Li G, Zheng M, Cai B, Wu L, Tang J. Study on inactivated vaccine in oily adjuvant against the egg drop syndrome. Journal of Nanjing Agricultural University. 1994;17(3):91–4.

21. Hess M, Blöcker H, Brandt P. The complete nucleotide sequence of the egg drop syndrome virus: an intermediate between mastadenoviruses and aviadenoviruses. Virology. 1997;238(1):145–56.

22. Raue R, Hess M. Hexon based PCRs combined with restriction enzyme analysis for rapid detection and differentiation of fowl adenoviruses and egg drop syndrome virus. J Virol Methods. 1998;73(2):211–7.

23. Zhang Z, Hu M. Detection of egg drop syndrome (EDS'76) virus by the polymerase chain reaction (PCR). Chinese Journal of Virology. 1996;12:156–61.

24. Zhen-Yuan MA, Gang LI, Wen-Chao LI, Guo YF. Development and application of TaqMan fluorescent real-time quantitative PCR for the detection of egg drop syndrome virus. Chinese Journal of Animal & Veterinary Sciences. 2012;5:767–72.

25. Schybli M, Sigrist B, Hess M, Van LB, Hoop RK, Vögtlin A. Development of a new real-time polymerase chain reaction assay to detect duck adenovirus a DNA and application to samples from Swiss poultry flocks. J Vet Diagn Investig. 2014;26(2):189–94.

26. Huang J, Tan D, Wang Y, Liu C, Xu J, Wang J. Egg drop syndrome virus enters duck embryonic fibroblast cells via clathrin-mediated endocytosis. Virus Res. 2015;210:69–76.

16

Geno- and seroprevalence of *Felis domesticus* Papillomavirus type 2 (FdPV2) in dermatologically healthy cats

Marco Geisseler[1,2,3], Christian E. Lange[1,2,4], Claude Favrot[2], Nina Fischer[2], Mathias Ackermann[1] and Kurt Tobler[1*]

Abstract

Background: Papillomaviruses can cause proliferative skin lesions ranging from benign hyperplasia to squamous cell carcinoma (SCC). However, asymptomatic infection is also possible. Several groups have detected *Felis domesticus* Papillomavirus type 2 (FdPV2) DNA in association with feline Bowenoid in situ carcinoma (BISC). Therefore, a causative connection has been suggested. However, the knowledge about FdPV2 epidemiology is limited. The aim of this study was to describe the genoprevalence and seroprevalence of FdPV2 in healthy cats.

For this purpose an FdPV2-specific quantitative (q)PCR assay was developed and used to analyse Cytobrush samples collected from 100 dermatologically healthy cats. Moreover, an ELISA was established to test the sera obtained from the same cats for antibodies against the major capsid protein (L1) of FdPV2.

Results: The genoprevalence of FdPV2 was to 98 %. Surprisingly, the quantities of viral DNA detected in some samples from the healthy cats exceeded the amounts detected in control samples from feline BISC lesions. The seroprevalence was much lower, amounting to 22 %. The concentrations of antibodies against FdPV2 were relatively low in healthy cats, whereas they were very high in control cats with BISC.

Conclusion: These observations suggest that FdPV2 is highly prevalent, even among healthy cats. However, cats that carry it on their skin mount in most instances no antibody response. It might be hypothesized that FdPV2 is only rarely productively replicating or its replication is only rarely exposed to the immune system.

Keywords: Cat, FdPV2, BISC, Papillomavirus, Prevalence

Background

Papillomaviruses (PVs) are small non-enveloped DNA viruses. They possess a double-stranded, circular genome of approximately 8 kilobasepairs (kbp), typically divided into an early (E) and late (L) region. The early regions encode viral regulatory proteins (E1, E2, E6, and E7) whereas the late regions encode the capsid proteins (L1 and L2). The capsid is of icosahedral shape and consists of major capsid protein L1, organised in 72 pentameric subunits, and minor capsid protein L2 [1].

PVs can be found in various higher vertebrates including mammals, birds and reptiles [2, 3]. Most host species can be infected by multiple different PV species and types but for few exceptions, PVs tend to be highly species specific [1, 4–6]. By 2007, seven PVs specific for *Felidae* were described. They were found in six different animal species, namely *Felis catus*, *Puma concolor* (Cougar), *Lynx rufus* (Bobcat), *Panthera leo* (Lion), *Neofelis nebulosa* (Clouded Leopard), and *Unica unica* (Snow Leopard) [7, 8]. All *Felidae* PVs that were actually sequenced at that time were classified into the genus *Lambdapapillomavirus* [2] and results from phylogenetic analysis proposed a coevolution of these viruses with their hosts [9].

The first partial sequences of the second feline PV were reported in 2006 [10]. After sequencing its whole genome in 2007, it was named feline PV type 2 (FdPV2) and classified into the newly created genus *dyo-Thetapapillomavirus* [2, 11]. The virus is also referred to as Felis catus PV type 2 (FcaPV2). Meanwhile, the number of

* Correspondence: kurt.tobler@uzh.ch
[1]Institute of Virology. Vetsuisse Faculty, University of Zurich
Winterthurerstrasse 266a, 8057 Zurich, Switzerland
Full list of author information is available at the end of the article

PVs specific for the domestic cat has increased to four [2, 12, 13]. Recently, BPV14 isolated from a domestic cat suffering from feline sarcoid was sequenced on its entire genome length [14]. A cross-species infection of BPV14 in cats was therefore suggested.

Clinically, PVs show a specific cellular tropism for squamous epithelial cells [1]. They can cause proliferative lesions ranging from benign warts to squamous cell carcinoma (SCC) [1, 2]. Initially, FdPV2 DNA had been solely detected in feline Bowenoid in situ carcinomas (BISC). BISC is a rare premalignant state of SCC [15, 16]. BISC is a non-painful, pigmented, plaque like lesion within the haired skin. It can occur at any site of the body and there are usually multiple ones. In some BISC was reported as partially alopecic and covered by crusts [15–18]. Histologically, the neoplastic cells are limited to the epidermis leaving the basement membrane still intact [16]. Surgical excision seems to be curative and no cases of metastasis have been reported so far. However, there are some reports of BISC that were left untreated and progressed to infiltrative SCC [16, 17].

The knowledge about the epidemiology of FdPV2 infections is still limited. Since its discovery, FdPV2 DNA has been found in BISC by various research groups with prevalence ranging from 18 % to 100 %. It was repeatedly amplified from viral plaques. Viral plaques are uncommon, non-neoplastic skin lesions that are clinically indistinguishable from BISC. Although complete regression has been reported, viral plaques are assumed to be precursor lesions of BISC [19, 20]. These studies overall support a causative role of FdPV2 in the development of viral plaques and BISC. However, most of these studies only include small numbers of cats. Furthermore, FdPV2 could also be found in other types of feline skin lesions. As in BISC, the determined prevalence rates of FdPV2 in these other lesions show a rather wide variety, comparable to those found in BISC lesions [4, 5, 18, 19, 21–23]. Only a few studies included samples from cats' normal skin. Amplification of PV DNA using broad range primers always failed. However, in one study, a set of specific primers was used. FdPV2 DNA could be amplified from 52 % of the samples [24] Furthermore, FdPV2 DNA prevalence in eleven queens and their kittens was reported to be 100 % and 91 %, respectively [25]. This study demonstrated the high prevalence of FdPV2 in the cat population.

There are no reports about the seroprevalence of FdPV2. Indeed, no assay for the measure of FdPV2 specific antibodies has been developed so far. We therefore established a Glutathione-S-Transferase (GST) capture ELISA [26] for detection of antibodies directed against the major capsid protein L1 of FdPV2. We have previously used this technique to determine the seroprevalence of

Canine Papillomavirus (CPV) 1 and CPV3 [27] and Equine Papillomavirus 2 (EcPV2) [28] in corresponding populations.

The aim of the present study was to determine the prevalence of FdPV2 in cats that do not suffer from any dermatological conditions. First, the aim was to determine the genoprevalence on skin. Secondly, we wanted to determine the seroprevalence of FdPV2. Finally, the genoprevalence and the seroprevalence of individual cats were compared.

Methods

Sampling of cats

With the owners consents, we sampled 125 cats, that were presented to the Clinic of Small Animals, Vetsuisse-Faculty, Zurich, Switzerland. In order to screen "healthy" cats with respect to PV infections. Furthermore, we included only cats without skin diseases nor any conditions impairing the immune system such as hypersensitivity, auto-immunity, neoplasia, immuno-modulatory treatments, FeLV, FIV and FIPV infections. From these 100 healthy cats, 60 were male (44 castrated) and 40 were female (20 spayed). Cat ranged in age from three months to 17 years with a median of seven years. Twenty-two cats were less than 1 year old. The age of 10 cats was unknown. Seventy-six cats were mixed breeds and 24 cats were purebred cats or descendants of two different purebred cats, respectively.

Skin cell samples were taken with a Cytobrush cell sampler (Deltalab; Barcelona, Spain). Two samples were taken from each cat. The first sample was taken from the haired skin around the mouth in the area where the left vibrissae are located. The second sample was taken from the right front paw, interdigitally between P3 and P4. If the described areas were not accessible for any reason (e.g. injury or bandage), the corresponding areas on the contralateral side were used for sampling. Briefly, a Cytobrush was wetted in 0.9 % sterile NaCl solution and rubbed with rotating movement for 30 s on the skin of the described area. The handle of the Cytobrush was then cut off and the brush part placed in a sterile 1.5 ml Eppendorf tube.

Serum samples were taken during routine diagnostics not related to our study or when a new intravenous catheter was placed. Animals with a known or suspected history of immunodeficiency or under treatment with immunosuppressive drugs were not included. If a complete blood count of a candidate was available, it was checked and cats suspected immunodeficiency were excluded.

Two cats with lesions that had been histologically confirmed as BISC served as positive controls. The Cytobrush samples were taken directly from the BISC lesions. One cat was sampled at two lesions on the neck whereas the other cat was sampled at one lesion on the forehead so that in a total of three samples were obtained. Serum samples were taken during routine diagnostics.

As a negative control, Cytobrush and serum samples were taken from 5 specific pathogen-free (SPF) cats [29]. The Cytobrush samples were taken from the same locations as described above. All serum and Cytobrush samples were stored at −20 °C until further analysis.

PCR

DNA was extracted from the Cytobrush samples using QIAamp® DNA Mini Kit (Qiagen; Basel, Switzerland) according to the manufacturer's protocol but with double amount of buffer ATL, proteinase K, buffer AL and ethanol. The extracted DNA was finally dissolved in 100 µl of buffer AE.

Quantitative real-time PCR (qPCR) was performed using the iCycler iQ™ Real-Time PCR Detection System (Bio-Rad; Hercules CA, USA). Reactions contained 10 µl iQ™ SYBR® Green Supermix (Bio-Rad; Hercules CA, USA), 0.6 µl forward primer (10 µM; fdpv2_qpcr_for: 5′-CAG CTC CCA GTC TCC TAA CG-3′), 0.6 µl reverse primer (10 µM; fdpv2_qpcr_rev: 5′-GCT GTG CCA TTA TCT GAG CA-3′), 3.8 µl sterile water and 5 µl template DNA. Negative controls contained no template DNA but additional 5 µl of sterile water. The following amplification conditions were used: 3 min at 95 °C, 41 cycles of 10 s at 95 °C and 30 s at 60 °C and 1 cycle of 1 min at 95 °C and 1 min at 55 °C. Afterwards temperature was raised by 0.5 °C per cycle during 84 cycles of 10 s to create the melt curve.

As a reference gene, feline glyceraldehyde-3-phosphate dehydrogenase (GAPDH) was chosen. A set of primers (gapdh_qpcr_for: 5′-GTG GAG GGA CTC ATG ACC AC-3′ and gapdh_qpcr_rev: 5′-GTG AGC TTC CCA TTC AGC TC-3′) was designed to amplify cat's GAPDH. qPCR was performed using the same protocol as described above.

Calibration curves were created with dilution series of plasmid DNA. For FdPV2, the plasmid containing the entire FdPV2 DNA was used. For GAPDH, an amplimer of a PCR reaction (with the primers gapdh_for: 5′-TCA TCA TCT CTG CCC CTT CT-3′ and gapdh_rev: 5′-GTG AGC TTC CCA TTC AGC TC-3′) was cloned, sequenced and then used as template DNA for calibration curve creation.

Antigen production for ELISA

The FdPV2 L1 coding sequence (CDS), lacking the first ten (5′) codons, was amplified by PCR from the cloned whole genome of FdPV2 [11] using Phusion™ High-Fidelity DNA Polymerase (Finnzymes; Espoo, Finland). Flanking BamHI sites at the ends of the amplimer,

introduced by the primers (fdpv2_L1_for: 5′-CGA CGG ATC CTT ATA TCT CCC ACC CTC CCC TG-3′ and fdpv2_L1_rev: 5′-AAT AGG ATC CTC ATT TGC GGG TGC GTT-3′), facilitated the cloning into the BamHI site of the pGEX-6P-1 vector (Pharmacia Biotech; Uppsala, Sweden). Protein expression in E.coli strain BL21(DE3), which express the T7 polymerase upon IPTG induction, was performed as described previously with minor modifications [30]. In brief, bacteria were grown in LB medium containing 100 µg/ml Ampicillin at 25 °C with shaking up to an OD_{600} of 0.3 when protein expression was induced by adding 0.25 mM isopropyl-β-D-thio-galactoside (IPTG) and incubated over night at 25 °C with shaking. Pelleted bacteria were resuspended in 1/10 of the culture volume of buffer L (40 mM Tris pH 8.0, 200 mM NaCl, 1 mM EDTA and 2 mM DTT) supplemented with Complete Protease Inhibitor Cocktail (Roche; Mannheim, Germany) and lysed by sonication. ATP (2 mM) and $MgCl_2$ (5 mM) were added and the lysate was incubated for 1 h at room temperature. Urea was slowly added over 5 min to a final concentration of 3.5 M. After incubation of 2 h at room temperature, the mixture was dialysed over night at 4 °C against buffer L using 7 K MWCO Slide-A-Lyzer® Dialysis Cassettes (Thermo Scientific; Rockford IL, USA). After centrifugation the obtained antigen mix was diluted 1:1 with glycerol and stored at −20 °C.

The protein expression procedure was simultaneously performed with three different E.coli strain BL21(DE3) cultures containing different pGEX-6P-1 vector derivatives. The first contained the FdPV2 L1 CDS fused to the GST CDS whereas the second contained a CPV1 L1 CDS fused to the GST CDS [27]. The third culture contained the GST CDS only. All ELISA assays reported in this study were performed with antigen from the same lot of antigen production.

GST capture ELISA

Throughout the protocol, plates were washed three times with PBS buffer supplemented with 0.3 % Tween 20 (PBS-T) between every incubation step. Polysorb 96-well plastic plates (Nunc; Roskilde, Denmark) were prepared for the ELISA. They were coated at 4 °C over night with 50 mM sodium carbonate buffer pH 9.6 containing 0.2 % glutathione casein (kindly provided by Martin Müller DKFZ, Heidelberg, Germany) and then blocked at 37 °C for 1 h with casein buffer (PBS-T containing 0.2 % casein). The GST tagged antigen, diluted 1:10 in casein buffer, was applied to the plates and incubated at 37 °C for 1 h.

Prior to ELISA, the sample sera had been diluted 1:500 in casein buffer, mixed with an equivalent of lysed untransformed E.coli strain BL21 (DE3) and incubated at 4 °C for 30 min to block reactions with contaminating bacterial proteins [26, 27, 30]. The sera, cleared by centrifugation, were applied to the plates and incubated at 37 °C for 1 h. Goat Anti-Feline IgG conjugated to Horseradish Peroxidase (HRP) (Southern Biotech; Birmingham AL, USA) diluted 1:1000 in casein buffer was added as secondary antibody and the plates were incubated again at 37 °C for 1 h. After six final washes with PBS-T, substrate (78 mM CH_3COOH, 24 mM CH_3COONa, 50 mM NaH_2PO_4, 2 mM ABTS [Roche; Rotkreuz, Switzerland] with 1.25 mM H_2O_2 applied shortly before use) was added. Absorbance was measured after 45 min at 405 nm in a Sunrise™ microplate reader (Tecan; Männedorf, Switzerland).

The cat sera were tested in triplicates against the antigen FdPV2 L1-GST and, as a negative control, against CPV1 L1-GST. For a subset of samples the ELISA was repeated. The according samples were then tested in duplicates against CPV1 L1 and against GST alone. In order to normalize the results of the different plates, the same positive and negative control sera were used on every plate. No serum was added in six wells serving as a plate control.

Data analysis and presentation

The C_q-values obtained from qPCR were converted into absolute numbers of copies of FdPV2 and GAPDH in each sample using the equation of the corresponding calibration curve. The C_q values of samples revealing no amplification within forty qPCR-cycles were set to 40 for further calculations. In order to obtain comparable results, in each sample the absolute number of FdPV2 copies was divided by the corresponding absolute number of GAPDH copies.

Serum samples were tested in triplicates in ELISA. To prevent outlier results from influencing the data, the median of the three observed values was used for further analysis. Plate to plate variability was compensated by dividing every value by the mean of the control sera values from the corresponding plate and multiplying the result by the mean of all control sera from all plates.

A cut-off value (COV) was set by the mean of all negative control samples plus two standard deviations. Figures were generated using R (Free Software Foundation; Boston, USA).

Results

Genoprevalence of FdPV2

In order to test the skin samples for the presence of FdPV2 specific DNA, a qPCR assay was developed. First, calibration curves for qPCR of FdPV2 and GAPDH DNA were determined (Additional file 1: Figure S1). Second, the cut-off-value (COV) for the ratio of FdPV2 to GAPDH was set as the mean of the negative control

samples plus two standard deviations of the mean from the negative control samples. Therefore, the COV was set at 0.367 and the $\log_2(COV)$ at −1.446, respectively. The newly developed qPCR assay was applied for the measurement of viral DNA in the Cytobrush samples collected from the healthy sample population. Eleven samples were excluded from the analysis because an unspecific by-product was amplified or the amplification of GAPDH failed and consequently the calculation of the copy number of FdPV2 per GAPDH was not possible. Out of the 200 DNA samples, 189 could therefore be used for the study representing all cats with at least one sample.

The genoprevalence within the sample population was evaluated considering the ratios of FdPV2 to GAPDH DNA copies as determined by the analysis of the qPCR-results. The log-transformed ratios are shown as box plots in Fig. 1. The medians of the negative controls were significantly lower than those from the positive controls The values of the negative and positive controls range from $2.0 \cdot 10^{-4}$ to $8.8 \cdot 10^{-2}$ and $3.7 \cdot 10^0$ to $1.7 \cdot 10^2$, respectively. The medians of the samples from the head as well from the paw were between the positive and the negative controls. Among the 189 samples used for further analysis, the calculated FdPV2 DNA copies per GAPDH varied from $4.0 \cdot 10^{-4}$ to $9.3 \cdot 10^4$ in the samples from the head and from $5.9 \cdot 10^{-3}$ to $2.0 \cdot 10^5$ in the samples from the paw.

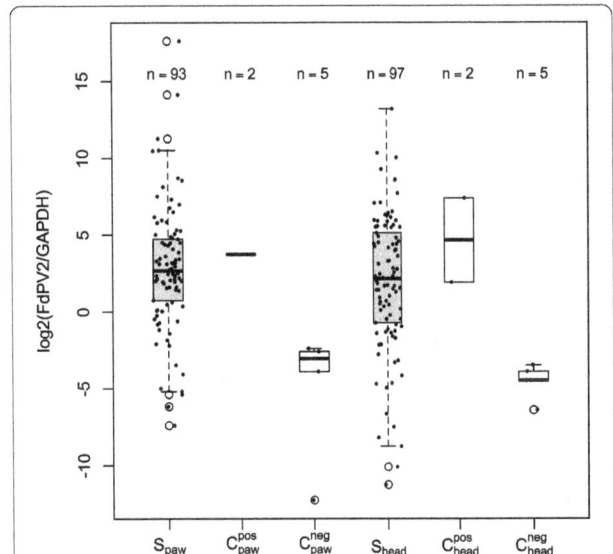

Fig. 1 Boxplot of the log-transformed ratios of FdPV2 to GAPDH molecule numbers. The positive controls (C_{paw}^{pos}, C_{head}^{pos}) and the negative controls (C_{paw}^{neg}, C_{head}^{neg}) are shown as white boxes and the samples (S_{paw}, S_{head}) as grey boxes. The solid bar represents the median, the box range from the first to the third quartiles, and the whiskers extend to the lowest and highest datum still within 1.5 times the interquartile range. Data not included within the whiskers are individually presented

No relationship between the copy numbers in head and paw samples could be found. A high number of copies in the head sample was not necessarily accompanied by a high number of copies in the corresponding paw sample and vice versa.

Of the 189 samples used in the analysis, 169 samples were above the COV whereas 20 samples remained below. Of these 20 samples, 13 were taken from the head and seven from the paw. Ninety-eight cats had at least one positive sample and were therefore counted as FdPV2 DNA positive. The two FdPV2 DNA negative cats both had two samples of sufficient quality (according to the exclusion criteria mentioned above) and did not have any relation to each other. Summarized, the DNA prevalence of FdPV2 in the studied population was determined to be 98 %.

Seroprevalence of FdPV2

In order to test the serum samples for the presence of FdPV2 specific antibodies, an ELISA was developed and applied. First, the antigen coating of the plates and the measurement of antibodies in the control sera were tested. Second, the antibody titres of all samples were determined and normalized. Third, the COV was set as the mean plus two standard deviations of the mean from the negative control samples. Fourth, the seroprevalence of the sample population was determined.

The serum samples were screened for antibodies against FdPV2 and, as a negative control, against CPV1 using a GST capture ELISA. The sample sera reacted against FdPV2 producing an OD ranging from 0.154 to 1.094 (mean = 0.301) and against CPV1 with an OD ranging from 0.162 to 1.096 (mean = 0.242). The COV was set at of 0.315 for FdPV2 and 0.472 for CPV1. The reactions against FdPV2 of 24 serum samples were above the COV and could thus be considered positive while 76 serum samples were counted as negative. Two serum samples showed a reaction against CPV1 above the according COV.

To incorporate the CPV1 control in the data analysis, the corrected OD values of the FdPV2 specific ELISA were plotted against the corrected OD values of the CPV1 specific ELISA (Fig. 2). As mentioned above, two of the 24 positive serum samples showed a reaction against CPV1 with an OD above the according COV (circles on the right side of the vertical and above or close to the diagonal line on Fig. 2). The ELISA was repeated with these samples. CPV1 L1 (tagged to GST) and GST alone were used as antigens. In both samples the reactions against GST alone were as strong as the reactions against CPV1 L1-GST. The samples were therefore categorised as negative for antibodies against CPV1 as well as against FdPV2. The seroprevalence of FdPV2 was corrected down to 22 %.

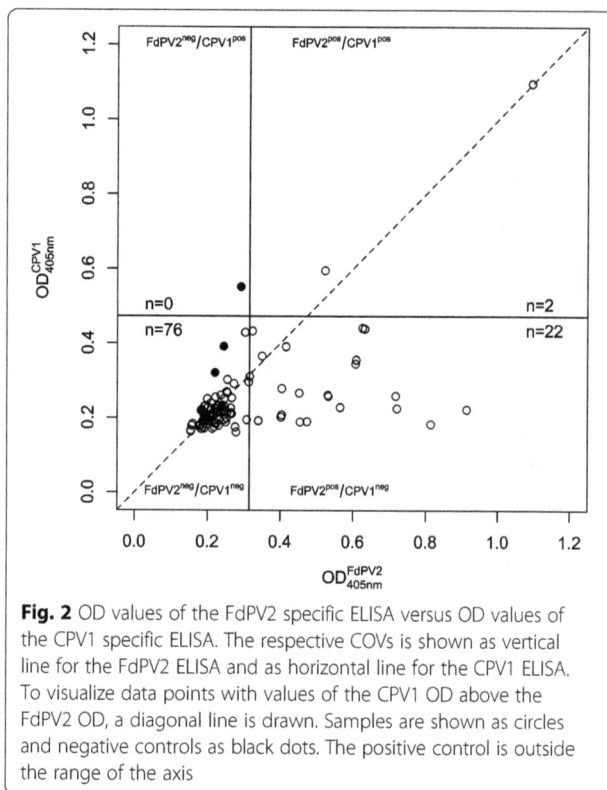

Fig. 2 OD values of the FdPV2 specific ELISA versus OD values of the CPV1 specific ELISA. The respective COVs is shown as vertical line for the FdPV2 ELISA and as horizontal line for the CPV1 ELISA. To visualize data points with values of the CPV1 OD above the FdPV2 OD, a diagonal line is drawn. Samples are shown as circles and negative controls as black dots. The positive control is outside the range of the axis

The OD of the negative control sera ranged from 0.184 to 0.291 (mean = 0.225) and the one of the positive control sera from 1.743 to 2.041 (mean = 1.892) in the FdPV2 specific ELISA. Bonferroni statistical test was used to compare the mean OD of the sample sera with those of the positive and negative control sera, respectively. The mean OD of the positive control sera was significantly higher than the mean OD of the sample sera ($p > 0.001$), whereas the mean OD of the negative control sera did not differ significantly from the mean OD of the sample sera ($p = 0.986$). Fisher's Exact Test was used to compare subgroups of the sample population. No difference in seropositivity could be found between purebred and mixed breed cats ($p = 0.386$), nor between male and female cats ($p = 0.448$), nor between intact individuals and neutered ones ($p = 0.061$). The seropositive cats had a median age of 12.0 years. This is significantly older (Univariate Analysis of Variance, $p > 0.001$) than the negative cats that had a mean age of 4.3 years. Yet, the age of three positive and seven negative cats was not known.

Correlation of genoprevalence to seroprevalence

Results obtained from ELISA and qPCR assay were compared. The log-transformed ratios of the Cytobrush samples isolated from the head were plotted against the ones from the paw and the size of the dot corresponded to the OD value of the FdPV2 specific ELISA (Fig. 3). Data

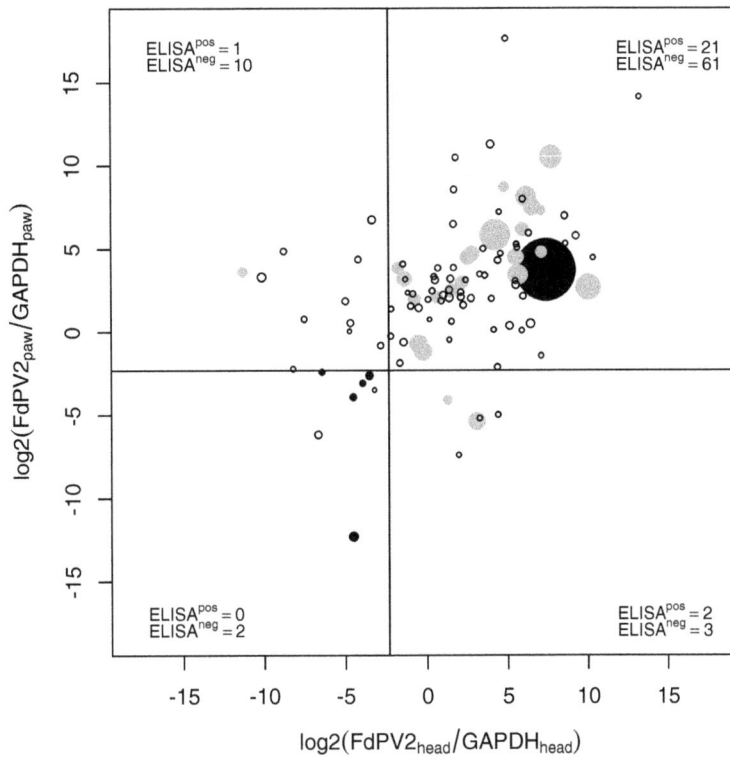

Fig. 3 Scatter plot of log-transformed qPCR copy numbers ratios of FdPV2 DNA to GAPDH DNA in head and paw samples and the according ELISA OD value from the serum sample. White circles represent FdPV2 ELISA negative cats from sample population, grey dots represent FdPV2 ELISA positive cats, and black dots represent positive and negative control cats. The drawing sizes of the circles/dots indicate the ELISA OD value. The vertical and horizontal lines are set at the level of the COV (0.202). The numbers of FdPV2 ELISA positive and negative cats in each of the four categories describing the genoprevalence of FdPV2 on head and paw are specified

points corresponding to cats with similar ratios of FdPV2 DNA to GAPDH DNA on paw and head would lie close to a diagonal line through the origin with a slope of 1. Such a correlation was not obvious. However there was a tendency of FdPV2 ELISA positive cats within the region of qPCR positive in paw- and head-samples and FdPV2 ELISA negative cats within the region of negative qPCR samples. Some cats had different ratios of FdPV2 and GAPDH DNA in the paw- and head-samples though. Cats tested positive in the ELISA showed a large variation in PCR results. The lowest observed copy number of FdPV2 per GAPDH was as low as $4.0 \cdot 10^{-4}$ whereas the highest value was $1.4 \cdot 10^{3}$. Two seropositive cats had less than one copy of FdPV2 per GAPDH in both samples and two other had one negative PCR sample each. None of the positive control samples had more than 200 copies of FdPV2 per GAPDH. The lowest FdPV2 DNA per GAPDH content in a sample from the seronegative cats was $9.0 \cdot 10^{-4}$, while the highest was $2.0 \cdot 10^{5}$. One seronegative cat had more than $9.0 \cdot 10^{3}$ copies of FdPV2 per GAPDH in both samples. The two cats that remained DNA negative were tested negative in ELISA as well. Overall the seronegative cats had on average almost 15 times as many copies of FdPV2 DNA (mean $1.4 \cdot 10^{3}$) than seropositive cats (mean $1.0 \cdot 10^{2}$).

Discussion

Several studies, which evaluate the genoprevalence of PVs in cats are published and listed in Table 1. Most of these papers address the presence of PV DNA in various lesions and only two were done on healthy skin samples. Consistently, the prevalence of DNA in SCC is high while it is hardly reported in unrelated lesions and healthy tissue. Instead of generating a Boolean data set for the genoprevalence, our study assessed the relative amount of FdPV2 DNA to genomic DNA. Such a quantification of viral DNA was previously proposed to rule out a possible causative role of FdPV2 in certain types of lesions since FdPV2 DNA was detected in normal feline skin [24]. Indeed, recently a qPCR-approach was used to quantify the FdPV2 DNA from swap samples of eleven queens and their kittens [25]. In our study we could amplify specific FdPV2 DNA above our threshold from 98 % of the healthy skin samples. This DNA prevalence was thus much higher than most previously reported ones. Only one study detected 52 % [24] and another one 91 % to 100 % FdPV2 DNA positive healthy skin

Table 1 Rates of reported PV DNA findings in skin samples from cats using PCR

Study	Primer set	BISC	Viral plaques	SCC	ISCC	OSCC	Other lesions	Normal skin
Kidney, 2001 [44]	NO1/NO2						0/50 (0 %)	
	E5+/E5−						0/50 (0 %)	
	IFNR-2/IDNT-2						0/50 (0 %)	
Antonsson, 2002 [41]	FAP59/FAP64							0/5 (0 %)
Nespeca, 2006 [10]	PapF/PapR	1/21 (5 %)			0/22 (0 %)		0/11 (0 %)	
	CP4/CP5/PPF1	5/21 (24 %)			4/22 (18 %)		0/11 (0 %)	
Munday, 2007 [15]	FAP59/FAP64	11/18 (61 %)					0/15 (0 %)	0/3 (0 %)
	IFNR-2/IDNT-2	9/11 (82 %)					0/15 (0 %)	0/3 (0 %)
		5/18 (28 %)						
Munday, 2008 [18]	a	20/20 (100 %)			17/20 (85 %)		3/17 (18 %)	
Munday, 2008 [33]	FAP59/FAP64		2/2 (100 %)					
	MY09/MY11		2/2 (100 %)					
	JMPF/JMPR		2/2 (100 %)					
Lange, 2009 [11]	A16/A37	3/3 (100 %)						
Munday, 2009 [31]	FAP59/FAP64					1/20 (5 %)	0/20 (0 %)	
	IFNR-2/IDNT-2					0/20 (0 %)	0/20 (0 %)	
	MY09/MY11					0/20 (0 %)	0/20 (0 %)	
Munday, 2009 [21]	FAP59/FAP64			1/1 (100 %)				
	JMPF/JMPR			1/1 (100 %)				
Anis, 2010 [4]	b	3/3 (100 %)			5/5 (100 %)	1/1 (100 %)	1/1 (100 %)	
		3/3 (100 %)			4/5 (80 %)			
Munday, 2009 [24]	MY09/MY11							0/44 (0 %)
	JMPF/JMPR							34/88 (39 %)c
Munday, 2010 [19]	MY09/MY11		4/14 (29 %)				0/14 (0 %)	
	FAP59/FAP64		4/14 (29 %)				0/14 (0 %)	
	JMPF/JMPR		14/14 (100 %)				1/14 (7 %)	
Munday, 2010 [45]	MY09/MY11						4/7 (57 %)	0/120 (0 %)
	jmpSA-F/jmpSA-R						6/7 (86 %)	0/120 (0 %)
Munday, 2011 [22]	MY09/MY11			7/70 (10 %)				
	JMPF/JMPR			33/70 (48 %)				
Munday, 2011 [23]	a	14/14 (100 %)	8/8 (100 %)		12/18 (67 %)		1/14 (7 %)	
Munday, 2011 [46]	FAP59/FAP64					0/30 (0 %)		
	MY09/MY11					0/30 (0 %)		
O'Neill, 2011 [5]	d	7/22 (32 %)			11/74 (15 %)		2/12 (17 %)	
		4/22 (18 %)			4/74 (5 %)			
Schwittick, 2011 [47]	a	1/1 (100 %)						
Munday, 2013 [12]	FAP59/FAP64	0/1 (0 %)						
	JMPF/JMPR	0/1 (0 %)						
	MY09/MY11	1/1 (100 %)						
	JMY2F/JMY2R	1/1 (100 %)						

Table 1 Rates of reported PV DNA findings in skin samples from cats using PCR *(Continued)*

Dunowska, 2014 [13]	FAP59/FAP64		0/1 (0 %)
	MY09/MY11		1/1 (100 %)
	Unnamed set		1/1 (100 %)
Thomson, 2015 [25]	E7SF/E7SR		*11/11 (100 %)*
			20/22 (91 %)

FdPV2 DNA detection are written in italic

BISC Bowenoid in situ carcinoma, *SCC* squamous cell carcinoma, *ISCC* infiltrative squamous cell carcinoma (when defined in study), other lesions = actinic keratosis, allergic dermatitis, apocrine gland cyst, apocrine gland cystadenoma, dermatophytosis, dysplasia, eosinophilic granuloma, eosinophilic plaques, feline leprosy, fibrosarcoma, glossitis, granulomatous dermatitis, hyperplastic tonsil, hypersensitivity dermatitis, mast cell tumour, melanoma, periodontal disease, plasmacytic stomatitis, sarcoids, trichoblastoma, ulcerative gingivitis

[a]nested PCR with FAP59/FAP64 and JMPF/JMPR
[b]HPV1, HPV2, HPV4, HPV7, HPV8, HPV10, HPV15
[c]equals 23/44 cats (52 %)
[d]FAP59/FAP64 and nested PCR with CP65/CP70 and CP66/CP69

samples [25]. The discrepancy between the various studies might have three primary reasons. First, several different primer sets were used to amplify PV specific DNA in the studies. As a consequence, it is difficult to directly compare the reported prevalences. In particular, only using FdPV2 specific primers but not degenerated ones allow the amplification of FdPV2 DNA [25, 31]. Second, DNA extraction in other studies was partly done from formalin fixed tissue samples. Formalin-fixation tends to degrade DNA, which might lead to a reduced sensitivity of detection [32]. Third, samples from healthy skin were typically collected using cotton-swabs. Indeed, Munday et al. showed that amplification of PV DNA from cotton-tipped swabs was a more sensitive method than amplification from formalin-fixed tissue [33]. However, we used Cytobrushes for the collection of cells from the surface of healthy skin. Cytobrushes are intended to take cell samples from mucosae. Chalvardjian et al. compared the uptake of cells of cotton-tipped swabs and Cytobrushes by performing endocervical sampling in women. The samples taken with a Cytobrush contained on average at least seventeen times more endocervical cells than the samples taken with a cotton-tipped swab in 87 % of the cases [34]. In our study, only 3 out of 200 Cytobrush samples were negative in qPCR for GAPDH and, therefore, 197 samples were apparently of sufficient quality. This demonstrates that Cytobrushes are a very useful and efficient tool for collecting cell samples even from the haired skin.

In a further set of experiments, we determined the seroprevalence of FdPV2. We found 22 % of the animals positive for FdPV2 specific antibodies in the investigated population of healthy cats. Similar ratios of seropositive animals were detected among healthy horses for antibodies against EcPV2 (28 %) [28], and healthy dogs with antibodies against CPV1 (22 %) or CPV3 (27 %) [27] and koalas (20 %) [35]. Likewise, antibodies against different types of HPV in an Australian human cohort was found

- although showing differences depending on age and gender - in the range of 0 % to 22 % positive individuals [36]. Neutralizing antibodies against EcPV2 were detected in 15 % of apparently healthy horses [37]. As negative controls, we used CPV1 L1 GST fusion protein and GST alone. Two of the cat sera reacted as strong to both of these proteins as to FdPV2 GST. GST is an enzyme that plays a key role in cellular detoxification. It is not only present in mammals but also in fungi, helminths and bacteria [38, 39]. Therefore it can be hypothesized that the mentioned two serum samples contained antibodies against GST. Consequently, these sera were not considered positive, neither for FdPV2 nor for CPV1.

Comparing the individual cat's results from the ELISA and the qPCR assay, a certain correlation might be expected. In humans, individuals with a high HPV DNA load are more prone to be also seropositive [40]. Likewise, contact with FdPV2 might as well induce the production of antibodies against it. Therefore, cats with a high virus load were expected to be more often seropositive. In our study, only a very weak correlation between virus load and seropositivity could be found. Indeed, significantly higher amounts of FdPV2 DNA were detected in some seronegative cats.

Asymptomatic PV infections are quite common. Up to 80 % of humans are infected asymptomatically with human PVs [18] and various animal species have also been investigated regarding such PV infections [25, 41–43]. Though, the prevalence varies from species to species. Interestingly, primates showed prevalence rates similar to that of humans [41]. Prior to the amplification of FdPV2 from the normal skin of cats, asymptomatic infections were not considered to be common. The "hit-and-run" carcinogenesis model was used to explain transformation of normal skin cells into neoplastic cells by transient PV infection [15]. It is not clear if the extracted PV DNA originates from infected epithelium cells or from virions that were attached to the skin's surface. Assuming the amount of asymptomatically infected animals based only on PCR results might therefore be inaccurate. However, it can be stated that 98 % of the cats had certainly been in contact with FdPV2. Considering the high tenacity of PV particles, it might be hypothesized that high amounts of virus are shed to build up an infectious virus reservoir on biological surfaces.

In our study the seropositive cats were significantly older than the seronegative cats. This correlates well with the existing data about BISC being more common in old cats [16, 17]. Seroconversion and disease development does not seem to be dependent on the virus load alone. This weakens again the hypothesis of a straightforward role of FdPV2 causing BISC. The virus might as well be a part of the normal skin flora of cats. The immune system would then not produce antibodies against it as long as the skin is intact. Suggesting that older cats have had more skin traumas in their lives, might further support this theory.

So far, the causative role of FdPV2 in the development of BISC has only been supported by repeatedly amplifying FdPV2 DNA from BISC samples. Here, we add one more argument for this, which may also be used as a clinical marker, namely the observation of an elevated antibody response going along with the development of malignancies. The very high genoprevalence found in our and in the Thompson study [25] implies that the virus is widespread in the cat population, irrespective of the health status of the cat. Thus, it is still possible that FdPV2 represents just a simple bystander of another yet unrecognized causative agent. Furthermore, we are not able to distinguish between intracellular DNA and viral nucleic acid just laying naked or within virions on the surface of the skin. Collection of different samples representing specifically the skin surface or skin tissue from the same patient is desirable. As such, multiple tests can be done and the results can be compared to reveal the replicative state of the viral genomes detected by PCR. Moreover, patients should be sampled repeatedly at different points in time to gain information about the interaction of the virus with the cat's body. Studies with large sample population are though needed in order to receive reliable data. Finally, this would allow to obtain a better comprehension of the virus' epidemiology.

Conclusion

The observed genoprevalence of FdPV2 in healthy cats was 98 %, while the seroprevalence was 22 %. Cats that carry the virus on their skin mount only rarely an antibody response. It might be hypothesized that though the virus is highly prevalent in the cat's population, it rarely leads to an immune response due to the fact that it is not productively replicating or the replication is sheltered from the exposure to the immune system.

Additional file

> **Additional file 1: Figure S1.** Calibration curves for qPCR. Dilution series of cloned feline GAPDH DNA amplimer and FdPV2 DNA were used as template and qPCR was performed using specific primer sets. The resulting equations of the calibration curves are shown. These equations were used to quantify the results of the qPCR using the same primer sets but the DNA extracted from the Cytobrush samples as templates. (PDF 572 kb)

Acknowledgements
We thank Prof. Dr. Martin Müller (DKFZ, Heidelberg, Germany) for providing the glutathione casein.

Funding
The authors declared that they received no financial support for their research.

Authors' contributions
MG collected the samples, performed the experiments, analysed the primary data, and wrote the manuscript draft. CL contributed to the design of the study, was involved in performing the experiments and reviewed the manuscript. NF was involved in collecting the samples and performing the experiments. CF made contributions to the design of the study and reviewed the manuscript. MA and KT contributed to the design of the study, were involved in analysing the data and reviewed the manuscript. All authors have approved the final version of the article.

Competing interests
The authors declare that they have no competing interests.

Consent for publication
Not applicable.

Author details
[1]Institute of Virology. Vetsuisse Faculty, University of Zurich Winterthurerstrasse 266a, 8057 Zurich, Switzerland. [2]Dermatology Department, Clinic for Small Animal Internal Medicine, Vetsuisse Faculty, University of Zurich, Winterthurerstrasse 260, 8057 Zurich, Switzerland. [3]Present address: Amt für Landwirtschaft und Natur des Kantons Bern, Veterinärdienst, Herrengasse 1, 3011 Berne, Switzerland. [4]Present address: Department of Microbiology and Immunobiology, Harvard Medical School, 77 Avenue Louis Pasteur, Boston, MA 02115, USA.

References
1. Howley PM, Schiller JT, Lowy DR. Papillomaviruses. In: Knipe DM, Howley PM, editors. Fields Virology. sixth. Philadelphia: Wolters Kluwer/Lippincott Williams & Wilkins; 2013. p. 1662–703.
2. Bernard H-U, Burk RD, Chen Z, van Doorslaer K, Hausen zur H, de Villiers E-M. Classification of papillomaviruses (PVs) based on 189 PV types and proposal of taxonomic amendments. Virology. 2010;401:70–9. doi:10.1016/j.virol.2010.02.002. Elsevier Inc.
3. de Villiers E-M. Cross-roads in the classi. Virology. 2013;445:2–10. doi:10.1016/j.virol.2013.04.023. Elsevier.
4. Anis EA, O'Neill SH, Newkirk KM, Brahmbhatt RA, Abd-Eldaim M, Frank LA, et al. Molecular characterization of the L1 geneof papillomaviruses in epithelial lesions of cats and comparative analysis with corresponding gene sequences of human and feline papillomaviruses. Am J Vet Res. 2010;71:1457–61.
5. O'Neill SH, Newkirk KM, Anis EA, Brahmbhatt R, Frank LA, Kania SA. Detection of human papillomavirus DNA in feline premalignant and

invasive squamous cell carcinoma. Vet Dermatol. 2011;22:68–74. doi:10.1111/j.1365-3164.2010.00912.x.

6. Munday JS, Hanlon EM, Howe L, Squires RA, French AF. Feline Cutaneous Viral Papilloma Associated with Human Papillomavirus Type 9. Vet Pathol. 2007;44:924–7.

7. Tachezy R, Duson G, Rector A, Jenson AB, Sundberg JP, Van Ranst M. Cloning and genomic characterization of Felis domesticus papillomavirus type 1. Virology. 2002;301:313–21.

8. Sundberg JP, Van Ranst M, Montali RJ, Homer BL, Miller WH, Rowland PH, et al. Feline Papillomas and Papillomaviruses. Vet Pathol. 2000;37:1–10. doi: 10.1354/vp.37-1-1.

9. Rector A, Lemey P, Tachezy R, Mostmans S, Ghim S-J, van Doorslaer K, et al. Ancient papillomavirus-host co-speciation in Felidae. Genome Biol. 2007;8: R57. doi:10.1186/gb-2007-8-4-r57.

10. Nespeca G, Grest P, Rosenkrantz WS, Ackermann M, Favrot C. Detection of novel papillomaviruslike sequences in paraffin-embedded specimens of invasive and in situ squamous cell carcinomas from cats. Am J Vet Res. 2006;67:2036–41. doi:10.2460/ajvr.67.12.2036.

11. Lange CE, Tobler K, Markau T, Alhaidari Z, Bornand V, Stöckli R, et al. Sequence and classification of FdPV2, a papillomavirus isolated from feline Bowenoid in situ carcinomas. Vet Microbiol. 2009;60–5. doi:10.1016/j.vetmic.2009.01.002. Elsevier B.V

12. Munday JS, Dunowska M, Hills SF, Laurie RE. Veterinary Microbiology. Vet Microbiol. 2013;165:319–25. doi:10.1016/j.vetmic.2013.04.006. Elsevier B.V.

13. Dunowska M, Munday JS, Laurie RE, Hills SFK. Genomic characterisation of Felis catus papillomavirus 4, a novel papillomavirus detected in the oral cavity of a domestic cat. Virus Genes. 2014;48:111–9. doi:10.1007/s11262-013-1002-3.

14. Munday JS, Thomson N, Dunowska M, Knight CG, Laurie RE, Hills S. Genomic characterisation of the feline sarcoid-associated papillomavirus and proposed classification as Bos taurus papillomavirus type 14. Vet Microbiol. 2015;177:289–95. doi:10.1016/j.vetmic.2015.03.019. Elsevier B.V.

15. Munday JS, Kiupel M, French AF, Howe L, Squires RA. Detection of papillomaviral sequences in feline Bowenoid in situ carcinoma using consensus primers. Vet Dermatol. 2007;18:241–5. doi:10.1111/j.1365-3164.2007.00600.x.

16. Baer KE, Helton K. Multicentric squamous cell carcinoma in situ resembling Bowen's disease in cats. Vet Pathol. 1993;30:535–43.

17. Favrot C, Welle M, Heimann M, Godson D. Clinical, Histologic, and Immunohistochemical Analyses of Feline Squamous Cell Carcinoma In Situ. Vet Path. 2009;46:25–33.

18. Munday JS, Kiupel M, French AF, Howe L. Amplification of papillomaviral DNA sequences from a high proportion of feline cutaneous in situand invasive squamous cell carcinomas using a nested polymerase chain reaction. Vet Dermatol. 2008;19:259–63. doi:10.1111/j.1365-3164.2008.00685.x.

19. Munday JS, Peters-Kennedy J. Consistent detection of Felis domesticus papillomavirus 2 DNA sequences within feline viral plaques. J Vet Diagn Invest. 2010;22:946–9.

20. Wilhelm S, Degorce-Rubiales F, Godson D, Favrot C. Clinical, histological and immunohistochemical study of feline viral plaques and bowenoid in situ carcinomas. Vet Dermatol. 2006;17:424–31. doi:10.1111/j.1365-3164.2006.00547.x.

21. Munday JS, Dunowska M, De Grey S. Detection of two different papillomaviruses within a feline cutaneous squamous cell carcinoma: case report and review of the literature. N Z Vet J. 2009;57:248–51. doi:10.1080/00480169.2009.36911.

22. Munday JS, Gibson I, French AF. Papillomaviral DNA and increased p16CDKN2A protein are frequently present within feline cutaneous squamous cell carcinomas in ultraviolet-protected skin. Vet Dermatol. 2011; 22:360–6. doi:10.1111/j.1365-3164.2011.00958.x.

23. Munday JS, French AF, Peters-Kennedy J, Orbell GMB, Gwynne K. Increased p16CDKN2A Protein Within Feline Cutaneous Viral Plaques, Bowenoid In Situ Carcinomas, and a Subset of Invasive Squamous Cell Carcinomas. Vet Pathol. 2011;48:460–5. doi:10.1177/0300985810374844.

24. Munday JS, Witham AI. Frequent detection of papillomavirus DNA in clinically normal skin of cats infected and noninfected with feline immunodeficiency virus. Vet Dermatol. 2009;21:307–10. doi:10.1111/j.1365-3164.2009.00811.x.

25. Thomson NA, Dunowska M, Munday JS. The use of quantitative PCR to detect Felis catus papillomavirus type 2 DNA from a high proportion of queens and their kittens. Vet Microbiol. 2015;175:211–7. doi:10.1016/j.vetmic.2014.11.028. Elsevier B.V.

26. Sehr P, Zumbach K, Pawlita M. A generic capture ELISA for recombinant proteins fused to glutathione S-transferase: validation for HPV serology. J Immunol Methods. 2001;253:153–62.

27. Lange CE, Tobler K, Favrot C, Müller M, Nöthling JO, Ackermann M. Detection of Antibodies against Epidermodysplasia Verruciformis-Associated Canine Papillomavirus 3 in Sera of Dogs from Europe and Africa by Enzyme-Linked Immunosorbent Assay. Clin Vaccine Immunol. 2009;16: 66–72. doi:10.1128/CVI.00346-08.

28. Fischer NM, Favrot C, Birkmann K, Jackson M, Schwarzwald CC, Müller M, et al. Serum antibodies and DNA indicate a high prevalence of equine papillomavirus 2 (EcPV2) among horses in Switzerland. Vet Dermatol. 2014; 25:210–e54. doi:10.1111/vde.12129.

29. Geret PC, Riond B, Cattori V, Meli LM, Hofmann-Lehmann R, Lutz H. Housing and care of laboratory cats: from requirements to practice. Schweiz Arch Tierheilkd. 2013;153:157–64. doi:10.1024/0036-7281/a000175.

30. Sehr P, Müller M, Höpfl R, Widschwendter A, Pawlita M. HPV antibody detection by ELISA with capsid protein L1 fused to glutathione S-transferase. J Virol Methods. 2002;106:61–70.

31. Munday JS, Howe L, French A, Squires RA, Sugiarto H. Detection of papillomaviral DNA sequences in a feline oral squamous cell carcinoma. Res Vet Sci. 2009;86:359–61. doi:10.1016/j.rvsc.2008.07.005.

32. Albini S, Zimmermann W, Neff F, Ehlers B, Hani H, Li H, et al. Identification and Quantification of Ovine Gammaherpesvirus 2 DNA in Fresh and Stored Tissues of Pigs with Symptoms of Porcine Malignant Catarrhal Fever. J Clin Microbiol. 2003;41:900–4. doi:10.1128/JCM.41.2.900-904.2003.

33. Munday JS, Willis KA, Kiupel M, Hill FI, Dunowska M. Amplification of three different papillomaviral DNA sequences from a cat with viral plaques. Vet Dermatol. 2008;19:400–4. doi:10.1111/j.1365-3164.2008.00710.x.

34. Chalvardjian A, De Marchi WG, Bell V, Nishikawa R. Improved endocervical sampling with the Cytobrush. CMAJ. 1991;144:313–7.

35. Antonsson A, McMillan NAJ. Papillomavirus in healthy skin of Australian animals. J Gen Virol. 2006;87:3195–200. doi:10.1099/vir.0.82195-0.

36. Newall AT, Brotherton JML, Quinn HE, McIntyre PB, Backhouse J, Gilbert L, et al. Population Seroprevalence of Human Papillomavirus Types 6, 11, 16, and 18 in Men, Women, and Children in Australia. Clin Infect Dis. 2008;46:1647–55. doi:10.1086/587895.

37. Schellenbacher C, Shafti-Keramat S, Huber B, Fink D, Brandt S, Kirnbauer R. Establishment of an in vitro equine papillomavirus type 2 (EcPV2) neutralization assay and a VLP-based vaccine for protectionof equids against EcPV2-associated genital tumors. Virology. 2015;486:284–90. doi:10.1016/j.virol.2015.08.016. Elsevier.

38. Sheehan D, Meade G, Foley VM, Dowd CA. Structure, function and evolution of glutathione transferases: implications for classification of non-mammalian members of an ancient enzyme superfamily. Biochem J. 2001;360:1–16.

39. Allocati N, Federici L, Masulli M, Di Ilio C. Glutathione transferases in bacteria. FEBS J. 2009;276:58–75. doi:10.1111/j.1742-4658.2008.06743.x.

40. Dillner J. The serological response to papillomaviruses. Semin Cancer Biol. 1999;9:423–30. doi:10.1006/scbi.1999.0146.

41. Antonsson A, Hansson BG. Healthy Skin of Many Animal Species Harbours Papillomaviruses Which Are Closely Related to Their Human Counterparts. J Virol. 2002;76:12537–42. doi:10.1128/JVI.76.24.12537-12542.2002.

42. Antonsson A, Forslund O, Ekberg H, Sterner G, Hansson BG. The ubiquity and impressive genomic diversity of human skin papillomaviruses suggest a commensalic nature of these viruses. J Virol. 2000;74:11636–41.

43. Lange CE, Zollinger S, Tobler K, Ackermann M, Favrot C. Clinically healthy skin of dogs is a potential reservoir for canine papillomaviruses. J Clin Microbiol. 2011;49:707–9. doi:10.1128/JCM.02047-10.

44. Kidney BA, Haines DM, Ellis JA, Burnham ML, Teifke JP, Czerwinski G, et al. Evaluation of formalin-fixed paraffin-embedded tissues from vaccine site-associated sarcomas of cats for papillomavirus DNA and antigen. Am J Vet Res. 2001;62:833–9.

45. Munday JS, Knight CG, Howe L. The Same Papillomavirus is Present in Feline Sarcoids from North America and New Zealand but Not in Any Non-Sarcoid Feline Samples. J Vet Diagn Invest. 2010;22:97–100. doi:10.1177/104063871002200119.

46. Munday JS, Knight CG, French AF. Research in Veterinary Science. Res Vet Sci. 2011;90:280–3. doi:10.1016/j.rvsc.2010.06.014. Elsevier Ltd.

47. Schwittlick U, Bock P, Lapp S, Henneicke K, Wohlsein P. Feline papillomavirus infection in a cat with Bowen-like disease and cutaneous squamous cell carcinoma. Schweiz Arch Tierheilkd. 2011;153:573–7. doi:10.1024/0036-7281/a000276.

17

Genetic characterization and mutation analysis of Qihe547 Aujeszky's disease virus in China

Cun Liu[1†], Yanhan Liu[2†], Ye Tian[2], Xuehua Wei[1], Yue Zhang[1] and Fulin Tian[1*] (iD)

Abstract

Background: Aujeszky's disease virus (ADV) can cause neurologic disease in young pigs, respiratory disease in older pigs and abortion or birth of mummified fetuses or stillborn neonates. The re-emergence of Aujeszky's disease (AD) in pig farms vaccinated with live vaccine (Bartha-K61) caused substantial economic losses to Chinese pig industry since late 2011. A field ADV, named Qihe547, was isolated from pigs that exhibited suspected AD clinical symptoms. To better understand the genetic characteristics and mutations of Qihe547 ADV, the whole genome was sequenced and analyzed.

Results: The genomic length of Qihe547 ADV was 143,404 bp, with 73.59% G + C contents. Phylogenetic analysis based on the whole genome of ADV strains revealed that Chinese ADV strains were located to one group with three subgroups. Qihe547 ADV was closely related to these novel ADV strains isolated in China since 2012. Qihe547 presented numerous hypervariable regions compared with oversea ADV strains. In 34 genes of Qihe547 ADV, amino acid (AA) insertion or deletion were observed. In addition, numerous AA mutations were found in the main protective antigen genes (gB, gC and gD genes). The differences of potential antigenic peptides in the main protective antigens between Qihe547 ADV and ADV Bartha were discovered in the dominant antigenic regions of gB (AA59-AA126, AA507-AA734),the extracellular region of gC and gD.

Conclusion: High diversity was observed between Qihe547 and foreign ADV isolates. The AA variations and the differences of potential antigenic peptides in the important functional regions of the main protective antigen (gB, gC and gD) of ADV Qihe547 may contribute to immune evasion of the virus and may be partial reason that the virus escapes from the vaccination of Bartha-K61 vaccine. In a word, the effect of the variations obviously requires further research.

Keywords: Aujeszky's disease virus, Qihe547, Phylogenetic analysis, Virus isolation

Background

Aujeszky's disease (AD), a devastating swine disease, also known as Pseudorabies (PR), is an economically important viral disease of swine worldwide [1–3]. AD is characterized by high morbidity in piglets, severe respiratory disorders in older pigs and sow's reproductive failure [2, 4]. AD occurs all over the world and caused great economic losses to the pig industry in many countries [5].

The causative agent responsible for AD is Aujeszky's disease virus (ADV) which is a member of the genus *Varicellovirus* of the subfamily *Alphaherpesvirinae* within the family *Herpesviridae* [6]. The genome of ADV is double-stranded DNA, approximately 150 kb, with G + C contents more than 70%. Swine are the natural host for ADV and the reservoir of the virus in nature [7], but a wide range of mammals can be infected with ADV [8].

Vaccination is widely used to control the AD. Attenuated live vaccine has been widely used for many years in many countries to diminish the economic losses caused by AD [9]. Due to large-scale compulsory vaccination with gE-deleted vaccines and the DIVA (differentiating infected from vaccinated animals) strategy, AD has been eradicated from domesticated pigs in many countries [3, 10]. In the 1970s, China imported the Bartha-K61 vaccine from Hungary [11]. Then, it was widely used in swine farmers, from the 1990s

* Correspondence: fulintianjn@163.com
†Cun Liu and Yanhan Liu contributed equally to this work.
[1]Shandong Provincial Center for Animal Disease Control and Prevention, Ji'nan 250022, Shandong, China
Full list of author information is available at the end of the article

until late 2011, AD well controlled. Since late 2011, however, AD broke out in China and has caused huge economic losses [11, 12]. It is great necessary to understand the genetical mutation of re-emergenced ADV strain.

In this study, we genetically characterized Qihe547 ADV strains from swine herd vaccinated with live vaccine by the Sanger dideoxy-chain termination method and phylogenetic analysis to get a better insight into the genetic relatedness between strains in both populations and to compare Qihe547 strains with strains circulating in neighboring countries. To understand the mutations of Qihe547 ADV and the causes of immune failure, comparative genomic analyses and phylogenetic analysis were performed.

Methods
Ethics and consents
Sampling was performed under the permission from the owner of the swine herd in Qihe, Shandong province, China and the ethical approval was granted by the Administration Bureau of Animal Husbandry and Veterinary Medicine of Shandong. Animal manipulation procedures adhered to the guidelines under the Animal Ethics Committee of the Administration Bureau of Animal Husbandry and Veterinary Medicine of Shandong.

Samples collection
In 2014, outbreak of AD occurred in a pig farm in Qihe, Shangdong province, China. In the farm, the pregnant sows showed high rates of abortion and some fattening pigs showed severe respiratory symptoms along with high fever and then died. The tissue samples, including brain, lung, kidney, liver, and lymph node, were collected from the clinical dead pigs for diagnostic tests and the corpses were harmlessly treated after anatomical examination. The clinically dead pigs were from the swine herd in Qihe area, Shandong province, China. The tissue samples were homogenated. The homologized samples were centrifuged at 8000 g for 5 min. Then, the supernatant was used to extracted DNA and virus isolation.

Virus isolation and identification
Nucleic acid was extracted from the supernatant using AxyPrep Body Fluid viral DNA/RNA Mini-prep kit according to the manufacturer's protocol (Corning Co., China). The nucleic acid was detected for Aujeszky's disease virus (ADV), porcine reproductive and respiratory syndrome virus (PRRSV), porcine circovirus type 2 (PCV2), swine fever virus (SFV) by the real-time PCR. The supernatant, ADV positive, was filtrated through 0.22 µm filter (Millipore, Milford, MA) and inoculated to monolayer culture of Monolayer culture of porcine kidney 15 (PK-15) cells cultured in dulbecco's modified eagle mediun (DMEM) (Hyclone) supplemented with 8% fetal bovine serum (Shanghai ExCell Biology Co.). PK-15

cells were maintained at 37 °C in an atmosphere of 5% CO2. The cytopathic effect (CPE) was examined daily.

Then, the virus was identified by indirect immuofluorescence assay (IFA). Briefly, inoculated cells were fixed with cold absolute ethanol for 1 h, blocked with 1% bovine serum albumin (Sigma, St. Louis, MO) for 30 min, and incubated with 200 times diluted Rabbit anti-ADV antibody (Abcam Ab3534) for 1 h, followed by a 2000 dilution of goat anti-rabbit IgG-FITC-conjugated secondary antibody (Abcam Ab6717) for 1 h. Cell staining was examined under a fluorescence microscope. Finally, the ADV was plaque-purified three times and propagated on PK-15 cells. Virus titer was determined by measuring the 50% tissue culture infective dose (TCID50).

PCR and sequencing
Primers for the amplification of Qihe547 ADV whole genome were designed based on ADV Becker (JF797219). The plaque-purified virus was used to extract DNA with the AxyPrep Body Fluid viral DNA/RNA Mini-prep kit according to the manufacturer's protocol (Corning Co., China). PCR was performed with the Takara LA Taq with GC buffer kit (Takara, Dalian), following the manufacturers' instructions. Cycling conditions for PCR was as follows: denaturation at 95 °C for 10 min, followed by 38 cycles of 95 °C for 1 min, gradiently annealing temperature 50 °C~ 70 °C for 50 s, and 72 °C for 1 min, and final extension of 72 °C for 10 min. The products were cloned into pMD18-T vector (Takara, Dalian). The recombinant plasmid, containing the target DNA fragments, was sent to BioSune Biotech Co. Ltd. (Jinan, China) for DNA sequencing by the Sanger dideoxy-chain termination method after restriction endonuclease digestion with EcoRI and Hind III.

Sequence assembling and alignments
Sequences of the target DNA fragments were assembled using SeqMan (DNASTAR, Madison, WI). Gene's annotation of Qihe547 ADV was based on ADV Becker and ADV JS-2012(KP257591). The alignment based on full-length genome sequences between Qihe547 ADV and reference ADV strains from GenBank was performed using the mVISTA genomics analysis tool with global LAGAN alignment [13].

Each gene's protein-coding sequence was filtered out from the whole genome sequence of Qihe547 ADV and foreign ADV strains (ADV32751/Italy2014, Kolchis, Bartha, Becker, and Kaplan). To study the deduced amino acid (AA) sequence mutations in each genes of Qihe547 ADV relative to foreign ADV isolates (ADV32751/Italy2014, Kolchis, Bartha, Becker, Kaplan), AA sequence alignments were performed. Alignments of the deduced amino acid sequence of each gene's protein-coding sequence were performed with MegAlign (DNASTAR, Madison, WI). The potential antigenic peptides in glycoprotein (g) B, gC, gD of

Qihe547 ADV and Bartha were predicted by Antigenic Peptide Prediction online tool (http://imed.med.ucm.es/Tools/antigenic.html).

Phylogenetic analyses

The phylogenetic analysis was performed by the neighbor-joining method. The phylogenetic tree based on full-length ADV genome sequences and gB, gC and gD genes sequences were constructed by MEGA 6.0 software with 1000 bootstrap replication.

Results

Virus isolation and identification

ADV-positive mixed tissue sample was confirmed by the real-time PCR and the supernatant inoculated onto PK-15 cells. A distinct CPE was observed (Fig. 1b) and identified with IFA (Fig. 1d). The isolated ADV was named Qihe547. The titers of Qihe547 ADV were $10^{8.0}$TCID$_{50}$/ml.

Genomic characterization of ADV Qihe547

The genome of Qihe547 ADV was sequenced and approximately 143 kb long. Its G + C contents was 73.59%. Its genome organization was similar to other ADV strains submitted to GenBank. The genomic annotation was performed according to the sequences information of ADV Becker and JS-2012. Genome sequence of Qihe547 ADV submitted to Genbank, the accession No. is KU056477. Interestingly, compared to other ADV isolates from GenBank, there are special polynucleotide sequences (5' CCCAGCTCTCCCCCGAGGGCCCAGC TCTCCCCC 3') at the genome 5'- terminal. Upstream gB gene, the unique repeats region, possible repeating unit (5' GGGGAGAGGGGAGACGAGA3') different from ADV Becker (5'GGGACGGAGGGGAGA3'), exists a gap. Multi-genome alignments based on ADV isolates indicated that numerous hypervariable regions were found between Qihe547 ADV and oversea ADV isolates (ADV32751/Italy2014, Batha, Becker), especially in intergenic sequences (Fig. 2a). However, between Qihe547 and other China ADV isolates (JS–2012, TJ, Ea, SC), the genetic diversity showed little differences.

Multiple sequences comparison

Compared with foreign ADV isolates (ADV32751/Italy2014, Kolchis, Bartha, Becker, Kaplan), a large amount of AA substitutions were observed in nearly each genes of Qihe547. Surprisingly, numerous unique AA indels were also found in Qihe547 genes (Table 1). The indels were observed mainly in genes associated with viral egress, viral replication and gene regulation. What's interesting is that the unique indels in subunits of primase complex, products of UL52, UL5 and UL8 genes and these variations closely related to repeating sequences. To the contrary, Qihe547 ADV showed high level identities with other Chinese ADV isolates in protein sequences. But, analysis of variation between Qihe547 and other Chinese ADV strains showed that some genes had a single AA insertion

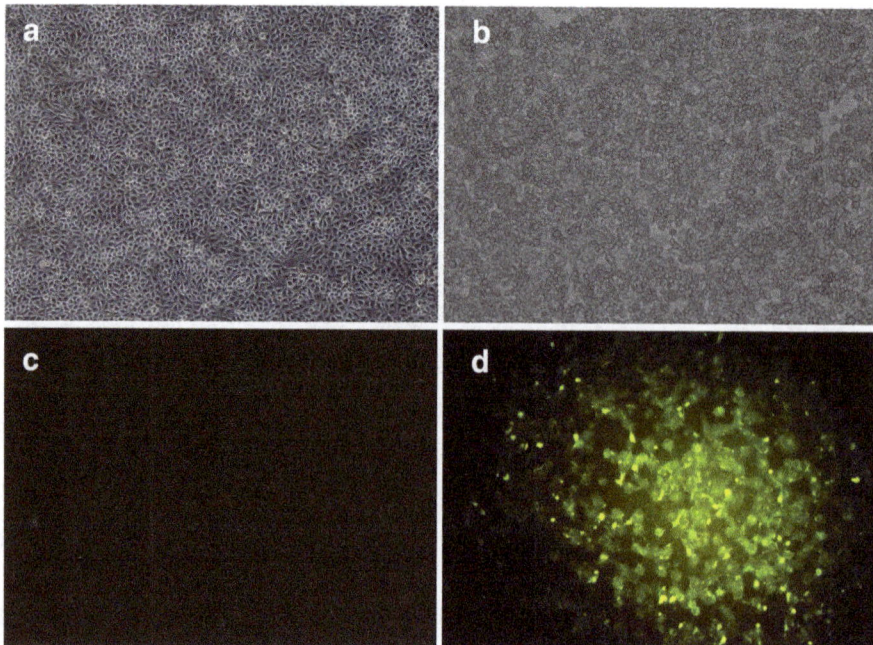

Fig. 1 The indentification of Qihe547 ADV. **a** and **c.** Negative control, PK-15 cells uninfected with Qihe547 ADV. **b.** Specific cytopathic effects of Qihe547 ADV (× 100). **d.** Indirect immunofluorescence assays of Qihe547 ADV in PK-15 cells (× 200)

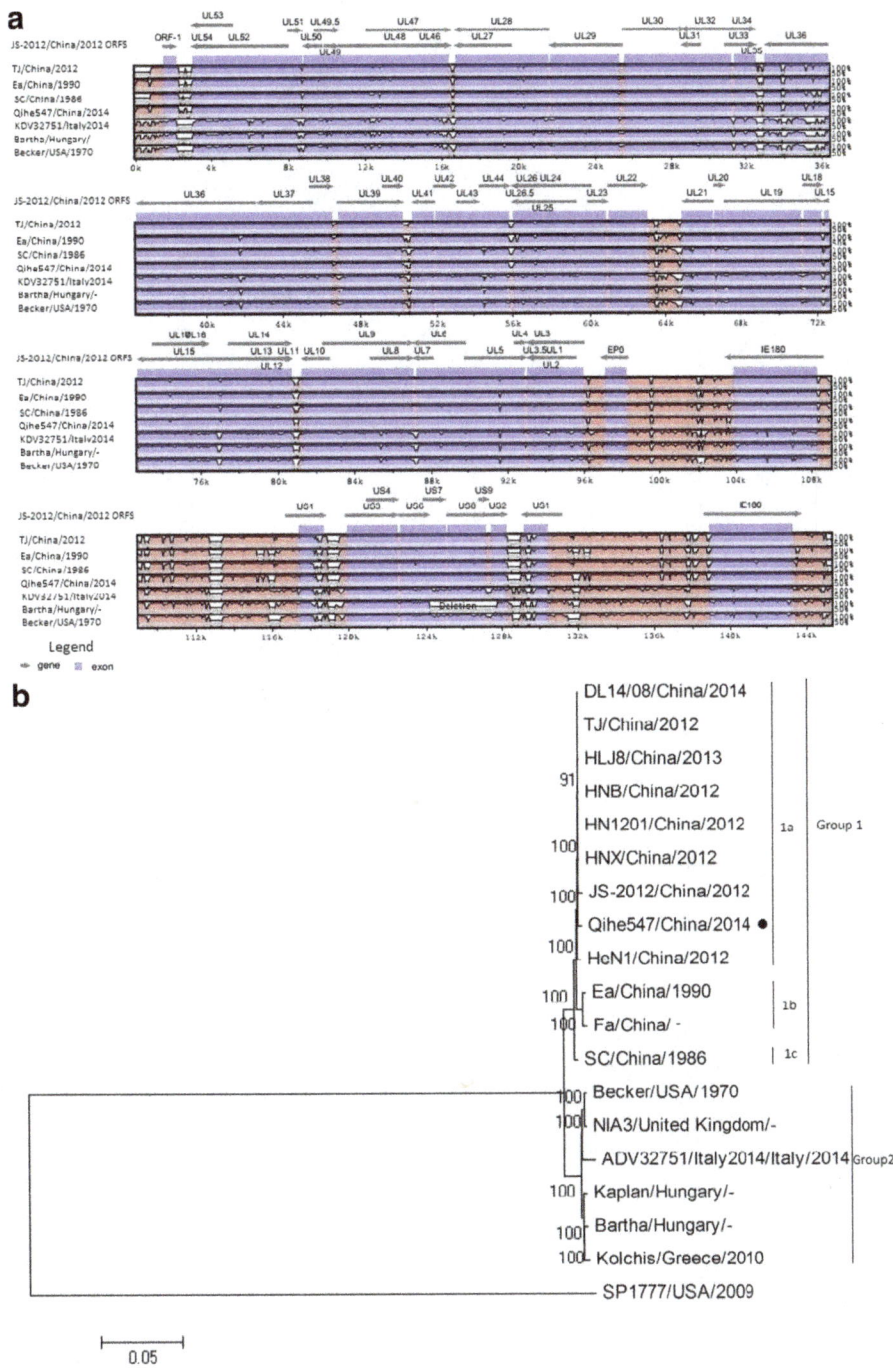

Fig. 2 Comparison and phylogenetic analysis based on the whole-genome sequences of ADV strains. **a.** The multiple sequence alignments showed the conserved regions within Qihe547, TJ, Ea, SC, ADV32751/Italy2014, Becker, Bartha strains. The conservation score was plotted in a sliding 100 bp window. **b.** Phylogenetic tree based on whole-genome sequences of PRV strains was constructed using MEGA 6.0 software. SP1777, a bovine herpesvirus strain, served as an outgroup. Bootstrap values obtained from 1000 replicates are shown at node points. Qihe547 isolated in this study was indicated by a solid dot

or deletion, IE180 gene sequence a insertion (456S) and a deletion (98Q), UL46 gene sequence one deletion (571E),UL3 gene sequence a insertion (101A). Compared with ADV strains (ADV32751/Italy2014, Kolchis, Bartha, Becker, Kaplan), numerous AA variations were discovered in gB, gC and gD of ADV Qihe547, there being also AA

with ADV Bartha, high diversity was observed in Qihe547 ADV genes (Fig. 3). The highest divergence showed in the US1 gene. Moreover, the divergence in UL36, UL49.5 and UL51 genes were more than 10%. Compared with foreign

Table 1 Amino acid Indels in Qihe547 ADV genes, compared to other ADV strains

Name	Indels	Name	Indels
UL52	428-432(+QAHSQ), 596-597(−AL), 622(+V), 663-665(+SSS)	UL51	196-200(+EADAE), 223-234(+KK)
UL50	21(+R)	UL49.5	5(+S)
UL47	71-75(−(E/R)(E/G)ER(M/T)), 128(−E), 224(+L)	UL46	502-503(-PP), 568-570(+EEE), 588-592(+GNAAD), 618-621(-GSFR)
UL44	63-69(+AAASTPA)	UL42	439(+S)
UL39	15-16(−(A/P)JP)	UL33	9-10(+GG)
UL32	221(+G)	UL29	345(+G), 1123(+G), 1140(+G)
UL28	429-430(+GA)	UL27	92(+I)
UL26	372-375(+GLPP),457-458(+AA)	UL26.5	126-129(+GLPP), 211-212(+AA)
UL25	242(+P), 318(+D)	UL21	364-368(+NGDGG),410-414(+PIVSA)
UL20	7(+V), 19-21(+AAV)	UL17	254-255(+PR)
UL13	28-34(+GGAIAAA)	UL9	285(+T)
UL8	121(+V), 134(+K), 142-144(+DED)	UL6	8(+A), 451(+R)
UL5	2(-A), 579-591(+PGGPGGPGAPAS)	UL3	95-98(+TTTT)
UL2	51-56(+GAGAGA)	EP0	187(+R)
IE180	27-28(+AA), 900-906(+SPGTKSG)	US7	178(-S), 238(+G)
US8	48(+D), 496(+D)		
UL15	168(+A), 189-190(-GP), 200-203(+RGES), 214-218(+GSGAK), 648-649(+PG)	US1	343-398(+EEDEEEEDEEEEDEEEEDEEEEDEEEEDEEEEDEEEEDEEEEDEEEEDEEEEDEEE)
UL36	249-257(+GAPAVAAVG), 562-563(-GA), 2256-2269(+APPAEAAPAAKPAP), 2339-2347(+RAPPPPQPQ), 2368-2392(+PPPQQQOQQQQORQEPPAATASKKA), 2449-2456(+TQPAAPAE), 2476-2478(+TAK), 2736(+S), 2911(+V), 2973-2974(+PA), 3040-3045(+RAEPAR), 3056(+E), 3110-3111(+PF), 3126-3128(+SPL), 3142(+E), 3211(+P)		

+: Amino acid insertion
-: Amino acid deletion

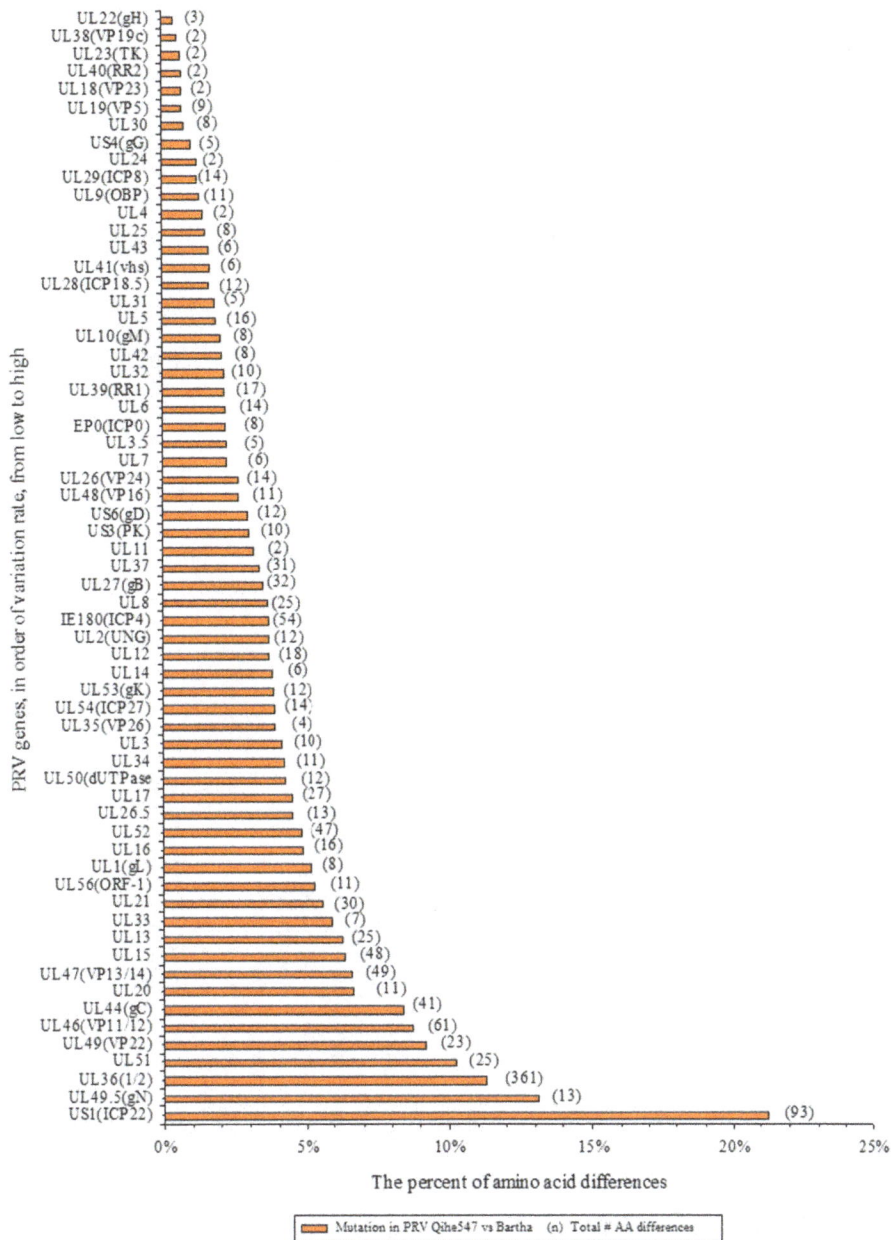

Fig. 3 Protein-coding variation in Qihe547 ADV versus Bartha ADV. The bars showed the percent of amino acid differences in Qihe547 ADV versus Bartha ADV. Protein names were listed on left. Four proteins (US7, US8, US9, and US2 proteins) were absent because of the deletion regions in the unique short region of Bartha ADV

consistent with ADV Becker or ADV ADV32751/Italy2014 (Table 2). Interestingly, there were AA differences between Chinese ancient ADV strains (Ea, Fa, SC) and novel ADV ADV strains (Qihe547, DL14/08, HeN1) isolated since 2012, such as R454K, H563Q, V740A in gB, G194E in gC and V338A in gD. Compared with Chinese ancient ADV isolates (Ea, Fa, SC), there were two AA deletion ([278](R/S) P[279]) in gD of three novel ADV isolates. In gD of ADV LA and MinA, another two Chinese ancient ADV strains, there are four AA deletion at the same position. Between

Qihe547 ADV and Bartha, the potential antigenic sites differences in the main protective antigen genes, gB, gC and gD genes, were found (Table 3). In AA59-AA126 of ADV Qihe547 gB gene, there were one more potential antigenic sites and three potential antigenic sites compared to ADV Bartha. In the N-terminal AA43-AA59 of Qihe547 ADV gC gene, two potential antigenic sites were predicted with the opposite to ADV Bartha a single potential antigenic site at the same position. Similarity, in the N-terminal AA62-AA88 of Qihe547 gD gene, there were two potential

Table 2 Amino acid mutation in gB, gC and gD of Qihe547 ADV

Name	g D	g B
	207,209,212,278-281,288,309,338,342,346,395	53 55 70 72-73 75-78, 81-83, 85, 87, 92-94, 96-97,102,361,454,507-508,553,563,571,674,740,843,850,852,898,911,917
Qihe547	SDK--RP ARAPTA	TTAGT--ADGFAEIDGVSDMKSPSQSGAAEDANP
DL14/08	SDK--RP ARAPTA	TTAGT---ADGFAEIDGVSDTKSPSQGGAAEDANP
HeN1	SDK--RP ARAPTA	TTAGT---ADGFAEIDGVSDTKSPSQGGAAEDANP
SC	SDKRPRP ARVPTA	TTAGT---ADGFTEIDGVSDTRSPSHGGTAEDVNP
Ea	SDKSPRP ARVPTA	TTAGT---ADGFTEIDGVSDTRSPSHGGTAEDVNP
Fa	SDKRPRP ARVPTA	TTAGT---ADGFTEIDGVSDTRSPSHGGTAEDVNP
ADV32751/Italy2014	SDK----GRAQPA	APTVP----INDVAE-DGVSGTRAAGQSSATDGASL
Kolchis	NER----VHAQPT	APTVPSPGLNDVAA-IEFTETRAAGQSSATDGASL
Becker	SDK--RP ARAQPA	APTGT---ANDVAA-IEFSETRSPGQSSATDGASL
Kaplan	NER----VHAQPT	APTVPSPGLNDVAA-IEFTETRSPGQSSATDGASL
Bartha	NER----VHAQPT	APTVPSPGLNDVAA-IEFTETRAAGQSSATDGASL

Name	g C
	16, 25, 34, 43, 52, 55, 57, 59-61, 63-69, 76, 87, 90, 99, 102, 130, 142, 187-192, 194, 240, 243, 431, 437, 449, 457, 461, 467, 485, 486, 487
Qihe547	TSNASEVGTTAAASTPAVQGKSVCSAWVEEVHMITTTASAL
DL14/08	TSNASEVGTTAAASTPAVQGKSVCSAWVEEVHMITTTASAL
HeN1	TSNASEVGTTAAASTPAVQGKSVCSAWVEEVHMITTTASAL
SC	ATNAPAAPEA------APNKAFYPWEDEGLSMITTTASAL
Ea	TSTASEVGTTAAASTPAVQGESVCSAWVEGVHMITTTASAL
Fa	TSTASEVGTTAAASTPAVQGESVCSAWVEGVHMITTTASAL
ADV32751/Italy2014	ASNEPAAPEA------APNQAFYP-VDVEGLSLVASVGAGP
Kolchis	ATNEPAAPEA------APNKAFYPALDDDGLSLVASVGAGP
Becker	ATNEPAAPEA------VPNKAFYP-VWEELSLVASVGAGP
Kaplan	ATNEPAAPEA------APNKAFYPAVEDEGLSLVASVGAGP
Bartha	ATNAPAAPEA------APNKAFYPWEDEGLSLVASVGAGP

Table 3 The differences of the potential antigenic peptide in gB, gC and gD between Qihe547 and Bartha ADV. In the N terminus of ADV gB, there are antigenically important regions, such as (AA59-AA126, AA214-AA279, AA507-AA734) [21]. The extracellular region (AA23-AA453) of ADV gC plays an important role in ADV adsorption, neutralizing antibody generation and virus neutralization. gD also is the main target of ADV antibody. Mutation occurred in these regions may result antigen drift. Numerous AA mutations and potential antigenic peptides differences were found in gB, gC and gD of Qihe547, this may lead to immune failure

Name	Qihe547[a]	Bartha[b]
gB	^{89}IDGAVSP^{95}	^{65}ASASPTPVPGSP^{76} ^{79}TPNDVSAE^{86} ^{433}EAIDAIYQ^{440}
gC	^{43}AGELSPSP^{50} ^{52}STPEPVSG^{59}	^{43}AGELSPSPPPTPAPA^{57}
gD	^{62}LEDPCGWALISDPQVDR^{79} ^{81}LNEAVAHR^{88}	^{62}LEDPCGVAALISDPQVDRLLSEAVAHR^{88}

[a]the potential antigenic peptide only in Qihe547
[b]the potential antigenic peptide only in Bartha

antigenic sites, one predicted at the same position of Bartha gD gene.

Phylogenetic analyses

Phylogenetic analysis based on completed genome sequence indicated that all ADV isolates used to construct phylogenetic tree were divided into two groups (group 1 and group 2) (Fig. 2b). All Chinese ADV isolates were located in group 1, and subdivided into group 1a, group 1b and group 1c. Group 1a comprised ADV strains isolated since 2012 (e.g., Qihe547, DL14/08, TJ, HeN1, HNB, HNX). Group 1b contained two Chinese traditional ADV isolates such as Ea and Fa. Another traditional ADV isolate SC belonged to group 1c. It indicated that Qihe547 ADV had close relationship with Chinese ADV strains isolated since 2012. Phylogenetic trees of gB, gC and gD indicated that nearly all Chinese ADV strains were located in one group (Fig. 4). In the phylogenetic trees of gC gene, six Chinese ADV isolates (SC, 783, SQ, SN-SICHUAN, HS, and SCZ) were clustered together with foreign isolates. This result showed that these strains were closely related to foreign ADV strains. Interestingly, phylogenentic relationship of SC ADV in the phylogennetic tree of gC was different from that in the phylogennetic tree based on completed genome sequences. In the phylogenentic of gD, Becker and NIA3 strains were clustered together with Chinese ADV isolates. Both genome-wide and gB, gC and gD genes phylogenetic analysis showed that Chinese ADV strains isolated since 2012 were not located in same subgroup with these strains isolated before 2012.

Discussion

AD caused substantial economic losses to the pig industry in many countries, especially in regions with dense pig populations [10]. Although it has been eradicated from domesticated pigs in many countries, AD remains one of the most important diseases of swine. Mounting evidences showed that ADV infections are more widespread in wild animals and it is a threat to domestic pigs [2, 14–16]. In China, tremendous progress has been made in controlling

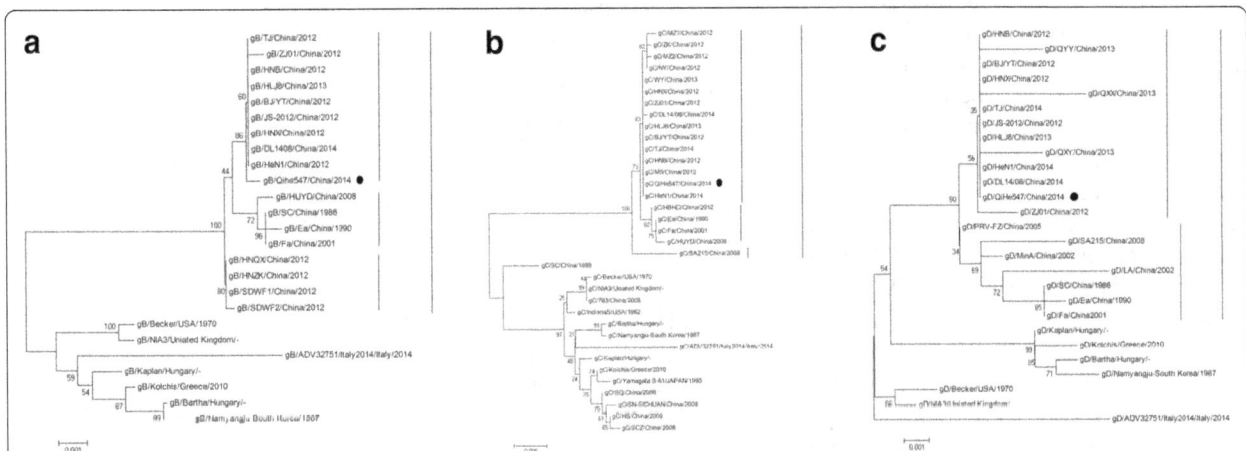

Fig. 4 Phylogenetic analysis of gB, gC and gD genes. Phylogenentic trees were constructed by Neighbour-joining method using MEGA 6.0 software. Bootstrap values obtained from 1000 replicates were shown at node points. Qihe547 ADV was indicated by a solid dot. **a.** Phylogenetic tree of gB gene. **b.** Phylogenetic tree of gC gene. **c.** Phylogenetic tree of gD gene

and eliminating ADV in herds, including large-scale vaccination with live vaccine from the 1990s until late 2011. Since late 2011, the re-emergence of AD in pig farms has attracted the attention of researchers. In this study, Qihe547 strain, a field Aujeszky's disease virus, was isolated and identified from herd vaccinated with live vaccine (Bartha-K61). For a better understanding of Qihe547 ADV genome characteristic, the full-length genome was sequenced.

Similar genome organization, with 73.59% G+C contents, were observed. The differences between Qihe547 and foreign ADV isolates distributed whole-genome-widely. Some interesting regions were found in Qihe547 ADV genome sequence, such as the unique upstream repeat region of gB gene which the possible repeat unit was different with Becker strain. Further study was needed to research the function of the repeat region for the virus. According to the phylogenetic analysis, Qihe547 ADV within subgroup 1a was closely related to these novel ADV strains isolated since 2012. Compared with foreign ADV strains (ADV32751/Italy2014, Kolchis, Bartha, Becker, Kaplan), unique AA indels were observed in 34 genes of Qihe547 ADV, mainly in proteins associated with viral egress, viral replication and genetic regulation. The highest divergence (more than 20%) was found in US1 protein acting as regulatory protein between Qihe547 ADV and Bartha. UL49.5 protein (gN) associated with immune evasion and UL36 and UL 51 proteins associated with viral egress also showed higher variation rates. Previous studies also draw similar conclusions [17]. Varying degrees of variation were also found in other proteins.

Mutations in antigen contribute to immune failure. Glycoprotein gB, gC and gD, the main protective antigen of ADV, could induce protective immune responses. The variability of gB, gC and gD was beneficial to ADV evading the host immune defense mechanism [18]. The differences displayed on glycoproteins of gB, gC and gD between ADV vaccine strain and prevalent ADV strains might contribute to immune failure. Vaccination or challenge experiments in sheep and pigs indicated that Bartha-K61 vaccine could not provide complete protection against the current prevalent ADV in China [19, 20]. In this study, numerous AA mutations were found in gB, gC and gD of Qihe547 ADV, especially in the dominant antigenic regions of gB (AA59-AA126 and AA507-AA734) [21].The heparan binding domains of gC (AA44-AA290) were also the antibody-binding domains [22]. As a result of AA variations, the potential antigenic peptides in gB, gC and gD were inconsistent between Qihe547 ADV and ADV Bartha. The differences of potential antigenic peptides in Qihe547 gB were mainly found in the dominant antigenic region (AA59-AA126, AA507-AA734). The extracellular region (AA23-AA453) in gC of ADV plays an important role in ADV adsorption, neutralizing antibody generation and virus neutralization. Many AA variations and the differences of potential antigenic peptides were observed in extracellular region in gC of ADV Qihe547. The differences of potential antigenic peptide were discovered in gD of ADV Qihe547. These antigenic peptide differences were found in the dominant antigenic region (AA59-AA126, AA507-AA734) of gB, the extracellular region (AA23-AA453) of gC, the dominant antigenic region of gD. The main protective antigen of Qihe547 ADV may be closely related with antigenicity of the virus, probably resulting neutralizing antibody differences between field strains and Bartha. And this may be the partial reason that the virus escapes from the vaccination of Bartha-K61 vaccine.

Conclusion

In this study, high diversity was observed between Qihe547 and other ADV isolates (ADV32751/Italy2014, Kolchis, Bartha, Becker, and Kaplan). AA mutations were found in nearly each gene and unique AA indels were observed in 34 genes of ADV Qihe547. The AA variations and the differences of potential antigenic peptides in the important functional regions of the main protective antigen (gB, gC and gD) of ADV Qihe547 may contribute to immune evasion of the virus. The effect of the variations obviously requires further research.

Funding

This study was funded by Agricultural Technology Innovation Projects of Livestock and Poultry Breeding.

Authors' contributions

CL carried out the studies and drafted the manuscript. YL helped to draft the manuscript and modified the manuscript. FT designed the study. All authors read and approved the final manuscript.

Competing interests

The authors declare that they have no competing interests.

Author details

[1]Shandong Provincial Center for Animal Disease Control and Prevention, Ji'nan 250022, Shandong, China. [2]College of Veterinary Medicine, China Agricultural University, Beijing 100193, China.

References

1. Ao JQ, Wang JW, Chen XH, Wang XZ, Long QX. Expression of pseudorabies virus gE epitopes in Pichia pastoris and its utilization in an indirect PRV gE-ELISA. J Virol Methods. 2003;114:145–50.
2. Cramer SD, Campbell GA, Njaa BL, Morgan SE, Smith SK, McLin WR, Brodersen BW, Wise AG, Scherba G, Langohr IM, et al. Pseudorabies virus infection in Oklahoma hunting dogs. J Vet Diagn Investig. 2011;23:915–23.
3. Pomeranz LE, Reynolds AE, Hengartner CJ. Molecular biology of pseudorabies virus: impact on neurovirology and veterinary medicine. Microbiol Mol Biol Rev. 2005;69:462–500.
4. Nauwynck HJ. Functional aspects of Aujeszky's disease (pseudorabies) viral proteins with relation to invasion, virulence and immunogenicity. Vet Microbiol. 1997;55:3–11.

5. Klupp BG, Hengartner CJ, Mettenleiter TC, Enquist LW. Complete, annotated sequence of the pseudorabies virus genome. J Virol. 2003;78:424–40.
6. Crandell RA. Pseudorabies (Aujeszky's disease). Vet Clin North Am Large Anim Pract. 1982;4:221–9.
7. Marcaccini A, Lopez Pena M, Quiroga MI, Bermudez R, Nieto JM, Aleman N. Pseudorabies virus infection in mink: a host-specific pathogenesis. Vet Immunol Immunopathol. 2008;124:264–73.
8. Müller TF, Teuffert J, Zellmer R, Conraths FJ. Experimental infection of European wild boars and domestic pigs with pseudorabies viruses with differing virulence. Am J Vet Res. 2005;62:252–8.
9. Ficinska J, Bienkowska-Szewczyk K, Jacobs L, Plucienniczak G, Plucienniczak A, Szewczyk B. Characterization of changes in the short unique segment of pseudorabies virus BUK-TK900 (Suivac a) vaccine strain. Arch Virol. 2003;148: 1593–612.
10. Muller T, Hahn EC, Tottewitz F, Kramer M, Klupp BG, Mettenleiter TC, Freuling C. Pseudorabies virus in wild swine: a global perspective. Arch Virol. 2011;156(10):1691–705.
11. An TQ, Peng JM, Tian ZJ, Zhao HY, Li N, Liu YM, Chen JZ, Leng CL, Sun Y, Chang D, et al. Pseudorabies virus variant in Bartha-K61-vaccinated pigs, China, 2012. Emerg Infect Dis. 2013;19(11):1749–55.
12. Wu R, Bai C, Sun J, Chang S, Zhang X. Emergence of virulent pseudorabies virus infection in northern China. J Vet Sci. 2013;14(3):363.
13. Brudno M, Do CB, Cooper GM, Kim MF, Davydov E, Green ED, Sidow A, Batzoglou S. LAGAN and multi-LAGAN: efficient tools for large-scale multiple alignment of genomic DNA. Genome Res. 2003;13(4):721–31.
14. Verin R, Varuzza P, Mazzei M, Poli A. Serologic, molecular, and pathologic survey of pseudorabies virus infection in hunted wild boars (Sus scrofa) in Italy. J wildlife Dis. 2014;50(3):559–65.
15. Verpoest S, Cay AB, Bertrand O, Saulmont M, De Regge N. Isolation and characterization of pseudorabies virus from a wolf (Canis lupus) from Belgium. Eur J Wildlife Res. 2013;60(1):149–53.
16. Hahn EC, Fadl-Alla B, Lichtensteiger CA. Variation of Aujeszky's disease viruses in wild swine in USA. Vet Microbiol. 2010;143(1):45–51.
17. Yu T, Chen F, Ku X, Fan J, Zhu Y, Ma H, Li S, Wu B, He Q. Growth characteristics and complete genomic sequence analysis of a novel pseudorabies virus in China. Virus Genes. 2016;52(4):474–83.
18. Takada A, Kida H. Induction of protective antibody responses against pseudorabies virus by intranasal vaccination with glycoprotein B in mice. Arch Virol. 1995;140(9):1629–35.
19. Wang CH, Yuan J, Qin HY, Luo Y, Cong X, Li Y, Chen J, Li S, Sun Y, Qiu HJ. A novel gE-deleted pseudorabies virus (ADV) provides rapid and complete protection from lethal challenge with the ADV variant emerging in Bartha-K61-vaccinated swine population in China. Vaccine. 2014;32(27):3379–85.
20. Luo Y, Li N, Cong X, Wang CH, Du M, Li L, Zhao B, Yuan J, Liu DD, Li S, et al. Pathogenicity and genomic characterization of a pseudorabies virus variant isolated from Bartha-K61-vaccinated swine population in China. Vet Microbiol. 2014;174(1-2):107–15.
21. Zaripov MM, Morenkov OS, Fodor N, Braun A, Schmatchenko VV, Fodor I. Distribution of B-cell epitopes on the pseudorabies virus glycoprotein B. J Gen Virol. 1999;80(3):537–41.
22. Ober BT, Teufel B, Wiesmüller KH, Jung G, Pfaff E, Saalmüller A, Rziha HJ. The porcine humoral immune response against pseudorabies virus specifically targets attachment sites on glycoprotein gC. J Virol. 2000;74(4):1752–60.

18

High sensitivity of domestic pigs to intravenous infection with HEV

Lisa Dähnert[1], Martin Eiden[1], Josephine Schlosser[2], Christine Fast[1], Charlotte Schröder[3], Elke Lange[3], Albrecht Gröner[4], Wolfram Schäfer[5] and Martin H. Groschup[1*] ⓘ

Abstract

Background: Hepatitis E virus (HEV) is one major cause of acute clinical hepatitis among humans throughout the world. In industrialized countries an increasing number of autochthonous HEV infections have been identified over the last years triggered by food borne as well as – to a much lower degree – by human to human transmission via blood transfusion. Pigs have been recognised as main reservoir for HEV genotype 3 (HEV-3), and zoonotic transmission to humans through undercooked/raw meat is reported repeatedly. The minimal infectious dose of HEV-3 for pigs is so far unknown.

Results: The minimum infectious dose of HEV-3 in a pig infection model was determined by intravenous inoculation of pigs with a dilution series of a liver homogenate of a HEV infected wild boar. Seroconversion, virus replication and shedding were determined by analysis of blood and faeces samples, collected over a maximum period of 91 days. A dose dependent incubation period was observed in faecal shedding of viruses employing a specific and sensitive PCR method. Faecal viral shedding and seroconversion was detected in animals inoculated with dilutions of up to 10^{-7}. This correlates with an intravenously (i.v.) administered infectious dose of only 6.5 copies in 2 ml (corresponding to 24 IU HEV RNA/ml). Furthermore the first detectable shedding of HEV RNA in faeces is clearly dose dependent. Unexpectedly one group infected with a 10^{-4} dilution exhibited prolonged virus shedding for more than 60 days suggesting a persistent infection.

Conclusion: The results indicate that pigs are highly susceptible to i.v. infection with HEV and that the swine model represents the most sensitive infectivity assay for HEV so far. Considering a minimum infectious dose of 24 IU RNA/ml our findings highlights the potential risk of HEV transmission via blood and blood products.

Keywords: HEV, Minimal infectious dose, Swine, In vivo

Background

Hepatitis E is caused by Hepatitis E virus (HEV) which is a major cause of acute hepatitis throughout the world with a total number of 44,000 HEV-related deaths in 2015 [1]. Hepatitis E virus is a small, quasi-enveloped, single-stranded RNA virus and a member of the *Hepeviridae* family. Novel taxonomic classification consists of the two genera *Piscihepevirus* and *Orthohepevirus* encompassing species A-D. All mammalian HEV isolates have been attributed to species *Orthohepevirus A* [2] and are further grouped into genotypes 1–8. Although displaying a highly diverse group on molecular level, all genotypes evidently belong to one serotype [3, 4].

Genotype 1 (HEV-1) and 2 (HEV-2) are restricted to humans and are the main cause of endemic outbreaks in developing countries in Asia, Africa and Central America [5, 6]. The transmission of these isolates mainly occurs by the faecal-oral route due to poor sanitation and contaminated water [3].

HEV genotypes 3 (HEV-3) and 4 (HEV-4) dominate in developed countries and are the main source for autochthonous human cases [7]. In Europe, North America, Australia and New Zealand HEV-3 is reported as causative agent for these autochthone HEV infections [8, 9]. In addition, infections with gt4 have been observed in China and Japan [10–12]. Besides humans various wild and domesticated animal species such as wild boar and

---
* Correspondence: martin.groschup@fli.de
[1]Institute of Novel and Emerging Infectious Diseases, Friedrich-Loeffler-Institut, Südufer 10, 17493 Greifswald, Insel Riems, Germany
Full list of author information is available at the end of the article

pigs [13] have been found to carry HEV-3 and HEV-4; HEV-3 has also been detected in deer [14, 15] and rabbits [16, 17]. HEV-3/4 therefore is a zoonosis and pigs as well as wild boar represent the main reservoirs [7, 18]. Domestic pig populations worldwide frequently include viraemic animals and high HEV seroprevalences on herd level [5, 19, 20]. The exact pathogenesis of the HEV infection in pigs has still to be clarified. Naturally the infection occurs by faecal-oral transmission route which has been experimentally shown by a number of studies [21–24], albeit the intravenous infection is used most frequently in experimental challenge studies [22, 24–29]. Infected animals have high viral loads of HEV RNA mainly in the liver accompanied by faecal virus shedding in high concentrations [22, 23, 25, 26, 28]. Remarkably, no clinical symptoms were observed in general [5, 22, 25, 26]. Domestic pigs are a suitable and sensitive model for HEV-3 infection studies since they are also susceptible to human HEV-3 isolates [27].

The HEV-3 transmission from animals to humans via the consumption of infected undercooked/raw meat has been documented in numerous cases [7, 9]. Furthermore, HEV has been detected in processed food products like sausages in Germany [30], Italy [31] and France [9]. Regional distinctions such as raw meat consumption, liver delicacies and close proximity to livestock can have an influence on the exposure towards HEV [8, 32]. Additionally, people with work-related exposure to reservoir animals such as veterinarians, slaughterhouse personal, hunters or stable hands show significantly higher seroprevalences compared to the general population [33, 34].

In general HEV causes a wide range of symptoms, from subclinical to acute hepatitis with icterus up to fulminant hepatic failure [13]: The strictly human pathogenic genotypes HEV-1/2 the main source of endemic outbreaks – cause frequently an acute self-limited hepatitis with icterus and affect in general younger patients [35]. HEV-1 infections in pregnant women are often associated with severe courses, especially in the third trimester [3, 13]. HEV-3 and HEV-4 strains can additionally trigger extrahepatic manifestations but exhibit significantly lower mortality [36]. In contrast to HEV-1, HEV-3 infections can cause chronic and persistent infections mainly in immunocompromised patients [37, 38]. So far the infectivity of HEV is not clearly understood and difficult to determine and due to the lack of a sensitive cell culture system. Since domestic pig are highly susceptible to HEV the porcine model was selected to perform an endpoint titration study based on the application of serial dilutions of a HEV-3 positive liver homogenate. Since swine share many similarities with humans in physiology and immunology the determination of the minimal infectious dose provides important indications also for HEV infectivity in humans.

Results
Inoculum
The infection studies were carried out using tenfold dilutions of the inoculum from 10^{-2} up to 10^{-9}. Corresponding ct-values, copy numbers and IU are summarized in Table 1. 3.7 IU correspond to 1 copy/µl RNA which was calculated from standard curves of both PCR assays used in this study [39] (Additional files 1 and 2). According to this calculation, a 10^{-2} dilution of the experimentally infected liver contained 9.4×10^5 copies (3.4×10^6 IU) in 2 ml respectively.

Clinical parameters and pathology
Through-out the whole observation period none of the animals showed a febrile response or any clinical signs consistent with hepatitis. Additionally, no signs typical for viral hepatitis were seen at gross examination after necropsy.

Virus detection in faeces and serum intra-vitam
Viral RNA was detected in the faeces of all groups up to a dilution of 10^{-7}. In the first experimental set up covering homogenate dilutions from 10^{-2} to 10^{-4} over an observation period of 27 days (Fig. 1) viral shedding was observed in all three groups. First detection of viral RNA was in the group 10^{-2} at 9 dpi (pig T1–14) followed by groups 10^{-3} at 15 dpi and 10^{-4} at 17 dpi. Due to this delayed onset it was decided to extend the observation period for the second experimental set up for up to 91 days. The second 10^{-4} group started with viral shedding also at 17dpi, followed by the group 10^{-5} at 27 dpi. Viruses started to be shed within a time frame of six days for all animals within all groups of the first experiment and these two groups in the second experiment. In the remaining groups inoculated with higher HEV dilutions, a more variable time frame of viral

Table 1 RT-qPCR Values of the inocula after infection

Exp. No.	Dilution Group	CT-Value	cop/µl RNA	cop/ml[t]	cop/dose[t]	IU/dose[t]
1	10-2	25,6	13,010	467.853,40	935.706,80	3.443.401,02
	10-3	28,7	139	49.642,46	99.284,92	365.368,51
	10-4	33,0	5,61	2.003,56	4.007,11	14.746,17
2	10-4	32,2	10	3.607,11	7.214,22	26.548,33
	10-5	35,5	0,91	325,71	651,42	2.397,23
	10-6	no ct	–	–	65,1[a]	239,70*
	10-7	no ct	–	–	6,5[a]	24,00*
	10-8	no ct	–	–	0,65[a]	2,40*
	10-9	no ct	–	–	0,065[a]	0,24*

Viral copy numbers were calculated from CT values determined by RT-qPCR (HEV copies/µl RNA), volume of dose applied was 2 ml, [t]calculated based on HEV copies/µl RNA (see Additional files 1 and 2)[a]extrapolated values

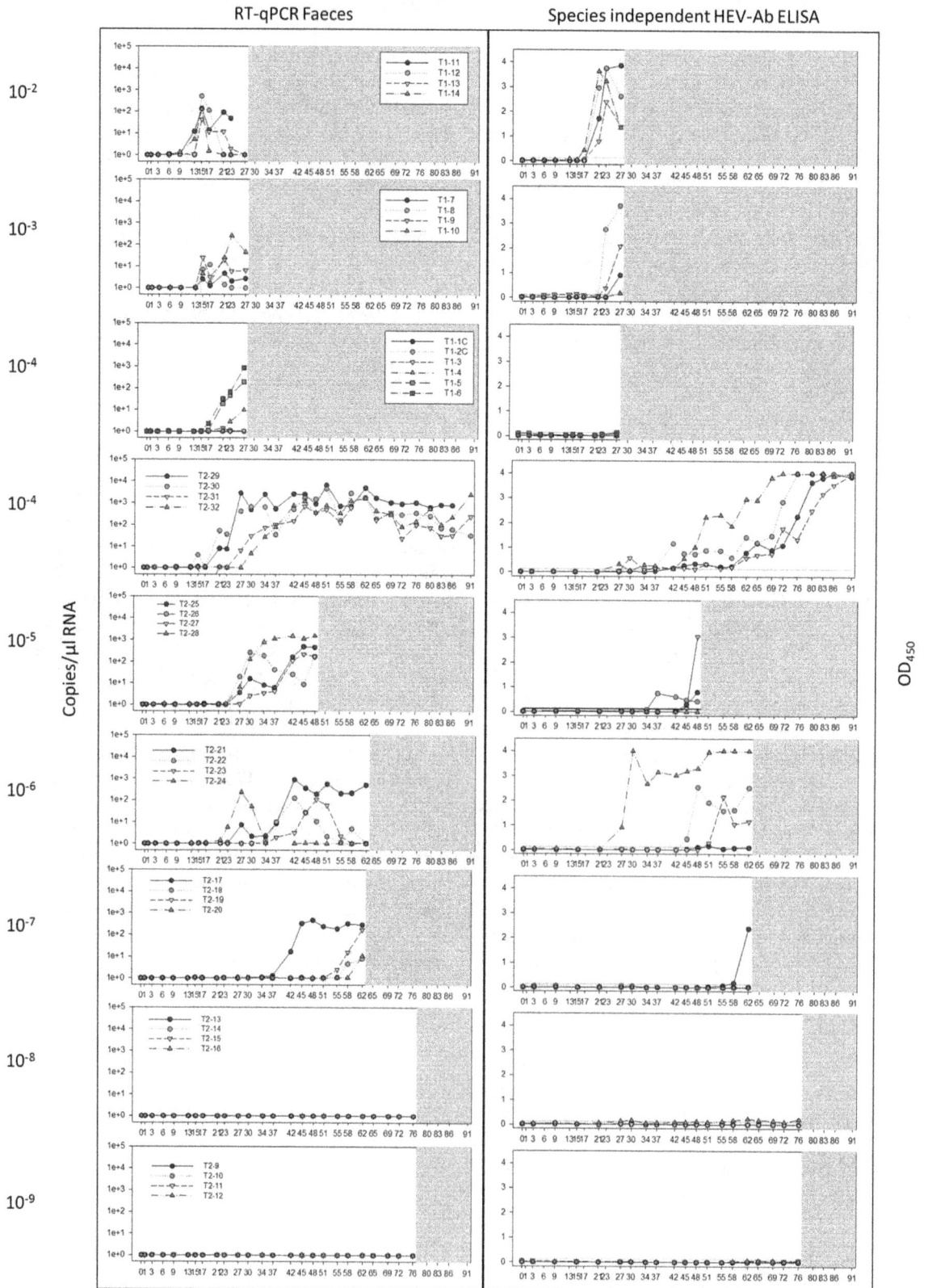

Fig. 1 Results of the RT-qPCR from faeces and the species independent HEV-Ab ELISA; the figures display the individual curves of viral RNA in faeces and the detection of total serum antibodies in correlation to days after infection

shedding was obvious. In group 10^{-6}, pig T2–24 shed from 21 dpi - 30 dpi. The remaining animals of this group started one by one after 27 dpi, 34 dpi, and 37 dpi. In group 10^{-7} pigs shed the virus until end of experiment; pig T2–17 started shedding virus at 37 dpi, while the other animals in this group started shedding at 55 dpi, 58 dpi and 62 dpi.

The majority of animals had between 1.3 and 51.7 HEV copies/µl RNA in the 10% faecal suspension initially and reached a plateau of about 10^3 HEV copies/µl RNA eventually. Only individual animals of the 10^{-4} and the 10^{-5} group of the second experiment (Fig. 1) occasionally excreted six fold higher viral loads (up to 6.6×10^3 HEV copies/µl RNA) in their faeces.

Viral clearance was detected in group 10^{-2} in which shedding of HEV RNA in faeces ceased in three out of four pigs (pig T1–12, pig T1–13, T1–14) after a mean of 7 d ($\pm$2.7). One individual (pig T1–11) remained positive till the end of the observation period on day 23. Unfortunately on day 27 (necropsy) no faeces was available from this animal for testing. In the 10^{-3} group only one animal (pig T1–08) cleared the virus before necropsy (27 dpi) 23 days dpi. No viral clearance was observed in the 10^{-4} group in both experiments even after 91 dpi. For group 10^{-5} an at least 18d period of faecal shedding was seen with no clearance till the end of the observation period of 49 dpi. In group 10^{-6} virus clearance occurred at very diverse intervals including 11d (pig T2–24), 21d (pig T2–23), 28d (pig T2–22) and more than 35d (pig T2–21), and in group 10^{-7} the animals shed virus for at least 36d (pig T2–17) or for 7, 4 or 1 days before being euthanized. Neither in the dilution 10^{-8} nor in the dilution 10^{-9} signs of virus replication or shedding were observed at all over a period of 76 dpi.

The detection of HEV RNA in serum was only sporadic and with low viral amounts (Additional files 2). Detection of viral RNA in serum samples at more than two consecutive sampling time points was only observed for groups 10^{-4}/2 (pig T2–29, pig T2–30, pig T2–31, pig T2–32), 10^{-5} (pig T2–25, pig T2–27, pig T2–28) and 10^{-7} (pig T2–17) after 24 dpi, 34 dpi and 58 dpi respectively.

Infectivity titre of liver

The infectivity in the liver tissue used for the inoculation can be quantified by the calculation of an ID_{50} according to Spearman and Kärber [40, 41] under particular assumptions: Restriction to an observation period of 27 days and in the case of longer incubation times the first positive animal within the group. The calculated titre was 6.3×10^5 ID_{50} per 1 ml liver tissue which represents 4.4×10^5 infectious units/ml. Setting this result in relation to the quantitation of HEV RNA in the liver

inoculum 773 IU HEV RNA correspond to one infectious unit (for calculations see Additional file 3).

Antibody detection

Seroconversion started in all animals after the excretion of viral RNA in faeces. Pigs seroconverted in group 10^{-2} at 17 dpi, group 10^{-3} at 23 dpi, group 10^{-4} at 27 dpi, 10^{-5} at 37 dpi, 10^{-6} at 27 dpi and 10^{-7} at 58 dpi. The interval between detection of HEV in faeces and seroconversion was 7.0 ± 1.0 days for group 10^{-2}, 11 ± 1.7 days for group 10^{-3}, 12.8 ± 12.8 days for group 10^{-4}, 16.0 ± 3.5 days for group 10^{-5}, 13.3 ± 7.2 days for group 10^{-6}, 31.0 ± 0.0 days for group 10^{-7}. In one case (pig T2–32) seroconversion was determined prior to virus detection which can be explained by the oscillation of corresponding OD values around the cut-off at sampling days 27 and 30 followed by a strong increase from day 34 on. HEV antibodies were assessed by the species independent HEV-Ab ELISA detecting total serum antibodies (e.g. IgG, IgM, IgA) in blood (Fig. 1). In the Priocheck HEV Ab porcine ELISA which only detects IgG, seroconversion was generally observed three to ten days later (Additional file 4). An exception was found for animals in the 10^{-7} group, where seroconversion was detected by both ELISA's at the same day. No seroconversion was observed in the 10^{-8} and 10^{-9} groups.

Virus detection in organs

All animals of the study were subjected to necropsy and blood, faeces, bile and tissue samples were taken and analysed by RT-qPCR for viral RNA (Tables 2 and 3). This included four different liver loci, the gallbladder, the hepatic lymph nodes, the spleen, the mesenteric lymph nodes and the mandibular lymph nodes. All animals which showed faecal shedding at necropsy were found HEV RNA positive in the bile and at least in two loci of the liver. The viral load of the gallbladder was considerably lower compared to bile itself. In pancreas no viral RNA was detected. In all other organs tested, individuals sporadically showed positive results. Corresponding data are shown in Tables 2 and 3.

Sequencing

The sequence of the hypervariable region of the inoculum (sequence and reference sequence KP294371 are shown in Additional file 5) was analysed by nested RT-PCR to monitor possible nucleotide exchanges after replication and passage through infected pigs. The original sequence displays a characteristic C/T wobble sequence at position 2200. Likewise virus sequences from isolates from pigs of dilution groups 10^{-2}, 10^{-3}, 10^{-4}, 10^{-5}, 10^{-6} and 10^{-7} displayed C/T polymorphisms at this site. In addition to the nucleotide 2200 polymorphism one isolate (derived from animal (T2–28))

Table 2 Results of RT-qPCR analysis of selected tissue samples groups 10^{-2} up to 10^{-4}

| | | 10^{-2} | | | | 10^{-3} | | | | 10^{-4} | | | | | | 10^{-4} | | | |
| | | 27dpi | | | | 27dpi | | | | 27dpi | | | | | | 91dpi | | | |
		T1-11	T1-12	T1-13	T1-14	T1-07	T1-08	T1-09	T1-10	T1-01c	T1-02c	T1-03	T1-04	T1-05	T1-06	T2-29	T2-30	T2-31	T2-32
Faeces	Ct	no Ct	no Ct	no Ct	no Ct	32.9	no Ct	31.7	29.0	no Ct	no Ct	no Ct	31.1	27.0	25.0	28.4	34.3	34.2	29.4
	copy					2.7		6.3	44.4				9.6	193.2	838.7	68.8	1.3	1.4	35.0
Bile	Ct	no Ct	no Ct	no Ct	no Ct	32.9	no Ct	30.7	26.6	no Ct	no Ct	31.2	26.7	25.1	23.0	24.4	31.5	30.1	24.1
	copy					2.6		13.2	267.0			9.5	243.2	777.6	3725.9	1007.5	8.7	21.8	1246.9
Liver 1	Ct	31.6	no Ct	no Ct	no Ct	30.8	no Ct	31.4	27.4	no Ct	no Ct	32.3	31.1	26.1	26.0	24.5	29.5	25.7	22.0
	copy	3.3				10.8		7.0	145.5			3.5	8.7	374.0	402.8	1884.9	38.6	727.6	13,142.2
Liver 2	Ct	32.8	no Ct	no Ct	no Ct	30.9	no Ct	31.0	27.8	24.9	no Ct	33.1	30.4	27.0	26.4	23.2	30.0	25.9	25.4
	copy	1.5				10.6		9.9	108.0	426.1		2.0	15.4	194.4	295.9	5187.7	25.0	637.3	898.7
Liver 3	Ct	32.9	25.2	no Ct	no Ct	29.5	no Ct	31.5	27.5	no Ct	no Ct	27.8	26.2	25.7	24.0	24.3	29.6	29.4	31.6
	copy	1.3	291.6			28.5		6.7	136.7			103.0	361.7	505.4	1753.2	2202.3	14.7	40.8	7.3
Liver 4	Ct	30.9	no Ct	33.1	no Ct	32.5	no Ct	32.6	26.8	no Ct	no Ct	no Ct	32.5	25.4	25.5	24.2	29.2	27.4	30.2
	copy	5.4		1.2		3.1		3.0	223.5				3.0	617.8	585.8	2354.4	46.2	191.2	21.5
Gall bladder	Ct	no Ct	no Ct	no Ct	no Ct	31.0	no Ct	32.5	28.2	no Ct	no Ct	32.2	31.1	28.7	25.8	30.0	no Ct	no Ct	32.1
	copy					9.5		3.2	76.4			3.9	8.9	52.6	485.5	25.2			5.1
Lnn. Hepat	Ct	no Ct	no Ct	no Ct	no Ct	no Ct	no Ct	33.4	no Ct	no Ct	no Ct	no Ct	no Ct	no Ct	32.1	31.9	no Ct	31.8	32.2
	copy							1.6							4.1	5.6		6.3	4.7
Spleen	Ct	no Ct	no Ct	no Ct	29.1	no Ct	no Ct	no Ct	no Ct	no Ct	no Ct	no Ct	no Ct	33.8	32.1	33.1	no Ct	32.8	32.2
	copy				18.9									1.2	4.1	2.3		2.9	4.5
Pancreas	Ct	no Ct	no Ct	no Ct	no Ct	no Ct	no Ct	no Ct	no Ct	no Ct	no Ct	no Ct	no Ct	no Ct	no Ct	no Ct	no Ct	no Ct	no Ct
	copy																		
Ln mes	Ct	no Ct	no Ct	no Ct	no Ct	no Ct	no Ct	no Ct	no Ct	no Ct	no Ct	no Ct	no Ct	no Ct	no Ct	30.6	32.8	no Ct	32.1
	copy															16.4	2.8		
Ln mand	Ct	no Ct	no Ct	no Ct	no Ct	no Ct	no Ct	no Ct	no Ct	no Ct	no Ct	no Ct	no Ct	no Ct	no Ct	no Ct	no Ct	33.4	no Ct
	copy																	1.8	4.9

No Ct $\geq$34.0 (=negative). Viral copy numbers were calculated from CT values determined by RT-qPCR (HEV copies/µl RNA). LN = lymph node

Table 3 Results of RT-qPCR analysis of selected tissue samples groups 10^{-5} up to 10^{-9}

| | | 10^{-5} | | | | 10^{-6} | | | | 10^{-7} | | | | 10^{-8} | | | | 10^{-9} | | | |
		T2-25 49dpi	T2-26	T2-27	T2-28	T2-21 63dpi	T2-22	T2-23	T2-24	T2-17 63dpi	T2-18	T2-19	T2-20	T2-13 77dpi	T2-14	T2-15	T2-16	T2-09 77dpi	T2-10	T2-11	T2-12
Faeces	Ct	n.s.	29.5	26.9	23.9	27.1	no Ct	no Ct	no Ct	26.3	32.3	27.1	32.3	no Ct	no Ct	no Ct	no Ct	no Ct	no Ct	no Ct	no Ct
	copy	n.s.	33.1	249.4	2473.3	163.2				283.7	3.5	163.0	3.6								
Bile	Ct	20.3	21.9	20.8	20.2	27.1	no Ct	no Ct	no Ct	24.3	27.5	24.6	25.1	no Ct	no Ct	no Ct	no Ct	no Ct	no Ct	no Ct	no Ct
	copy	36,698.1	10,918.1	25,355.3	39,844.7	160.0				1269.3	124.9	1006.4	700.5								
Liver 1	Ct	23.5	23.7	29.4	23.8	23.3	no Ct	no Ct	no Ct	24.4	30.3	no Ct	30.1	no Ct	no Ct	no Ct	no Ct	no Ct	no Ct	no Ct	no Ct
	copy	4016.9	3422.8	41.9	3226.2	1572.7				686.2	7.0		7.8								
Liver 2	Ct	23.0	23.5	28.4	25.2	26.0	no Ct	no Ct	no Ct	23.5	30.6	no Ct	32.2	no Ct	no Ct	no Ct	no Ct	no Ct	no Ct	no Ct	no Ct
	copy	5946.9	4025.0	86.6	1115.6	203.2				1383.5	5.2		1.6								
Liver 3	Ct	25.3	23.1	29.1	25.2	24.7	no Ct	no Ct	no Ct	23.2	28.0	25.9	no Ct	no Ct	no Ct	no Ct	no Ct	no Ct	no Ct	no Ct	no Ct
	copy	1017.0	5475.4	50.6	1120.3	534.1				1704.9	40.2	219.8									
Liver 4	Ct	24.3	22.3	28.2	26.4	22.2	no Ct	no Ct	no Ct	24.1	29.7	26.9	30.3	no Ct	no Ct	no Ct	no Ct	no Ct	no Ct	no Ct	no Ct
	copy	2191.1	10,005.2	103.4	423.1	3978.3				882.4	11.3	99.3	7.0								
Gall bladder	Ct	29.5	30.6	28.1	25.9	29.4	no Ct	no Ct	no Ct	26.2	29.9	24.9	29.1	no Ct	no Ct	no Ct	no Ct	no Ct	no Ct	no Ct	no Ct
	copy	39.0	15.5	110.0	634.2	14.3				171.0	9.5	478.6	17.0								
Lnn. Hepat	Ct	31.4	29.2	32.6	31.2	30.1	no Ct	no Ct	no Ct	30.9	no Ct	31.5	no Ct	no Ct	no Ct	no Ct	no Ct	no Ct	no Ct	no Ct	no Ct
	copy	8.5	45.9	3.3	10.0	7.8				4.3		2.6									
Spleen	Ct	no Ct	29.3	33.4	31.9	29.1	no Ct	no Ct	no Ct	28.1	no Ct	no Ct	no Ct	no Ct	no Ct	no Ct	no Ct	no Ct	no Ct	no Ct	no Ct
	copy		43.9	1.8	5.7	17.4				39.0											
Pancreas	Ct	no Ct	no Ct	no Ct	no Ct	no Ct	no Ct	no Ct	no Ct	no Ct	no Ct	no Ct	no Ct	no Ct	no Ct	no Ct	no Ct	no Ct	no Ct	no Ct	no Ct
	copy																				
Ln mes	Ct	33.2	32.3	no Ct	no Ct	27.2	no Ct	no Ct	no Ct	30.2	no Ct	no Ct	no Ct	no Ct	no Ct	no Ct	no Ct	no Ct	no Ct	no Ct	no Ct
	copy	2.1	4.4			74.5				7.4											
Ln mand	Ct	no Ct	33.2	no Ct	31.4	31.3	no Ct	no Ct	no Ct	no Ct	no Ct	no Ct	no Ct	no Ct	no Ct	no Ct	no Ct	no Ct	no Ct	no Ct	no Ct
	copy		2.0		8.2	3.2															

No Ct ≥34.0 (=negative). Viral copy numbers were calculated from CT values determined by RT-qPCR (HEV copies/µl RNA). LN = lymph node

had a C to T exchange at position 2322. The alignment was done with a previously published full genome (KP294371). The wobble sequence translates into a proline/serine exchange. The point mutation in animal T2–28 was silent.

Horizontal controls

The two horizontal controls (T1-01c, T1-02c) showed no clinical or molecular evidence for a HEV infection and faecal samples were always negative for HEV RNA. However, one liver lobe of pig T1-01c gave a positive RT-qPCR result (shown in Table 2).

Discussion

An increasing number of human hepatitis E genotype 3 (HEV-3) infections has been observed in developed countries in the last decade. This is accompanied by a substantial proportion of HEV RNA positive blood products and a growing number of transfusion transmitted HEV infections [8]. The high incidence represents a major health concern especially for immunosuppressed patients or people with pre-existing liver diseases. This raises the question of the minimal infectious dose for intravenously applied HEV-3. For this purpose - based on the similarities of pig and human physiology and the assumption of pigs as reliable animal models for human HEV infection, a comprehensive dose-titration study was carried out in domestic pigs to determine the minimum infectious dose. The titration was based on serial dilutions of a HEV RNA positive liver homogenate of an experimentally infected wild boar [23]. It ranged from a 10^{-2} dilution up to a 10^{-9} dilution and was applied into groups of 4 pigs each. Seroconversion and viral shedding was observed up to a 10^{-7} dilution which corresponds to a minimal infection dose of 6.5 copies in 2 ml total volume (according to 24 IU HEV RNA) for swine. In contrast to pigs, the infectious dose for humans is reportedly significantly higher: the transfusion of a platelet concentrate with residual plasma containing 7056–8892 IU HEV RNA was infectious [59]. In another study, the minimal infectious dose of HEV through transfusion was 3.6×10^4 IU [42]. A minimal infectious dose for humans in the order of 10,000 IU HEV RNA was also reported by *F. Rossi*, IPFA [43]. However, a direct correlation is not possible, since our pig study evaluated liver derived HEV in contrast to human blood (plasma) products. The reduced infectivity of HEV positive plasma samples has been recently demonstrated in human liver chimeric mice [44]. In monkeys, intravenous HEV infection studies were performed as well, however generally using inocula from faecal samples. The used dosages encompassed titers of 6.4 $\log_{10}$ copies/ml of a HEV-3 isolate [45] as well as 2.45×10^5 IU/ml for a HEV-4 and 7.51×10^5 IU/ml for a HEV-1 isolate [46]. Similar

dosages were used in a recent study [47] using different HEV-1 isolates (3.5–6.4 $\log_{10}$ IU/ml) and HEV-3 isolates (2.5–9.5 log 10 IU/ml). In each case cynomolgus monkeys could be productively infected.

All infected animals remained asymptomatic and showed no signs of illness such as fever, reduced alertness nor weight loss throughout the experiment which is in line with previous studies in domestic pigs [24, 25, 28]. So far, only singular cases of icterus in HEV infected pigs [27] as well as the elevation of liver specific enzymes in infected wild boar [23] have been reported. This study established faeces as appropriate source to determine the course of virus replication in pigs via RT-qPCR due to weak and sporadic detection of viral RNA in serum with a general lower viral load. This is in accordance with previous reports [22, 48]. It is of special significance that the calculated copy number of 5 copies /ml in the 10^{-7} dilution lies below the limit of detection of both PCR assays but is sufficient to infect pigs by the intravenous route. This is in accordance with a previous pig titration study where infectivity was observed below the RT-PCR detection limit [28]. In addition a clear correlation between applied dosage and incubation period was determined: A 10^{-2} dilution of the inoculum induced viral shedding at 9 dpi, followed by virus shedding at 15 dpi (10^{-3} dilution), 17 dpi (10^{-4} dilution) and 27 dpi (the 10^{-5} dilution) in at least one pig per group, on average 30 dpi (10^{-6} dilution) and on average 51 days (10^{-7} dilution). The subsequent dilutions also induce growing individual differences in onset of virus shedding varying from 6 to 9 days (10^{-4} group), 10–18 days (10^{-6} group) up to 16 days (10^{-7} group). The increasing extension of incubation period raises the question regarding replication site and persistence of the virus. Further dilutions (10^{-8}, 10^{-9}) of the inoculum induced no HEV infection. In contrast, in the aforementioned study none of the animals inoculated with dilutions below 10^{-4} were infected [28]. However a direct comparison of both studies was not possible, since the viral load was determined by Genomic Equivalents (GE/ml) and due to the use of faecal material for intravenous inoculation. Both sources are able to induce virus replication as reported in multiple studies. A comparison of these studies regarding onset and duration of virus shedding and seroconversion is difficult because applied doses are incomparable and inoculum sources were different: After i.v. application of HEV positive faeces (10% suspension) virus detection in faeces started 3 dpi [25, 26] or 7–14 dpi [28] and took 1–3 weeks. Using inoculum from a HEV positive pig derived bile led to virus shedding from 3 dpi on [22]. Using high titre liver homogenates i.v. infected pigs started to excrete virus between 2 and 7 days and continued 12–52 days [24, 29]. In another study faecal RNA samples of infected pigs were detected at day 9 to 19 and faecal HEV excretion lasted 21–30 days [21]. Interestingly both sources harbour significant differences because blood as well as hepatocyte/liver derived HEV particles are enveloped and covered with cellular membranes in contrast to non-enveloped HEV from bile or faeces [49–51]. This envelope has been also detected in cell-culture produced HEV particles [50, 52] and appears to modulate infectivity in vitro as well as in vivo. The generation of quasi-enveloped HEV particles by hepatocytes and hepatoma cell lines [51] strongly indicates that HEV from liver homogenate harbors an envelope as well, but experimental data are still lacking. Cell culture derived enveloped HEV revealed an almost tenfold lower infectivity compared to non-enveloped faeces derived HEV in cell culture [53], which may be a consequence of inefficient cell attachment [54]. However, the findings are complicated by the fact, that human liver homogenate as well as faeces from HEV positive patients is capable to infect humanized chimeric mice in contrast to human plasma or cell-culture derived HEV [55]. Therefore the question how biochemical properties of HEV particles modulate infectivity needs further investigation.

A presumably faecal-oral transmission of HEV was observed by one of the horizontal controls (pig T1-01c) within the 10^{-4} group. The animal exhibited a HEV positive liver as determined by necropsy at 27dpi. The first onset of viral shedding in this group was on 17dpi from pig T1–06 which leads to a maximum incubation period of 10 days. The reported periods between oral intake and HEV particles shed in faeces can range from 7.2d [24] up to 22d [22] and possibly even longer.

Antibody response and seroconversion were assessed by two ELISAs with different specifications: The AXIOM ELISA was a multispecies test for detecting total serum antibodies against HEV (including IgM, IgA and IgG) whereas the PrioCheck ELISA only detected porcine HEV specific IgG antibodies. Specific porcine IgM ELISAs were not available. In general the serum samples were confirmed positive in the AXIOM ELISA three to ten days before the porcine specific IgG ELISA which is consistent with a primary IgM related immune response followed by IgG response. However, no reliable conclusions can be drawn about the specific occurrence of IgM or IgA or of a mixture of both during the primary immune response. In any case IgA antibodies play an important role in the human immune response against HEV [55, 56].

Virus clearance could be observed in different groups examined: Three animals in group 10^{-2}, one individual from 10^{-3} and three from the 10^{-6} dilution group cleared the infection within 7d (±2,7d), 8d and 9d up to 27d respectively. All remaining animals were subjected to necropsy prior to a possible virus clearance. All animals of the second 10^{-4} group displayed a prolonged infection where HEV RNA was detectable in faeces for more than 62 days. Necropsy samples of this group harboured high viral load in bile and liver samples. The 10^{-5} group exhibited faecal viral shedding for 22d and individual animals from group 10^{-6} and 10^{-7} group for 35d and 36d, respectively. Again, all animals showed high viral loads in bile and liver at necropsy. In general, HEV is considered to be a transient infection and viral clearance usually occurs within three weeks after first occurrence in faeces [22, 24]. The long-term viral shedding for more than 32 up to 62 days is therefore uncommon and possibly indicates a chronic or persistent infection. Only rare data from similar studies are available: HEV shedding was significantly increased by co-infection with Porcine Reproductive and Respiratory Syndrome Virus (PRRSV) and extended from 9.7 to 48.6 days [57]. However, all animals in this study had been pre-tested to be negative for PRRSV as well as Porcine circovirus 2 (PCV2). This analysis was continued with faeces samples from the second 10^{-4} group encompassing the whole observation period, to exclude a newly or internal acquired PRRSV and PCV2 infection. Again all samples were negative for both viruses. All animals showing prolonged viral shedding belonged to the dilutions 10^{-4} or higher, indicating a correlation between inoculated virus load and prolonged viral shedding. The 10–4 g.

Another research study showed extended virus shedding in wild boars for more than 16 weeks [48]. Interestingly the animals showed high antibody titers as well, with apparently no influence on virus shedding and replication. The presence of neutralising antibodies within the context of a HEV infection has been shown in vivo [58, 59] and in vitro [58–60]. Other factors such as age and individual immune status should be taken into account as well.

It may be of interest that in humans various cases of chronic HEV-3 infections were reported in immunosuppressed patients [61]. For all other genotypes such a correlation has not been made so far.

Conclusion

In conclusion, this study determined an extremely low minimum infectious dose of HEV to elicit a productive infection in pigs. Additionally our work demonstrated a dose dependent incubation time in pig resulting in a delayed onset of virus replication and shedding. Moreover the study emphasises the potential risk of HEV transmission in humans via blood and blood products. Finally further investigations should be undertaken to determine and compare the infectivity through oral transmission.

Methods

Inoculum

The inoculum used in this study was prepared from a highly HEV RNA positive liver of an experimentally infected wild boar from a previous HEV-3 infection study [23].

20 g of liver tissue was grounded with mortar, pestle and sterile sea sand. The homogenate was then diluted 1:5 with sterile 1x phosphate buffered saline (1xPBS) to obtain a 20% dilution and centrifuged at 4400 g for 15 min at 4 °C. The supernatant was eventually sterile-filtered through 0.22 μm MILLEX-GP Syringe Filter Unit (Millipore, Ireland), aliquoted and stored at − 80 °C. The inoculum was titrated with the Logarithmic dilutions in sterile 1xPBS starting from a 10^{-2} dilution (referring to the original liver tissue) to 10^{-9}. The 10^{-2} dilution was sterile filtered prior to inoculation and further dilutions. The corresponding inocula were prepared shortly before inoculation and kept on ice until use and administered intravenously (i.v.) to 4 animals/group.

Experimental design

The competent authority of the Federal State of Mecklenburg Western-Pomerania approved all described animal experiments based on European Directive 2010/63/EU and associated national regulation (reference number in Germany LALLF M-V/TSD/7221.3–2.1.014/10, LALLF M-V/TSD/7221.3–2.1-017/13).

For this study 38 domestic pigs (Large White breed) from a commercial breeder (animal husbandry, 18,196 Dummerstorf, Germany) were acquired and housed under containment level 3** conditions. Healthy animals of compatible sizes and ages were allocated randomly to groups by animal technicians. Social incompatibilities were taken into account in few instances for animal welfare reasons. As the study was a titration experiment with a clear cut yes/no readout, the number of animals per group was set as described together with a biostatistician. The experimental design was part of the individualized animal experimental license (reference number in Germany LALLF M-V/TSD/7221.3–2.1.014/10, LALLF M-V/TSD/7221.3–2.1-017/13 according to the Germany animal welfare law (§7 paragraph 1 phrase 2 TierSchG) which was applied for and given after scientific and ethical assessment by an independent advisory board. After entry, the animals were divided into the experimental groups and held in quarantine for 2 weeks. Individuals of each group were housed together whereas each group was held in a separate stable unit. Complete change of clothes was mandatory before entry of each of the rooms. All animals were pretested for HEV and yielded negative PCR findings in faeces and serum as well negative HEV ELISA results. Due to facility limitations the study was divided in two experiments. The groups were randomly formed consisting of 4 animals each as described in Table 4. Two horizontal transmission control animals were included and kept with the 10^{-4} group. A blinded experimental and analysis design was not possible as researchers and animal technicians were supposed to carry out husbandry and experimental work in a defined way in order to mitigate (cross-) contamination risks: every day animal manipulations started in the low dose challenged animals, followed by intermediate dose challenged pigs and in high-dose animals eventually. Moreover, as only standardized quantitative data were obtained (e.g. body temperatures, qPCR (Ct-values and copy numbers) and ELISA derived data), blinding was not considered necessary.

After at least two weeks of acclimatisation negative HEV RNA and Anti-HEV-antibody results were reconfirmed by Rt-qPCR and ELISA. Additionally samples were examined by a novel multiplex PCR [62] (Results not shown) to exclude an influence of co-infection with other viral diseases (PRRSV, PCV2). The inoculation was done with 2.0 ml inoculum intravenously given into the *Vena cava cranialis*.

Table 4 Experimental set up for the titration of highly HEV positive liver tissue in log steps in the porcine model

Exp. No.	Dilution = Group	Animal identity	Animals		Sex	Age at inoculation in weeks	Observation period
			infected	horizontal control			
1	10^{-2}	Pig T1-11, −12, −13, −14	4	–	f/m	10	27dpi
	10^{-3}	Pig T1-07, −08, −09, −10,	4	–	f/m	10	27dpi
	10^{-4}	Pig T1-01c -02c,-03, −04, −05, −06	4	2	f/m	10	27dpi
2	10^{-4}	Pig T2-29, −30, −31, −32	4	–	m	11	91dpi
	10^{-5}	Pig T2-25, −26, −27, −28	4	–	f/m	11	49dpi
	10^{-6}	Pig T2-21, −22, −23, −24	4	–	f	11	63dpi
	10^{-7}	Pig T2-17, −18, −19, −20	4	–	f/m	11	63dpi
	10^{-8}	Pig T2-13, −14, −15, −16	4	–	m	11	77dpi
	10^{-9}	Pig T2-09, −10, −11, −12	4	–	f	11	77dpi

The dilution step of the homogenate as assigned the group designation, f-female, m-male, dpi-days post infection, -- none

Animal behaviour and rectal body temperatures were checked daily. As described in previous studies body temperatures over 40 °C for at least two consecutive days were considered a febrile response [23, 48]. Depression, diarrhoea, vomitus, icterus, ascites and neurological symptoms were considered signs of acute hepatitis and would have led to an immediate removal and euthanasia of the animal.

During the experiments blood and faecal samples were taken regularly every two to three days. Blood was allowed to clot for 30 min at room temperature and then centrifuged at 2300 g for 12 min. Serum was then collected, aliquoted and stored at – 20 °C. From the faecal samples a 10% faecal suspension was made using 0.89% NaCl-solution. After vortexing and centrifugation (4400 g, 4 °C, 20 min) the supernatant was sterile filtrated using a sterile 0.22 μm MILLEX-GP Syringe Filter Unit (Millipore, Ireland) and stored at – 20 °C. This solution was the starting point for RNA extraction.

At the end of the observation period all animals were slaughtered (electro stunning followed by exsanguination). Animals were euthanized by a veterinarian following EU and German animal welfare regulations and carcasses necropsied by a trained veterinary pathologist assisted by a necropsy technician. Necropsies were performed and samples from blood, faeces and bile as well as different tissue samples were taken for RNA extraction and stored at – 20 °C. Retrieved tissue samples were immediately immersed in 4% neutral buffered formalin.

RNA and antibody detection

A sample volume of 140 μl of serum, bile and faecal filtrates were extracted manually with the QIAmp® Viral RNA Mini Kit (QIAGEN GmbH, Hilden Germany) following the manufactures instructions and eluated in 50 μl buffer. Manual RNA extraction from tissue samples was performed with the RNEasy® Mini Kits (QIAGEN GmbH, Hilden Germany). For this purpose 10 mg of a tissue sample was homogenized in 600 μl RLT buffer using TissueLyser II ® (Qiagen). After centrifugation the supernatant was used for RNA extraction according to manufacturer's instructions. A heterologous internal control [63] was added to each extraction sample. Obtained RNA was stored at – 80 °C until further use.

To monitor the course of the infection a diagnostic quantitative real-time RT-PCR (RT-qPCR) targeting a fragment of ORF3 was performed. Primer, probes and protocol were used as previously reported [23, 39]. Each reaction containing 25 μL had a final primer concentration of 0.8 μM and of 0.1 μM probe and 5 μl RNA. RT-qPCR was carried out using the Quanti Tect Probe RT-PCR Kit (QIAGEN GmbH). The CFX96™Real-Time System (Bio-Rad Laboratories GmbH, Munich, Germany) was set to 50 °C for 30 min for reverse transcription followed by

denaturation/activation 95 °C for 15 min. DNA amplification was performed in 45 cycles consisting each of 95 °C for 10 s, 55 °C for 25 s and 72 °C for 25 s in immediate succession. The quantification of RNA (HEV copies/μl RNA) was performed by a standard curve based on serial dilutions of a HEV standard, which was included in each qRT-PCR run (Additional file 1). The copy number of HEV standards was calculated by a synthetic calibrator which consists of the qRT-PCR amplicon (81 nucleotides) and a T7-Promotor sequence at the 5′-end to allow in vitro transcription [39]. International Units/ml (IU) were calculated from the WHO International standard for HEV. This standard was provided by the Paul-Ehrlich-Institut (PEI), Langen, Germany (PEI code 6329/10). All RT-qPCR results are given in HEV copies/μl RNA (originating from 50 μl elution volume). The corresponding cop/ml (originating from 140 μl fluid sample) were calculated by conversion factor of 357, 14 (elution volume divided by fluid sample volume) and is depicted in Additional file 2. The limit of detection of about 1 cop/μl is reached at ct-values of ~ 35. The limit of quantification corresponds to the 10^{-4} HEV standard at 7 copies/μl. Viral loads below the lowest HEV standard were retested by an alternative HEV specific qRT-PCR [64].

To monitor the immunological responses, the serum samples were tested with two commercially available HEV ELISA kits. One was the species independent HEV-Ab ELISA (AXIOM, Bürstadt, Germany) [23] detecting total serum anti-HEV-antibodies. The other was the porcine specific Priocheck HEV Ab porcine ELISA (Mikrogen GmbH, Neuried, Germany), which is specific for serum Anti-HEV-IgG. Both were carried out and interpreted as described by the manufacturer.

Sequencing

To monitor and verify the identity of the inoculated compared to the replicated/excreted virus genome a 348 nucleotide long partial sequence of the hypervariable region (HVR) was recovered from corresponding faecal samples. SYBR Green-RT-qPCR followed by a SYBR Green nested PCR both with melting curve analysis was performed as described [39]. The resulting cDNA was sequenced by a commercial provider (Eurofins Genomics GmbH, Ebersberg, Germany).

Additional files

Additional file 1: Amplification curve of HEV specific qRT-PCR. A) Amplification curve of HEV specific qRT-PCR targeting the HEV standard (red line) and the RNA extracts from the inocula (blue line). B) Standard curves were obtained by Ct values plotted against the log of starting quantity. C) Obtained Ct values and determined copy numbers (DOCX 219 kb)

Additional file 2: Determination of the conversion factor from copies per μl RNA to IU of the WHO standard (DOCX 16 kb)

Additional file 3: Calculation of the Infectivity titer of the Liver Used for Inoculation (DOCX 16 kb)

Additional file 4: Results of the RT-qPCR from serum and the porcine IgG HEV-Ab ELISA; The graphs display the individual curves of viral RNA in serum and the detection of Anti-HEV-IgG antibodies in correlation to days after infection (DOCX 462 kb)

Additional file 5: Alignment of the hypervariable region. Alignment done with Geneious version 10.2 created by Biomatters. Available from https://www.geneious.com (DOCX 244 kb)

Abbreviations
Ab: Antibody; dpi: Days post inoculation; ELISA: Enzyme-linked immunosorbent assay; f: Female; gt: Genotype; HEV: Hepatitis E Virus; HVR: Hypervariable region; i.v.: Intravenous; Ig: Immunglobulin; IU: International Unit; m: Male; PBS: Phosphate buffered saline; PCR: Polymerace chain reaction; PCV2: Porcine circovirus 2; PRRSV: Porcine reproductive and respiratory syndrome virus; RNA: Ribonucleic acid

Acknowledgments
We thank Hanan Sheik Ali for their profound help. In addition, we thank Birke Boettcher and Tobias Winterfeld, for their excellent technical support, Svenja Wiechert and Erika Hilbold for their excellent help during their internship and all animal caretakers for the excellent animal husbandry. This study was partially funded by a research grant from CSL Behring Marburg.

Funding
This project has received funding from the European Union's Horizon 2020 research and innovation program VetBioNet under grant agreement No 731014.

Authors' contributions
Conceived and designed the experiments: JS, LD, ME, MHG. Performed the experiments and necropsy: LD, JS, CF, EL, CS, ME, AG. Analysed the Data: LD, JS, ME, CF, MHG. Wrote the paper: LD, JS, ME, MHG, WS. All authors have read and approved the manuscript.

Consent for publication
Not applicable.

Competing interests
W. Schäfer is an employee and A. Gröner is a former employee of CSL Behring. The authors declare that they have no competing interests

Author details
[1]Institute of Novel and Emerging Infectious Diseases, Friedrich-Loeffler-Institut, Südufer 10, 17493 Greifswald, Insel Riems, Germany. [2]Department of Veterinary Medicine, Institute of Immunology, Freie Universität Berlin, Robert-von-Ostertag-Straße 7-13, 14163 Berlin, Germany. [3]Department of Experimental Animal Facilities and Biorisk Management, Friedrich-Loeffler-Institut, 17493 Greifswald, Insel Riems, Germany. [4]PathoGuard Consult, Fasanenweg 6, 64342, Seeheim-Jugenheim, Germany. [5]CSL Behring Biotherapies for Life™, P.O. Box 1230, 35002 Marburg, Germany.

References
1. WorldHealthOrganization. Hepatitis E, Fact sheet [updated July 2017]. Available from: http://www.who.int/mediacentre/factsheets/fs280/en/en/.
2. International Committee on Taxonomy of Viruses Hepeviridae Study G, Smith DB, Simmonds P, Jameel S, Emerson SU, Harrison TJ, et al. Consensus proposals for classification of the family Hepeviridae. The Journal of general virology. 2014;95(Pt 10):2223–32.
3. Balayan MS. Epidemiology of hepatitis E virus infection. J Viral Hepat. 1997; 4(3):155–65.
4. Arankalle VA, Chadha MS, Chobe LP, Nair R, Banerjee K. Cross-challenge studies in rhesus monkeys employing different Indian isolates of hepatitis E virus. J Med Virol. 1995;46(4):358–63.
5. Meng X, Purcell R, Halbur P, Lehman J, Webb D, Tsareva T, et al. A novel virus in swine is closley related to the human HEV. Proc Natl Acad Sci U S A. 1997;94:6.
6. Khuroo MS, Khuroo MS. Hepatitis E: an emerging global disease - from discovery towards control and cure. J Viral Hepat. 2016;23(2):68–79.
7. Christou L, Kosmidou M. Hepatitis E virus in the Western world--a pork-related zoonosis. Clin Microbiol Infect. 2013;19(7):600–4.
8. Aspinall EJ, Couturier E, Faber M, Said B, Ijaz S, Tavoschi L, et al. Hepatitis E virus infection in Europe: surveillance and descriptive epidemiology of confirmed cases, 2005 to 2015. Euro Surveill. 2017;22(26):13–22.
9. Colson P, Borentain P, Queyriaux B, Kaba M, Moal V, Gallian P, et al. Pig liver sausage as a source of hepatitis E virus transmission to humans. J Infect Dis. 2010;202(6):825–34.
10. Matsubayashi K, Kang JH, Sakata H, Takahashi K, Shindo M, Kato M, et al. A case of transfusion-transmitted hepatitis E caused by blood from a donor infected with hepatitis E virus via zoonotic food-borne route. Transfusion. 2008;48(7):1368–75.
11. Zhang W, He Y, Wang H, Shen Q, Cui L, Wang X, et al. Hepatitis E virus genotype diversity in eastern China. Emerg Infect Dis. 2010;16(10):1630–2.
12. Dai X, Dong C, Zhou Z, Liang J, Dong M, Yang Y, et al. Hepatitis E virus genotype 4, Nanjing, China, 2001-2011. Emerg Infect Dis. 2013;19(9):1528–30.
13. Kamar N, Bendall R, Legrand-Abravanel F, Xia N-S, Ijaz S, Izopet J, et al. Hepatitis E. Lancet. 2012;379(9835):2477–88.
14. Takahashi K, Kitajima N, Abe N, Mishiro S. Complete or near-complete nucleotide sequences of hepatitis E virus genome recovered from a wild boar, a deer, and four patients who ate the deer. Virology. 2004;330(2):501–5.
15. Rutjes SA, Lodder-Verschoor F, Lodder WJ, van der Giessen J, Reesink H, Bouwknegt M, et al. Seroprevalence and molecular detection of hepatitis E virus in wild boar and red deer in the Netherlands. J Virol Methods. 2010; 168(1–2):197–206.
16. Zhao C, Ma Z, Harrison TJ, Feng R, Zhang C, Qiao Z, et al. A novel genotype of hepatitis E virus prevalent among farmed rabbits in China. J Med Virol. 2009;81(8):1371–9.
17. Hammerschmidt F, Schwaiger K, Dähnert L, Vina-Rodriguez A, Hoper D, Gareis M, et al. Hepatitis E virus in wild rabbits and European brown hares in Germany. Zoonoses Public Health. 2017;64(8):612–22.
18. Thiry D, Mauroy A, Pavio N, Purdy MA, Rose N, Thiry E, et al. Hepatitis E Virus and Related Viruses in Animals. Transbound Emerg Dis. 2015;64(1):37–52.
19. Krumbholz A, Joel S, Neubert A, Dremsek P, Durrwald R, Johne R, et al. Age-related and regional differences in the prevalence of hepatitis E virus-specific antibodies in pigs in Germany. Vet Microbiol. 2013;167(3–4):394–402.
20. Hsieh SY, Meng XJ, Wu YH, Liu ST, Tam AW, Lin DY, et al. Identity of a novel swine hepatitis E virus in Taiwan forming a monophyletic group with Taiwan isolates of human hepatitis E virus. J Clin Microbiol. 1999;37(12): 3828–34.
21. Andraud M, Dumarest M, Cariolet R, Aylaj B, Barnaud E, Eono F, et al. Direct contact and environmental contaminations are responsible for HEV transmission in pigs. Vet Res. 2013;44:102.
22. Casas M, Pina S, de Deus N, Peralta B, Martin M, Segales J. Pigs orally inoculated with swine hepatitis E virus are able to infect contact sentinels. Vet Microbiol. 2009;138(1–2):78–84.
23. Schlosser J, Eiden M, Vina-Rodriguez A, Fast C, Dremsek P, Lange E, et al. Natural and experimental hepatitis E virus genotype 3-infection in European wild boar is transmissible to domestic pigs. Vet Res. 2014;45:121.
24. Bouwknegt M, Rutjes SA, Reusken CB, Stockhofe-Zurwieden N, Frankena K, de Jong MC, et al. The course of hepatitis E virus infection in pigs after contact-infection and intravenous inoculation. BMC Vet Res. 2009;5:7.
25. Halbur PG, Kasorndorkbua C, Gilbert C, Guenette D, Potters MB, Purcell RH, et al. Comparative pathogenesis of infection of pigs with hepatitis E viruses recovered from a pig and a human. J Clin Microbiol. 2001; 39(3):918–23.
26. Lee YH, Ha Y, Ahn KK, Chae C. Localisation of swine hepatitis E virus in experimentally infected pigs. Vet J. 2009;179(3):417–21.
27. Balayan. Brief Repor, Experimental Hepatitis E Infection in Domestic Pigs. 1990.
28. Meng XJ, Halbur PG, Shapiro MS, Govindarajan S, Bruna JD, Mushahwar IK, et al. Genetic and experimental evidence for cross-species infection by

swine hepatitis E virus. J Virol. 1998;72(12):9714–21.

29. Bouwknegt M, Frankena K, Rutjes SA, Wellenberg GJ, de Roda Husman AM, van der Poel WH, et al. Estimation of hepatitis E virus transmission among pigs due to contact-exposure. Vet Res. 2008;39(5):40.

30. Szabo K, Trojnar E, Anheyer-Behmenburg H, Binder A, Schotte U, Ellerbroek L, et al. Detection of hepatitis E virus RNA in raw sausages and liver sausages from retail in Germany using an optimized method. Int J Food Microbiol. 2015;215:149–56.

31. Di Bartolo I, Angeloni G, Ponterio E, Ostanello F, Ruggeri FM. Detection of hepatitis E virus in pork liver sausages. Int J Food Microbiol. 2015;193:29–33.

32. Lapa D, Capobianchi MR, Garbuglia AR. Epidemiology of hepatitis E virus in European countries. Int J Mol Sci. 2015;16(10):25711–43.

33. Krumbholz A, Mohn U, Lange J, Motz M, Wenzel JJ, Jilg W, et al. Prevalence of hepatitis E virus-specific antibodies in humans with occupational exposure to pigs. Med Microbiol Immunol. 2012;201(2):239–44.

34. Meng XJ, Wiseman B, Elvinger F, Guenette DK, Toth TE, Engle RE, et al. Prevalence of antibodies to hepatitis E virus in veterinarians working with swine and in Normal blood donors in the United States and other countries. J Clin Microbiol. 2002;40(1):117–22.

35. Aggarwal R. Clinical presentation of hepatitis E. Virus Res. 2011;161(1):15–22.

36. Ankcorn MJ, Tedder RS. Hepatitis E: the current state of play. Transfus Med. 2017;27(2):84–95.

37. Kamar N, Selves J, Mansuy JM, Ouezzani L, Peron JM, Guitard J, et al. Hepatitis E virus and chronic hepatitis in organ-transplant recipients. N Engl J Med. 2008;358(8):811–7.

38. Dalton HR, Bendall RP, Keane FE, Tedder RS, Ijaz S. Persistent carriage of hepatitis E virus in patients with HIV infection. N Engl J Med. 2009;361(10):1025–7.

39. Vina-Rodriguez A, Schlosser J, Becher D, Kaden V, Groschup MH, Eiden M. Hepatitis E virus genotype 3 diversity: phylogenetic analysis and presence of subtype 3b in wild boar in Europe. Viruses. 2015;7(5):2704–26.

40. Kärber G. Beitrag zur kollektiven Behandlung pharmakologischer Reihenversuche. Naunyn-Schmiedeberg's Arch Exp Pathol Pharmakol. 1931;162(4):480–3.

41. Spearman C. The method of "right and wrong cases" (constant stimuli) without Gauss's formula. Br J Psychol. 1908;2:227–42.

42. Satake M, Matsubayashi K, Hoshi Y, Taira R, Furui Y, Kokudo N, et al. Unique clinical courses of transfusion-transmitted hepatitis E in patients with immunosuppression. Transfusion. 2017;57(2):280–8.

43. Rossi F. Reflection paper on viral safety of plasma-derived medicinal products with respect to hepatitis E virus. Committee for Medicinal Products for Human Use (CHMP). 2015(EMA/CHMP/BWP/723009/2014).

44. van de Garde MD, Pas SD, van der Net G, de Man RA, Osterhaus AD, Haagmans BL, et al. Hepatitis E virus (HEV) genotype 3 infection of human liver chimeric mice as a model for chronic HEV infection. J Virol. 2016;90(9):4394–401.

45. Roques P, Gardinali NR, Guimarães JR, Melgaço JG, Kevorkian YB, FDO B, et al. Cynomolgus monkeys are successfully and persistently infected with hepatitis E virus genotype 3 (HEV-3) after long-term immunosuppressive therapy. PloS one. 2017;12(3):e0174070.

46. Geng Y, Zhao C, Huang W, Harrison TJ, Zhang H, Geng K, et al. Detection and assessment of infectivity of hepatitis E virus in urine. J Hepatol. 2016;64(1):37–43.

47. Choi YH, Zhang X, Tran C, Skinner B. Expression profiles of host immune response-related genes against HEV genotype 3 and genotype 1 infections in rhesus macaques. J Viral Hepat. 2018;25(8):986–95.

48. Schlosser J, Vina-Rodriguez A, Fast C, Groschup MH, Eiden M. Chronically infected wild boar can transmit genotype 3 hepatitis E virus to domestic pigs. Vet Microbiol. 2015;180(1–2):15–21.

49. Nagashima S, Takahashi M, Kobayashi T, Tanggis, Nishizawa T, Nishiyama T, et al. Characterization of the Quasi-Enveloped Hepatitis E Virus Particles Released by the Cellular Exosomal Pathway. J Virol. 2017;91(22):e00822–17.

50. Takahashi M, Yamada K, Hoshino Y, Takahashi H, Ichiyama K, Tanaka T, et al. Monoclonal antibodies raised against the ORF3 protein of hepatitis E virus (HEV) can capture HEV particles in culture supernatant and serum but not those in feces. Arch Virol. 2008;153(9):1703–13.

51. Okamoto H. Culture systems for hepatitis E virus. J Gastroenterol. 2013;48(2):147–58.

52. Qi Y, Zhang F, Zhang L, Harrison TJ, Huang W, Zhao C, et al. Hepatitis E virus produced from cell culture has a lipid envelope. PLoS One. 2015;10(7):e0132503.

53. Chapuy-Regaud S, Dubois M, Plisson-Chastang C, Bonnefois T, Lhomme S, Bertrand-Michel J, et al. Characterization of the lipid envelope of exosome encapsulated HEV particles protected from the immune response. Biochimie. 2017;141:70–9.

54. Yin X, Ambardekar C, Lu Y, Feng Z. Distinct entry mechanisms for nonenveloped and quasi-enveloped hepatitis E viruses. J Virol. 2016;90(8):4232–42.

55. Tian DY, Chen Y, Xia NS. Significance of serum IgA in patients with acute hepatitis E virus infection. World J Gastroenterol. 2006;12(24):3919–23.

56. Zhang S, Tian D, Zhang Z, Xiong J, Yuan Q, Ge S, et al. Clinical significance of anti-HEV IgA in diagnosis of acute genotype 4 hepatitis E virus infection negative for anti-HEV IgM. Dig Dis Sci. 2009;54(11):2512–8.

57. Salines M, Barnaud E, Andraud M, Eono F, Renson P, Bourry O, et al. Hepatitis E virus chronic infection of swine co-infected with porcine reproductive and respiratory syndrome virus. Vet Res. 2015;46:55.

58. Emerson SU, Clemente-Casares P, Moiduddin N, Arankalle VA, Torian U, Purcell RH. Putative neutralization epitopes and broad cross-genotype neutralization of hepatitis E virus confirmed by a quantitative cell-culture assay. The Journal of general virology. 2006;87(Pt 3):697–704.

59. Tang ZM, Tang M, Zhao M, Wen GP, Yang F, Cai W, et al. A novel linear neutralizing epitope of hepatitis E virus. Vaccine. 2015;33(30):3504–11.

60. Takahashi M, Hoshino Y, Tanaka T, Takahashi H, Nishizawa T, Okamoto H. Production of monoclonal antibodies against hepatitis E virus capsid protein and evaluation of their neutralizing activity in a cell culture system. Arch Virol. 2008;153(4):657–66.

61. Kamar N, Rostaing L, Legrand-Abravanel F, Izopet J. How should hepatitis E virus infection be defined in organ-transplant recipients? Am J Transplant. 2013;13:2.

62. Wernike K, Hoffmann B, Beer M. Single-tube multiplexed molecular detection of endemic porcine viruses in combination with Background screening for Transbopundery diseases. J Clin Microbiol. 2013;51(3):7.

63. Hoffmann B, Depner K, Schirrmeier H, Beer M. A universal heterologous internal control system for duplex real-time RT-PCR assays used in a detection system for pestiviruses. J Virol Methods. 2006;136(1–2):200–9.

64. Jothikumar N, Cromeans TL, Robertson BH, Meng XJ, Hill VR. A broadly reactive one-step real-time RT-PCR assay for rapid and sensitive detection of hepatitis E virus. J Virol Methods. 2006;131(1):65–71.

19

Novel protein chip for the detection of antibodies against infectious bronchitis virus

Liping Yan[1,2,3*] [iD], Jianhua Hu[1,2], Jing Lei[1,2], Zhiyu Shi[1,2], Qian Xiao[1,2], Zhenwei Bi[1,2], Lu Yao[1,2], Yuan Li[1,2], Yuqing Chen[1,2], An Fang[1,2], Hui Li[1,2], Suquan Song[1], Min Liao[4,5] and Jiyong Zhou[1,2,3,4,5*]

Abstract

Background: Infectious bronchitis (IB) caused by the IB virus (IBV) can cause acute damage to chickens around the world. Therefore, rapid diagnosis and immune status determination are critical for controlling IBV outbreaks. Enzyme-linked immunosorbent assays (ELISAs) have been widely used in the detection of IBV antibodies in the early infection and continuous infection of IB because they are more sensitive and quicker than other diagnostic methods.

Results: We have developed two indirect microarray methods to detect antibodies against IBV: a chemiluminescent immunoassay test (CIT) and a rapid diagnostic test (RDT). IBV nonstructural protein 5 (nsp5) was expressed, purified from *Escherichia coli*, and used to spot the initiator integrated poly(dimethylsiloxane), which can provide a near "zero" background for serological assays. Compared with the IDEXX IBV Ab Test kit, CIT and RDT have a sensitivity and specificity of at least 98.88% and 91.67%, respectively. No cross-reaction was detected with antibodies against avian influenza virus subtypes (H5, H7, and H9), Newcastle disease virus, Marek's disease virus, infectious bursal disease virus, and chicken anemia virus. The coefficients of variation of the reproducibility of the intra- and inter-assays for CIT ranged from 0.8 to 18.63%. The reproducibility of RDT was consistent with the original results. The application of the IBV nsp5 protein microarray showed that the positive rate of the CIT was 96.77%, that of the nsp5 ELISA was 91.40%, and that of the RDT was 90.32%. Furthermore, the RDT, which was visible to the naked eye, could be completed within 15 min. Our results indicated that compared with nsp5 ELISA, the CIT was more sensitive, and the RDT had similar positive rates but was faster. Furthermore, the two proposed methods were specific and stable.

Conclusions: Two microarray assays, which were rapid, specific, sensitive, and relatively simple, were developed for the detection of an antibody against IBV. These methods can be of great value for the surveillance of pathogens and monitoring the efficiency of vaccination.

Keywords: Infectious bronchitis, Protein chip, Antibody detection

Background

Infectious bronchitis (IB) is an acute, highly contagious, and economically important respiratory disease in chickens; it is caused by the IB virus (IBV), which is a significant respiratory pathogen that causes considerable economic losses in the commercial poultry industry worldwide [1]. The IBV genome is a single-stranded, positive-sense RNA that is 27.6 kb in size [2]. It encodes four major structural proteins, namely, glycosylated spike protein (S), membrane protein (M), phosphorylated nucleoprotein (N), and envelope protein (E) [3], and 15 nonstructural proteins (nsp2–nsp16). Generally, nonstructural proteins are present in infected cells but not in the virus, and they only play a role in the process of virus infection and replication [4]. Chickens immunized with an inactivated vaccine will produce no antibodies or low levels of antibodies against viral nonstructural proteins. Thus, nonstructural proteins have the potential application in differentiating natural infection from inactivated vaccine immunity [5].

IB diagnosis is complicated due to the continual emergence of new serotypes [6] and the difficulty in differentiating IB from other upper respiratory diseases [7]. Virus isolation is regarded as the gold standard for the diagnosis of IBV infection, but it is time-consuming and costly [8]. The agar gel precipitation test is used in IBV antibody

* Correspondence: yanliping@njau.edu.cn; jyzhou@njau.edu.cn
[1]MOE Joint International Research Laboratory of Animal Health and Food Safety, Institute of Immunology and College of Veterinary Medicine, Nanjing Agricultural University, Nanjing 210095, People's Republic of China
Full list of author information is available at the end of the article

detection; however, this method has low sensitivity. Hemagglutination inhibition (HI) assays are suitable for the rapid diagnosis of IB, which requires a series of methods to treat the antigen; however, the HI titer is not related to protection. The virus neutralization test correlates with protection and has the highest specificity among IB diagnostic methods, but it is tedious and laborious [9].

Compared with these methods, enzyme-linked immunosorbent assay (ELISA) has been widely used for testing IBV early infection and continuous infection, and this technique can be used for both antigenic and antibody detection. The immunogenicity of the coating antigen is one of the crucial factors when performing an ELISA test for antibody detection. An inactivated whole virus is the most commonly used coating antigen in commercial diagnosis kits for IBV diagnosis. Recombinant antigenic protein expressed using prokaryotic, yeast, or baculovirus systems has been widely used in preparing specific coating antigens for ELISA kits [10–13]. ELISAs based on purified recombinant protein may have higher specificity and sensitivity as the target antigen is immune-dominant and devoid of any nonspecific immune responses [14]. ELISAs based on whole virus particles as well as recombinant S1 (spike protein 1 subunit) and N proteins (nucleoproteins) can provide a rapid and large-scale detection method for IBV infection. However, few IBV detection methods have been developed based on nonstructural proteins (nsps). Our laboratory has established an nsp5 ELISA to detect IBV infection [4]. The nsp5 antibodies detected are likely to be non-neutralizing and exist in lower numbers than the ones generated by other proteins. Based on previous studies, we developed a rapid, highly sensitive protein microarray and a visible detection method to detect IBV nsp5 antibodies for epidemiological investigation and antibody level monitoring.

Methods
Reagents
Initiator integrated poly(dimethylsiloxane) (iPDMS) membrane 26 (15×15 mm^2) was obtained from BS Company (Zhejiang, China). 1-Ethyl-3-(3-dimethylaminopropyl) carbodiimide (EDC) and N-hydroxysuccinimide (NHS) were purchased from Medpep (Shanghai, China). Chicken IgY was purchased from SouthernBiotech (Birmingham, UK). Horseradish peroxidase-labeled goat anti-chicken IgY (HRP-IgG) was obtained from KPL (Dianova, USA). Peroxidase conjugate stabilizer/diluent and chemiluminescent substrate (SuperSignal ELISA Femto Maximum Sensitivity Substrate) were purchased from Thermo Fischer (Massachusetts, USA). Tetramethylbenzidine (TMB) chromogenic reagent was purchased from Nanjing Jiancheng Bioengineering Institute, China. Marek's disease virus (MDV) was purchased from Harbin National Engineering Research Center of Veterinary Biologics Corp (Harbin, China).

Serum samples
In this study, 328 clinical serum samples were collected from a chicken farm. Forty-two negative sera were obtained from different ages of specific-pathogen-free (SPF) chickens raised in SPF isolators in Zhejiang University.

Three-month-old SPF chickens, which were purchased from Shennong Company (Zhejiang, China) and reared in SPF isolators, were used to prepare negative serum and positive serum. We prepared standard positive serum samples from chickens infected with H5, H7, and H9 avian influenza virus (AIV); Newcastle disease virus (NDV); IBV; infectious bursal disease virus (IBDV); and chicken anemia virus (CAV).

Protein chip microarray preparation
A microarray was prepared in a 100,000-grade clean room. Proteins were first dissolved with 30% acetonitrile solution (v/v, in Milli-Q water) to 1 mg/mL stock solution and then diluted into the optimized concentration (200 µg/mL) with printing buffer (0.3 M phosphate buffer, 0.2% glycerin, 0.01% Triton X-100, and 1.5% mannitol) for further printing. iPDMS membranes were first activated with 0.1 M EDC and 0.1 M NHS mixtures for 30 min, rinsed with Milli-Q water, and immediately used for printing. To determine the optimal antigen concentration, the protein was diluted with 0.3 M phosphate buffer to different concentrations. Each dilution of protein was printed on iPDMS using a protein microarray (SCIENION, Germany). Once the antigen concentration was determined, the optimized concentration of nsp5 was achieved by dilution with printing buffer and printed on iPDMS for subsequent experiments in triplicate. The protein microarray was prepared using the SmartArrayer 48 contact printer (Capitalbio, China) with approximately 0.6 nL of printing solution for each sample. Each subarray had a positive control with chicken-IgY at a concentration of 0.1 mg/mL and negative control with printing buffer.

Establishment of the chemiluminescent immunoassay test (CIT)
The procedure for the CIT is shown in Fig. 1. Serum samples were first diluted with serum-dilution buffer (1% bovine serum albumin, 1% casein, 0.5% sucrose, 0.2% polyvinylpyrrolidone, 0.5% Tween 20 in 0.01 M phosphate-buffered saline, pH = 7.4). In total, 100 µL of the diluted serum samples was then added into each protein microarray and incubated for 30 min on a shaker (Thermo Fischer, USA) at 500 rpm and 37 °C. Microarrays incubated with serum-dilution buffer were used as negative controls. Each microarray was then rinsed thrice with washing buffer and incubated with 100 µL of 1 mg/mL HRP-IgG diluted 1:20,000 in peroxidase conjugate stabilizer/diluent for another 30 min on the shaker (500 rpm, 37 °C), followed by the same washing

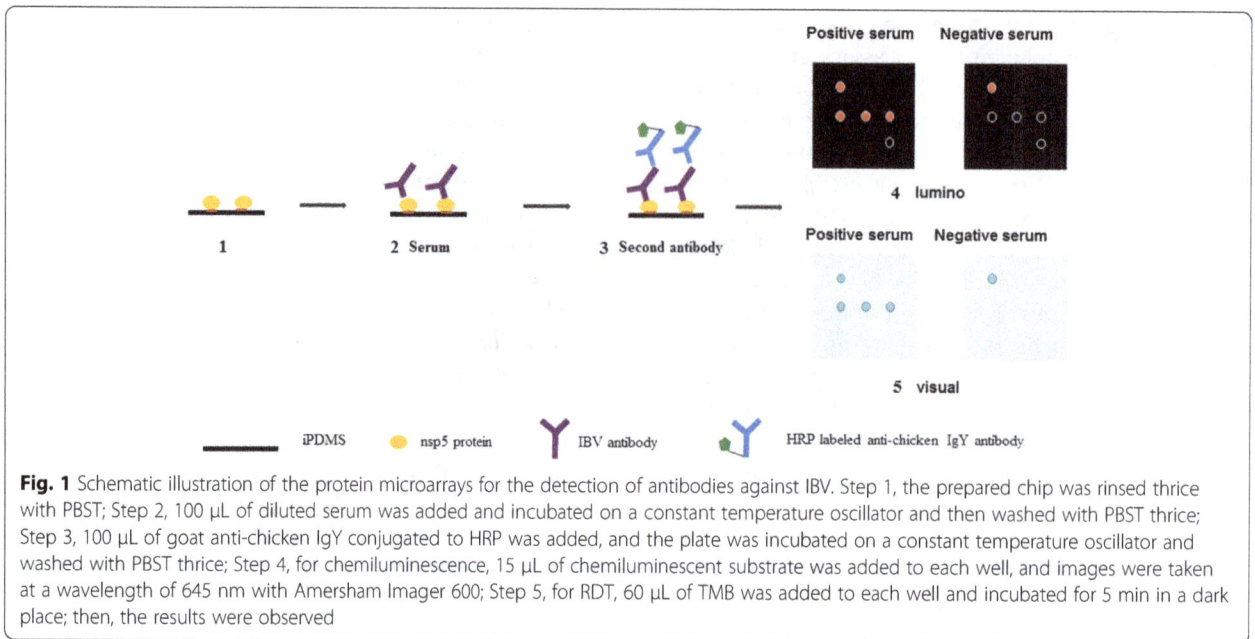

Fig. 1 Schematic illustration of the protein microarrays for the detection of antibodies against IBV. Step 1, the prepared chip was rinsed thrice with PBST; Step 2, 100 μL of diluted serum was added and incubated on a constant temperature oscillator and then washed with PBST thrice; Step 3, 100 μL of goat anti-chicken IgY conjugated to HRP was added, and the plate was incubated on a constant temperature oscillator and washed with PBST thrice; Step 4, for chemiluminescence, 15 μL of chemiluminescent substrate was added to each well, and images were taken at a wavelength of 645 nm with Amersham Imager 600; Step 5, for RDT, 60 μL of TMB was added to each well and incubated for 5 min in a dark place; then, the results were observed

steps described above. A total of 15 μL of the chemiluminescent substrate was added to the microarray, and images were taken at a wavelength of 645 nm with the Amersham Imager 600 (GE, USA). Chemiluminescent signals were acquired using GenePix Pro 6.0 software, and the signal-to-noise ratio (SNR) was calculated.

The purified recombinant nsp5 was printed on the iPDMS membrane to form a microarray with a concentration of 0.05, 0.1, 0.2, and 0.4 mg/mL. Subsequently, the serum samples were added to microplates at the following dilutions: 1:100, 1:200, 1:400, 1:600, 1:800, 1:1600, 1:3200, and 1:6400. To identify the optimal time of exposure, the images were taken at an exposure time of 30 s, 1 min, 2 min, 3 min, and 4 min.

To determine the CIT threshold, a total of 184 serum samples, including 142 positive samples and 42 negative samples, identified by the IDEXX IBV Ab Test kit were tested according to the optimal working conditions. Results were then compared with those obtained using the IDEXX IBV Ab Test kit. Finally, receiver operating characteristic (ROC) curve analysis was conducted to determine the accuracy of the IBV protein microarray test.

The specificity of the CIT was evaluated by detecting the positive sera against AIV (H5, H7, and H9), NDV, MDV, IBDV, and CAV.

The evaluation of the CIT reproducibility within and between runs was carried out as described by Jacobson [15]. Thirteen field serum samples (nine IDEXX positive samples and four IDEXX negative samples) were selected for the reproducibility experiments. For intra-assay reproducibility, three replicates of each serum sample were analyzed within the same plate. For inter-assay reproducibility, three replicates of each sample were run in different plates. The mean SNR, standard deviation (SD), and coefficient of variation (CV) were then calculated.

Development of the rapid diagnostic test (RDT)

The procedure of the RDT is also shown in Fig. 1. Serum was first diluted 1:100 with serum-dilution buffer, and 100 μL of the diluted serum sample was added into each protein microarray and incubated for 5 min on a shaker (500 rpm, 37 °C). The microarray incubated with serum-dilution buffer was used as a negative control. The microarray was then rinsed thrice with washing buffer and incubated with 100 μL of 1 mg/mL HRP-IgG diluted 1:2000 in peroxidase conjugate stabilizer/diluent for another 5 min on a shaker (500 rpm, 37 °C), followed by the same washing steps described above. A total of 60 μL of TMB was added to the microarray and incubated for 5 min in the dark; then, the results were observed.

To confirm the concentration of the nsp5 protein in the RDT, the purified recombinant nsp5 was printed on an iPDMS membrane to form a microarray with concentrations of 0.05, 0.1, 0.2, and 0.4 mg/mL. The specificity of the RDT was evaluated by detecting the positive sera against AIV (H5, H7, and H9), NDV, MDV, IBDV, and CAV. The sensitivity experiments of the RDT were conducted by detecting the IBV positive serum with different titers. Then, the results were observed, and the detection limit was determined.

Application of the CIT and the RDT

To further evaluate the CIT and RDT, 186 clinical serum samples were detected by the CIT, RDT, and nsp5 ELISA antibody test kit [4]. Subsequently, the positive rate of each method was determined.

Statistical analysis

Chemiluminescent signals were acquired using Gene-Pix, and the SNR was calculated as follows: SNR = (Signal intensity − Background)/Background. Graph-Pad Prism 6 and Microsoft Excel were used for the statistical analysis of all data, including the determination of the threshold and the calculation of the SNR value, means, SDs, and CVs. The ROC curve was obtained using GraphPad Prism 6. Sensitivity and specificity were calculated according to the following formulas: Sensitivity = True positive/(True positive + False negative) × 100%; Specificity = True negatives/(False positives + True negatives) × 100%. The area under the curve (AUC) was used to validate the diagnostic application of the CIT. The area under the ROC curve quantifies the overall ability of the test to discriminate between those individuals with the disease and those without the disease. A truly useless test (one no better at identifying true positives than flipping a coin) has an AUC of 0.5, whereas a perfect test (one that has zero false positives and zero false negatives) has an AUC of 1.

Results

Establishment of the CIT

For the CIT, the optimal antigen concentration was 0.2 mg/mL (Fig. 2a, b), and the dilution for the serum samples was 1:600 (Fig. 2c, d), on the assumption that the SNR between the positive and the negative sera was the highest. The dilution of the HRP-conjugated goat anti-chicken antibody was defined as 1:20,000. When the exposure time was more than 2 min, the SNR of the negative serum rose rapidly; thus, we set the exposure time to 2 min (Fig. 3).

ROC analysis showed that the IBV nsp5 microarray had high selectivity ($p < 0.0001$) between the positive and the negative samples, and the AUC was 0.9993 (Fig. 4a). Based on the ROC analysis of the IBV nsp5 microarray, the SNR value of the IDEXX-negative serum samples varied from a minimum of 0.01 to a maximum of 1.964, whereas the SNR value of the IDEXX-positive serum samples was from a minimum of 1.82 to a maximum of 23.59 (Fig. 4b). A threshold SNR value of 2 for IBV nsp5 microarray was found to provide optimal results, with a

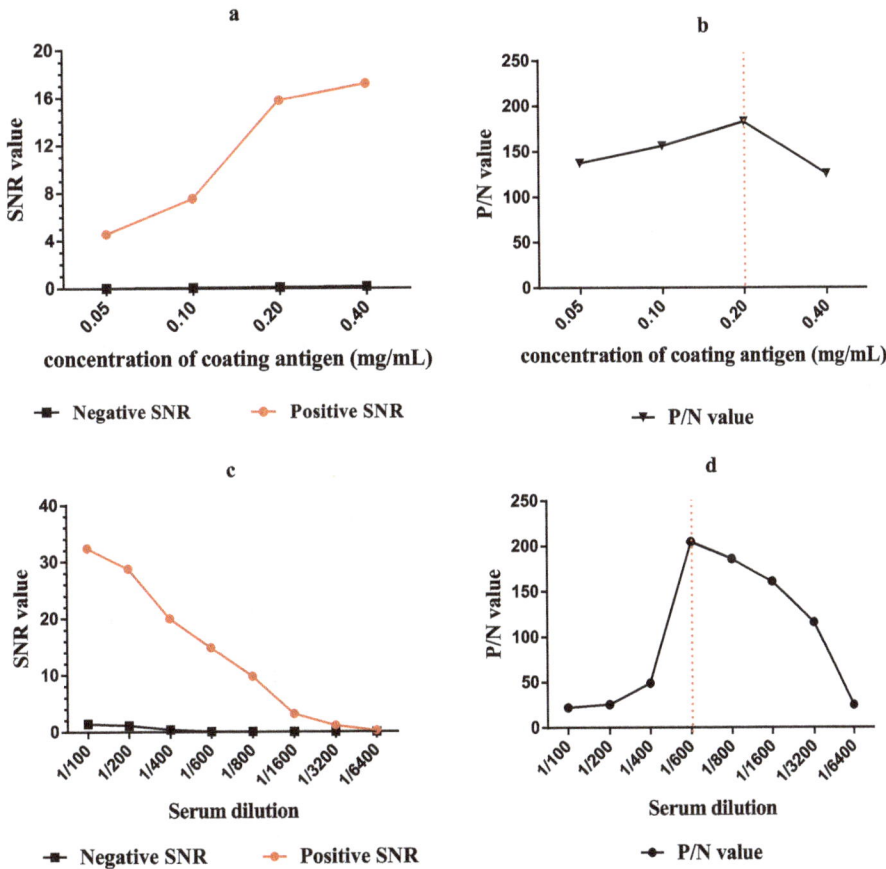

Fig. 2 Optimization of the microarray working conditions. a SNR variation across different IBV nsp5 concentrations (0.05, 0.1, 0.2, and 0.4 mg/mL) of the coating antigen. b P/N value between the positive and negative SNRs with different IBV nsp5 concentrations. c SNR variation across different serum dilutions. d P/N value between the positive and the negative SNRs with the optimal dilution of the serum sample (1:600)

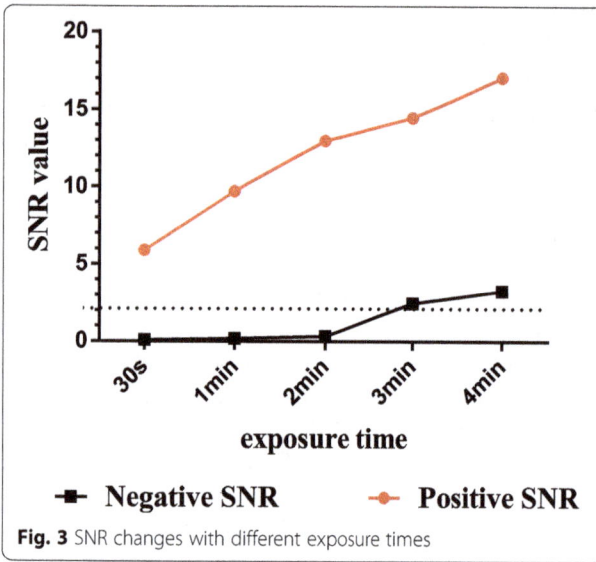

Fig. 3 SNR changes with different exposure times

sensitivity of 98.59%, a specificity of 100%, and an accuracy of 98.91% compared with the results of other thresholds (Table 1). Thus, the samples with SNR < 2 were considered negative, whereas those with SNR ≥ 2 were considered positive.

The specificity of the CIT was evaluated by detecting the cross-reactivity of the antibodies against AIV (H5, H7, and H9), NDV, MDV, IBDV, and CAV. The SNRs of all sera from the previously mentioned viruses were all below the threshold of 2. These data revealed that no cross-reactivity occurred between the IBV GST-fused nsp5 antigen and antibodies against other avian viruses. This result demonstrated that the antigen has a high specificity.

The reproducibility of the CIT detection was determined by comparing the SNR value of each clinical serum sample from the below tests. The within-plate CVs of nine positive and four negative serum samples tested ranged from 0.8 to 18.63% (Table 2), whereas the between-run CVs of these serum samples ranged from 1.89 to 18.01% (Table 3). These results showed that the CIT detection results were reproducible and had low and acceptable variation.

Development of the RDT

One hundred and forty-four clinical serum samples (130 samples were positive for antibodies against IBV, and 14 samples were negative as confirmed by the IDEXX IBV Ab Test kit) were subjected to visual rapid detection following the procedure described above. The data showed that 130 serum samples were positive for antibodies against IBV, and 14 samples were negative, similar to the results of the IDEXX IBV Ab Test kit with the nsp5 concentration of 0.2 mg/mL (Table 4). If IBV antibodies exist in the serum, the spot with the IBV antigen turns

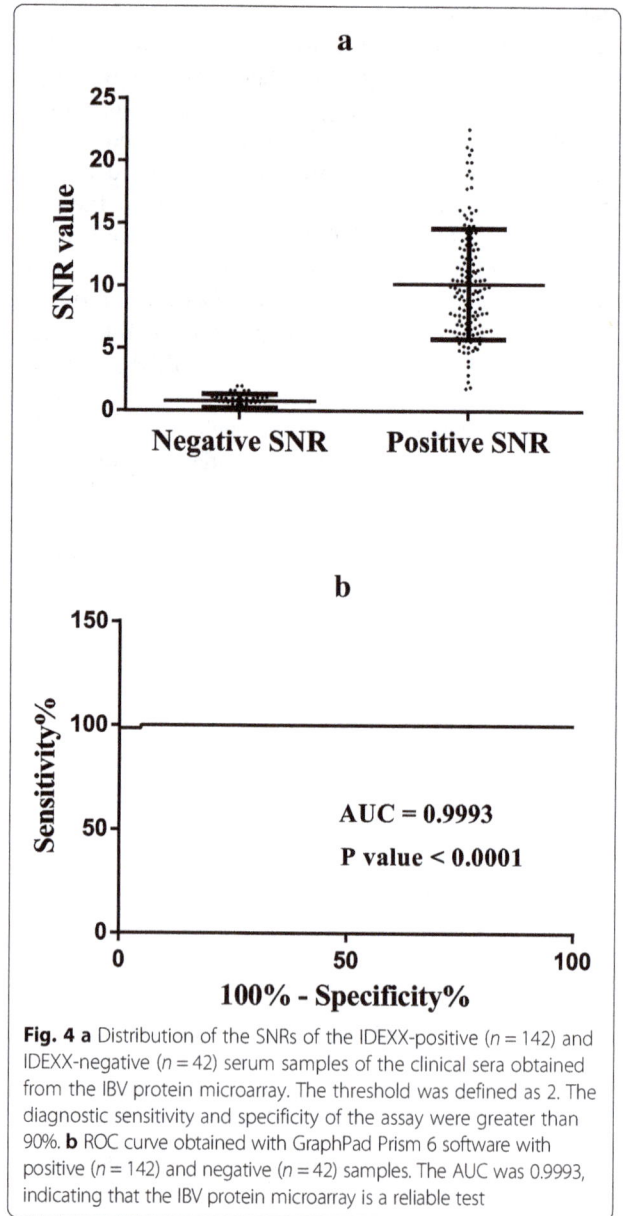

Fig. 4 a Distribution of the SNRs of the IDEXX-positive (*n* = 142) and IDEXX-negative (*n* = 42) serum samples of the clinical sera obtained from the IBV protein microarray. The threshold was defined as 2. The diagnostic sensitivity and specificity of the assay were greater than 90%. **b** ROC curve obtained with GraphPad Prism 6 software with positive (*n* = 142) and negative (*n* = 42) samples. The AUC was 0.9993, indicating that the IBV protein microarray is a reliable test

blue, thereby allowing us to determine the concentration of nsp5 as 0.2 mg/mL.

The specificity of the RDT was evaluated by detecting the cross-reactivity of antibodies against AIV (H5, H7, and H9), NDV, MDV, IBDV, and CAV. The specific experiments of the RDT showed that no cross-reaction

Table 1 Evaluation of the IBV protein microarray with selected thresholds

Threshold	Sensitivity (%)	Specificity (%)	Accuracy (%)
> 1.940	98.59	95.24	98.91
> 1.962	98.59	97.62	98.91
> 2	98.59	100	98.91
> 2.635	97.89	100	98.3

Table 2 CVs of positive sera within the same run

No.	I (SNR value)	II (SNR value)	III (SNR value)	X (Mean)	SD	CV (%)
1	0.69	1.02	0.72	0.81	0.15	18.63
2	1.64	1.80	1.30	1.58	0.21	13.22
3	0.26	0.30	0.29	0.28	0.02	6.80
4	0.01	0.00	0.00	0.00	0.00	17.78
5	9.40	9.94	7.43	8.92	1.08	12.09
6	14.58	17.49	18.49	16.85	1.66	9.84
7	15.43	15.42	15.69	15.51	0.12	0.80
8	14.57	14.46	15.09	14.70	0.28	1.87
9	21.88	19.47	19.34	20.23	1.17	5.78
10	14.16	13.74	14.00	13.96	0.17	1.24
11	21.52	23.23	22.00	22.25	0.72	3.24
12	17.42	15.34	15.70	16.15	0.91	5.61
13	19.11	20.15	20.55	19.94	0.61	3.06

occurred between the IBV GST-fused nsp5 antigen and the antibodies against other avian viruses. The sensitivity experiments demonstrated that when the positive serum was diluted 1:1000, the spot still turned blue (Fig. 5).

Application of the IBV nsp5 protein microarray

To further evaluate the IBV nsp5 microarray, 186 clinical serum samples were detected by the IBV nsp5 microarray and the nsp5 ELISA antibody test kit. The results showed that of the 186 samples, 170 samples were positive for antibodies against IBV, and 16 samples were negative according to the nsp5 ELISA kit. A total of 180 out of the 186 samples were positive, and 6 samples were negative according to the CIT. A total of 167 positive samples and 19 negative samples were detected

by the RDT. The positive rate of the CIT was 96.77%, that of the nsp5 ELISA was 91.40%, and that of the RDT was 90.32% (Table 5).

Discussion

Most serological assays, including the IDEXX ELISA kit, use viral particles of IBV as an antigen for the detection of antibodies against IBV. However, the preparation of purified virions for use as an antigen is time-consuming and expensive. In the present study, recombinant nonstructural proteins expressed in *Escherichia coli* antigen-based protein microarray was evaluated for the first time in the serological diagnosis of IB [4]. Protein microarrays have high sensitivity and good reproducibility in quantitative and qualitative assays, and they are a valuable asset when analyzing complex biological samples [16]. In clinical

Table 3 CVs of positive and negative sera between runs

No.	I (SNR value)	II (SNR value)	III (SNR value)	X (Mean)	SD	CV (%)
1	0.70	0.69	0.72	0.70	0.01	1.89
2	2.00	1.80	1.70	1.83	0.12	6.73
3	0.30	0.30	0.34	0.31	0.02	6.06
4	0.01	0.01	0.01	0.01	0.00	14.48
5	3.08	2.07	2.24	2.46	0.44	18.01
6	15.94	17.49	16.86	16.76	0.64	3.80
7	13.16	11.57	14.70	13.14	1.28	9.73
8	14.57	15.09	13.69	14.45	0.58	4.00
9	21.88	22.33	19.47	21.22	1.26	5.92
10	11.88	13.61	14.38	13.29	1.04	7.86
11	17.56	15.89	14.24	15.90	1.35	8.52
12	11.83	13.78	14.42	13.34	1.10	8.25
13	13.96	17.50	14.80	15.42	1.51	9.80

Table 4 Comparison of the detection results at different antigen concentrations for the RDT

Antigen concentration	0.4 mg/mL	0.2 mg/mL	0.1 mg/mL	0.05 mg/mL
Positive number	134	130	129	127
Negative number	10	14	15	17

sample testing, many factors, including time, cost, accuracy, sensitivity, and throughput, determine the performance and usefulness of an immunoassay. In this study, a new solidly supported material, iPDMS membrane, which has a near "zero" background for identification, was used. It achieved high sensitivity in detecting antibodies in serum [17]. These unique features of iPDMS not only simplify data analysis but also reduce nonspecific interactions [18]. ELISA detection has been widely used in the detection of IBV antibodies in early infection and continuous infection of IB and vaccine-immune, and no diagnosis method is more sensitive and quicker than ELISA. In this study, two microarray methods (CIT and RDT) were established. Except for the method of the observation, the reaction processes of the two methods are akin to the detection process of ELISA. However, unlike ELISA, the established methods only require 2 ng of antigen coating on each spot, and the amount of HRP-IgG required for each reaction well is only 5 ng. The antigen and HRP-IgG used in both methods were less than those used in ELISA, thereby reducing the cost of detection. In addition, the CIT can detect antibodies against IBV nsp5 quantitatively and is more sensitive than the IBV nsp5 ELISA kit. The RDT was developed to detect antibodies against IBV visually, and the results can be obtained within 15 min with great sensitivity and specificity. Compared with ELISA, RDT has a shorter detection time and better detection efficiency. In this study, we only used one antigen of IBV for testing and verification. In the future, we will apply antigens of different diseases to iPDMS to achieve high-throughput test results.

For the establishment of the IBV nsp5 protein chip, we first optimized the procedure and determined the CIT threshold as 2 with the IDEXX IBV antibody detection kit. With the threshold of 2, the CIT showed high sensitivity (98.59%), specificity (100%), and accuracy (98.91%) in the antibody detection of the samples compared with those of other thresholds (Table 1). The RDT demonstrated a high success rate compared with the commercial IDEXX IBV Ab Test kit, suggesting that the RDT is a reliable assay for the detection of IBV infection. Clinical serum samples were also subjected to rapid detection. Furthermore, the RDT has higher sensitivity than the commercial IDEXX IBV Ab Test kit. It is also simpler and faster than ELISA methods. To further evaluate IBV nsp5 protein chip, 186 clinical serum samples were detected by the IBV nsp5 protein chip and the nsp5 ELISA antibody test kit. The positive rates of the CIT, nsp5 ELISA, and RDT were 96.77%, 91.40%, and 90.32%, respectively. Compared with nsp5 ELISA, the CIT was more sensitive, and the RDT had similar positive rates but was faster.

Protein chips are a high-throughput monitoring system that monitors the interaction among protein molecules through the interaction between a target molecule and a capture molecule. Although protein chips have been produced in the context of proteomics research, its application is not limited to proteomics alone. With the development of protein chip technology, researchers have gradually applied this technology to other fields, such as food inspection, disease diagnosis, drug screening, agriculture, forestry, animal husbandry, and forensic science. At present, this technology is rarely studied and applied in veterinary medicine. High throughput is an

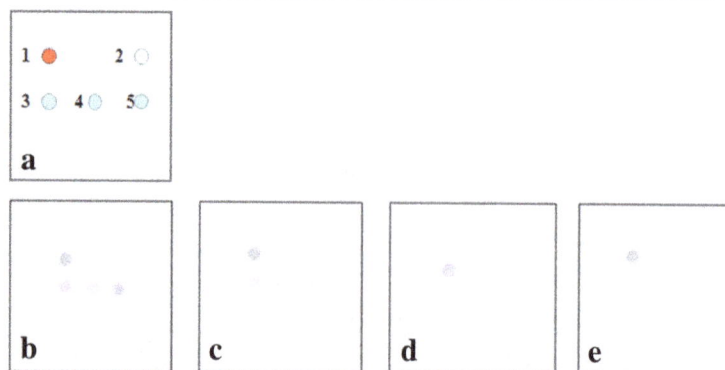

Fig. 5 Sensitivity experiments of the RDT. **a** Array of the protein chip: 1, positive control (chicken IgY); 2, negative control; 3, 4, and 5, IBV nsp5 spots. **b** IBV positive serum diluted 1:100. **c** IBV positive serum diluted 1:1000. **d** IBV positive serum diluted 1:10,000. **e** Negative SPF chicken serum

Table 5 Comparison of IBV protein microarray and the nsp5 ELISA kit with 186 clinical serum samples

Samples	nsp5 ELISA	CIT	RDT
Positive	170	180	168
Negative	16	6	18
Positive rates	91.40%	96.77%	90.32%

important feature of protein chips. Antibodies against several diseases can be detected from only a single serum, and this factor is especially important for clinical research, which uses precious samples from rare and wild animals. Substrate selection and surface modification, as well as new substrate research and development, have become major research foci in the field of protein chips. Our present work indicates that iPDMS can provide a matrix for the detection of antibodies in chicken serum. In addition, the high sensitivity and specificity of protein microarrays render them powerful tools in disease detection [19, 20] and enable their use for determining antibody responses to infectious diseases [21]. In the future, we will print recombinant antigenic proteins of different avian viruses to achieve high-throughput detection results with the same serum.

Conclusion

The nsp5 protein chips were developed for the detection of antibodies against IBV. These assays are comparable to the commercial IDEXX IBV Ab Test kit in terms of sensitivity and specificity. The RDT can generate results within 15 min and may be a suitable alternative to screen for the presence of IBV in chickens.

Abbreviations
AIV: Avian influenza virus; CAV: Chicken anemia virus; CIT: Chemiluminescent immunoassay test; EDC: 1-Ethyl-3-(3-dimethylaminopropyl) carbodiimide; ELISA: Enzyme-linked immunosorbent assay; HRP-IgG: Horseradish peroxidase-labeled goat anti-chicken IgY; IB: Infectious bronchitis; IBDV: Infectious bursal disease virus; IBV: IB virus; iPDMS: initiator integrated poly(dimethylsiloxane); MDV: Marek's disease virus; NDV: Newcastle disease virus; NHS: N-hydroxysuccinimide; nsp5: Nonstructural protein 5; RDT: Rapid diagnostic test; SPF: Specific-pathogen-free; TMB: Tetramethylbenzidine

Funding
The design of the protein chip method was supported by National Key Research and Development Program of China (2016YFD0500800). The sampling serum of chicken was supported Key Program of Science and Technology Planning of Guangdong Province (2017B020202010). The analysis and interpretation of data was supported by the Agricultural Science & Technology Independent Innovation Fund of Jiangsu Province (CX(15)1065) and National Natural Science Foundation of China (31402150). The writing of the manuscript was supported by the Fundamental Research Funds for the Central Universities (Y0201600146) and the Priority Academic Program Development of Jiangsu Higher Education Institutions.

Authors' contributions
LPY, JHH, and JL performed the experiments, interpreted the results, and drafted the manuscript. LPY, ZYS, and QX analyzed the data. ZWB, LY, YL, YQC, AF, and HL collected the clinical samples. SQS and ML analyzed the data and revised the manuscript. LPY and JYZ designed the study, analyzed the data, and revised the manuscript. All authors reviewed the results and approved the final version of the manuscript.

Ethics approval and consent to participate
This study was performed in accordance with the recommendations in the Guide for the Care and Use of Laboratory Animals of the Ministry of Health, China. The protocol of the current study was reviewed and approved by the Institutional Animal Care and Use Committee of Nanjing Agricultural University (approval no. SYXK 2017–0007). Written informed consent to use 328 clinical serum samples, which were collected from a chicken farm, were obtained from the owner of the animals. All efforts were made to minimize animal suffering during sample collection.

Consent for publication
Not applicable

Competing interests
The authors declare that they have no competing interests.

Author details
[1]MOE Joint International Research Laboratory of Animal Health and Food Safety, Institute of Immunology and College of Veterinary Medicine, Nanjing Agricultural University, Nanjing 210095, People's Republic of China. [2]Jiangsu Engineering Laboratory of Animal Immunology, Institute of Immunology and College of Veterinary Medicine, Nanjing Agricultural University, Nanjing 210095, People's Republic of China. [3]Jiangsu Detection Center of Terrestrial Wildlife Disease, Institute of Immunology and College of Veterinary Medicine, Nanjing Agricultural University, Nanjing 210095, People's Republic of China. [4]Key Laboratory of Animal Virology, Ministry of Agriculture, Zhejiang University, Hangzhou 310058, People's Republic of China. [5]Collaborative Innovation Center for Diagnosis and Treatment of Infectious Diseases, The First Affiliated Hospital, Zhejiang University, Hangzhou 310058, People's Republic of China.

References
1. Jordan B. Vaccination against infectious bronchitis virus: a continuous challenge. Vet Microbiol. 2017;206:137–43.
2. Boursnell MEG, Brown TDK, Foulds IJ, Green PF, Tomley FM, Binns MM. Completion of the sequence of the genome of the coronavirus avian infectious bronchitis virus. J Gen Virol. 1987;68:57–77.
3. Lai MMC, Cavanagh D. The Molecular Biology of Coronaviruses. Adv Virus Res. 1997;48:1–100.
4. Lei J, Shi TT, Sun DN, Mo KK, Yan Y, Jin YL, Liao M, Zhou JY. Development and application of nsp5-ELISA for the detection of antibody to infectious bronchitis virus. J Virol Methods. 2017;243:182–9.
5. Elnekave E, Shilo H, Gelman B, Klement E. The longevity of anti NSP antibodies and the sensitivity of a 3ABC ELISA - a 3 years follow up of repeatedly vaccinated dairy cattle infected by foot and mouth disease virus. Vet Microbiol. 2015;178(1–2):14–8.
6. Lim TH, Lee HJ, Lee DH, Lee YN, Park JK, Youn HN, Kim MS, Lee JB, Park SY, Choi IS, et al. An emerging recombinant cluster of nephropathogenic strains of avian infectious bronchitis virus in Korea. Infect Genet Evol. 2011;11(3):678–85.
7. Han ZX, Zhao F, Shao YH, Liu XL, Kong XG, Song Y, Liu SW. Fine level epitope mapping and conservation analysis of two novel linear B-cell epitopes of the avian infectious bronchitis coronavirus nucleocapsid protein. Virus Res. 2013;171(1):54–64.
8. Dawson PS, Gough RE. Antigenic variation in strains of avian infectious bronchitis virus. Archiv fur die gesamte Virusforschung. 1971;34(1):32–9.
9. Singh NK, Dey S, Mohan CM, Kataria JM, Vakharia VN. Evaluation of four enzyme linked immunosorbent assays for the detection of antibodies to infectious bursal disease in chickens. J Virol Methods. 2010;165(2):277–82.

10. Gibertoni AM, Montassier Mde F, Sena JA, Givisiez PE, Furuyama CR, Montassier HJ. Development and application of a Saccharomyces cerevisiae-expressed nucleocapsid protein-based enzyme-linked immunosorbent assay for detection of antibodies against infectious bronchitis virus. J Clin Microbiol. 2005;43(4):1982–4.

11. Chen H, Coote B, Attree S, Hiscox JA. Evaluation of a nucleoprotein-based enzyme-linked immunosorbent assay for the detection of antibodies against infectious bronchitis virus. Avian Pathol. 2003;32(5):519–26.

12. Chan KW, Hsieh HH, Wang HC, Lee YJ, Sung MH, Wong ML, Hsu WL. Identification, expression and antigenic analysis of recombinant hemagglutinin proteins of canine distemper virus. J Virol Methods. 2009;155(1):18–24.

13. Lin KH, Lin CF, Chiou SS, Hsu AP, Lee MS, Chang CC, Chang TJ, Shien JH, Hsu WL. Application of purified recombinant antigenic spike fragments to the diagnosis of avian infectious bronchitis virus infection. Appl Microbiol Biotechnol. 2012;95(1):233–42.

14. Dey S, Mohan CM, Kumar TM, Ramadass P, Nainar AM, Nachimuthu K. Recombinant LipL32 antigen-based single serum dilution ELISA for detection of canine leptospirosis. Vet Microbiol. 2004;103(1–2):99–106.

15. Jacobson. Validation of serological assays for diagnosis of infectious diseases. Rev Sci Technol. 1998;17(2):469–526.

16. Sutandy FX, Qian J, Chen CS, Zhu H. Overview of protein microarrays. Curr Protoc Protein Sci. 2013; Chapter 27:Unit 27 21.

17. Ma HW, Wu YZ, Yang XL, Liu X, He JA, Fu L, Wang J, Xu HK, Shi Y, Zhong RQ. Integrated poly(dimethysiloxane) with an intrinsic nonfouling property approaching "absolute" zero background in immunoassays. Anal Chem. 2010;82(15):6338–42.

18. Huang M, Ma Q, Liu X, Li BA, Ma HW. Initiator integrated poly(dimethysiloxane)-based microarray as a tool for revealing the relationship between nonspecific interactions and irreproducibility. Anal Chem. 2015;87(14):7085–91.

19. Smith L, Watson MB, O'Kane SL, Drew PJ, Lind MJ, Cawkwell L. The analysis of doxorubicin resistance in human breast cancer cells using antibody microarrays. Mol Cancer Ther. 2006;5(8):2115–20.

20. Joos T, Bachmann J. Protein microarrays: potentials and limitations. Front Biosci. 2009;14:4376–85.

21. Mezzasoma L, Bacarese-Hamilton T, Di Cristina M, Rossi R, Bistoni F, Crisanti A. Antigen microarrays for serodiagnosis of infectious diseases. Clin Chem. 2002;48(1):121–30.

Specific detection of Muscovy duck parvovirus infection by TaqMan-based real-time PCR assay

Chunhe Wan*[iD], Cuiteng Chen, Longfei Cheng, Hongmei Chen, Qiuling Fu, Shaohua Shi, Guanghua Fu, Rongchang Liu and Yu Huang*

Abstract

Background: Muscovy duck parvovirus (MDPV) causes high mortality and morbidity in Muscovy ducks, with the pathogenesis of the virus still unknown in many respects. Specific MDPV detection is often rife with false positive results because of high identity at the genomic nucleotide level and antigenic similarity with goose parvovirus (GPV). The objective of this study was to develop a sensitive, highly specific, and repeatable TaqMan-based real-time PCR (qPCR) assay for facilitating the molecular detection of MDPV.

Results: The specific primers and probe were designed based on the conserved regions within MDPVs, but there was a variation in GPVs of the nonstructural (NS) genes after genetic comparison. After the optimization of qPCR conditions, the detection limit of this qPCR assay was 29.7 copies/μl. The assay was highly specific for the detection of MDPV, and no cross-reactivity was observed with other non-targeted duck-derived pathogens. Intra- and inter-assay variability was less than 2.21%, means a high degree of repeatability. The diagnostic applicability of the qPCR assay was proven that MDPV-positive can be found in cloacal swabs samples, Muscovy duck embryos and newly hatched Muscovy ducklings.

Conclusions: Our data provided incidents that MDPV could be possible vertically transmitted from breeder Muscovy ducks to Muscovy ducklings. The developed qPCR assay in the study could be a reliable and specific tool for epidemiological surveillance and pathogenesis studies of MDPV.

Keywords: MDPV, NS gene, Specific detection, TaqMan-based real-time PCR assay, Vertical transmission

Background

The family *Parvoviridae* is comprised of two subfamily members, *Densovirinae* and *Parvovirinae*. The subfamily *Parvovirinae* contains eight distinct genera: *Amdoparvovirus*, *Aveparvovirus*, *Bocaparvovirus*, *Copiparvovirus*, *Dependoparvovirus*, *Erythroparvovirus*, *Protoparvovirus*, and *Tetraparvovirus*. At present, the International Committee on Taxonomy of Viruses (ICTV) has classified Muscovy duck parvovirus (MDPV) and goose parvovirus (GPV) as a single species (namely, *Anseriform dependoparvovirus* 1), which is classified into the genus

* Correspondence: chunhewan@126.com; huangyu_815@163.com
Fujian Provincial Key Laboratory for Avian Diseases Control and Prevention, Fujian Animal Diseases Control Technology Development Center, Institute of Animal Husbandry and Veterinary Medicine of Fujian Academy of Agricultural Sciences, Xi-feng Road No.100, Jiantian village, Jin'an district, Fuzhou 350013, China

Dependoparvovirus in subfamily *Parvovirinae* due to similar genetic properties and evolutionary origins (https://talk.ictvonline.org/taxonomy/).

Reportedly, the genomes of MDPVs and GPVs contain a single copy of the linear, single-stranded DNA genome of approximately 5100 nucleotides in length. The genomes of these viruses are flanked by identical inverted terminal repeats (ITR) at both the 5′- and 3′-terminus. The ITR can fold on itself to form a palindromic hairpin structure. A terminal resolving site (TRS), Rep protein binding site (RBS), and transcription factor binding sites can be found in ITR, which were involved in viral replication, packaging and transcription. There are two major open reading frames (ORFs) in both MDPVs and GPVs genome. The left ORF that encodes for the nonstructural (NS) protein, which is involved in viral replication and regulatory function. However, the right ORF that produces three capsid proteins (VP1, VP2 and VP3), which plays important roles in virus tropism, host range, and pathogenicity. In addition, VP2 and VP3 contain the same carboxyl-terminal portion of VP1 of the viruses, which was generated by differential alternative splicing of mRNA [1–8].

Normally, GPVs have been found in goslings, Muscovy ducklings [5, 8], swans [9], and *Anser cygnoides* [10], whereas MDPV has only been discovered in Muscovy ducklings. MDPV infection was initially described by Professor Lin in our laboratory in the early 1980s [11]. Muscovy ducklings infected by MDPV were characterized by watery diarrhea, wheezing, and locomotor dysfunction. MDPV is mainly observed in Muscovy ducklings less than three-week-old, with the mortality rate reaching as high as 80% depending on age [3, 7, 11].Then the disease was widespread in China, leading huge economic loss to waterfowl husbandry due to the high mortality and morbidity.

A previous study revealed that MDPV and GPV are also pathogenic to Muscovy ducklings, even in the same Muscovy duck flocks [8, 12]. Compared with the GPV virulent strain B and MDPV virulent strain FM, they share more than 80.0% nucleotide similarity at the genome level. In addition, these two viruses exhibit nucleotides and amino acids identities of 83.0 and 90.6% at NS gene level, and 81.5% and 87.6% at VP1 gene level, respectively [1–3]. The high identity at the amino acids level of the VP1 protein indicates potential immunogenic cross-reactivity between MDPVs and GPVs [13–15]. Therefore, differentiating between MDPV and GPV in Muscovy ducklings is essential. Nevertheless, the high homologies in nucleotide identities and immunogenic cross-reactivity between MDPVs and GPVs, increases the risk of omissive and mistaken diagnoses for the specific detection of MDPV.

Recently, a TaqMan-based real-time quantitative PCR (qPCR) assays have gained wide acceptance due to their rapid nature, sensitivity, reproducibility, and the reduced risk of carry-over contamination as a result of the specific TaqMan probe, which had been widely used for viral epidemiological surveillance and pathogenesis studies [16–19]. Thereby, the aim of this study was to design a fully validated, reliable, and highly specific TaqMan-based real-time quantitative PCR assay for precise detection of MDPV infection based on specific primers and probe, designed by targeting the conserved region of the MDPV NS gene after bioinformatics analysis.

Methods
Viruses and bacteria strains
Avian influenza virus (H9N2 AIV), avian Tembusu virus (ATmV), duck hepatitis virus type 1 and 3 (DHAV-1 and DHAV-3), Muscovy duck reovirus (MDRV), duck adenovirus A (DAdV-A), duck enteritis virus (DEV), Muscovy duck origin goose parvovirus (GPV), novel goose parvovirus (N-GPV), *Escherichia coli* (*E. coli.*), Pasteurellamultocida (*P.M.),* Rimerella anatipstifer (*R.A.*) and Salmonella spp. (*S.S.*) were isolated and kept in our laboratory, which was decribed the same as previously reported [12, 20, 21].

Nucleic acids extraction and cDNAs preparation
Viral RNAs (RNA viruses, i.e. AIV, ATmV, DHAV-1 and DHAV-3, MDRV) were extracted using EasyPure Viral DNA/RNA Kit (Tiangen, Beijing, China), viral DNAs (DNA viruses, i.e. DAdV-A, DEV, GPV and N-GPV) were extracted using EasyPure Micro Genomic DNA Kit (Transgen Bioteck, Beijing, China), according to instructions provided by the manufacturer. cDNAs (AIV, ATmV, DHAV-1 and DHAV-3, MDRV) were prepared with viral RNAs (with approximate 100 ng of viral RNA) using TransScript II All-in-One First-Strand cDNA Synthesis SuperMix (One-Step gDNA Removal) (Tiangen, Beijing, China). Bacteria genomic DNAs (bacteria pathogens, i.e. E. coli., P.M., R.A. and S.S.) were extracted using EasyPure Bacteria Genomic DNA Kit (Transgen, Beijing, China). DNAs and cDNAs were then quantified using a NANODROP 2000 spectrophotometer (Thermo Scientific, Waltham, MA, USA). All extracted DNAs and cDNAs templates were stored at – 80 °C until use.

Primers and probe selection and design
Primer and probe selection was performed on the evolutionarily most conserved regions of the NS gene of MDPV. Briefly, a total of 15 MDPVs and 37 GPVs NS gene sequences were downloaded from the GenBank database (https://www.ncbi.nlm.nih.gov/nucleotide/), and these 52 NS gene sequences were aligned using the Lasergene package MegAlign program by ClustalW method. The identification of the conserved region, which was highly conserved in MDPVs, but there was a characteristics variation in GPVs. The obvious different region between MDPVs with GPVs, was selected for the primers and TaqMan probe design. The forward primer MDPV-qF (5′-TACGAATGAACAAACCAA-3′), the reverse primer MDPV-qR (5′- CGCTCTTAATATCTCCTCTA-3′), and the TaqMan probe MDPV-qP (FAM-5′- TGAA CGAGCGAATGAGCCTTCC-3′-Eclipse) were designed using Primer Premier Software version 5.0 (Premier Biosoft, Palo Alto, CA, USA). The length of amplicon was 118 base pairs (bp). Primers (MDPV-qF and MDPV-qR) and probe (MDPV-qP) were verified by Basic Local Alignment Search Tool (BLAST, https://blast.ncbi.nlm.nih.gov/Blast.cgi) for specificity analysis, then these verified primers and probe were synthesized by a commercial company (TaKaRa, Dalian, China).

Construction of recombinant plasmid containing the NS gene of MDPV

The NS gene (1884 bp) of MDPV (strain FJM5) [8] was amplified by PCR, with the primer sets of forward primer (NSF) 5'-ATGGCATTTTCTAGGCCTCTTCA-3' and reverse primer (NSR) 5'-TTATTGTTCATTCT CCATATCAT-3'. A conventional PCR was performed in a Thermal Cycler Dice (TaKaRa, OTSU, SHIGA, Japan), the isolated DNAs were used as template in a reaction volume of 50 µl reaction mixture containing 25 µl ThermoScientific DreamTaq Green PCR Master Mix (2×) (Thermo Fisher Scientific Inc., Shanghai, China), 1 µl primers NSF and NSR (20 µM each), 1 µl DNA template, and 22 µl Nuclease-free water. The PCR reaction was conducted using a program that included with an initial denaturation at 94 °C for 5 min, followed by 35 cycles of denaturation at 94 °C for 50 s, annealing at 53 °C for 35 s, elongation at 72 °C for 120 s, and then a final extension at 72 °C for 10 min.

The amplified PCR products were then subjected to electrophoresis on 1.0% agarose gels for analysis. The expected PCR amplicons were purified and then T-A cloned using the pMD18-T Vector Cloning Kit (TaKaRa, Dalian, China). Then the transformants were identified by a commercial company (Sangon, Shanghai, China) for nucleotide sequencing. The selected recombinant plasmid, pMD18-NS, used as the standard plasmid. The plasmid pMD18-NS was then quantified using a NANODROP 2000 spectrophotometer (Thermo Scientific, Waltham, MA, USA). According to the formula described by Yun et al. [22], the copy numbers of the plasmid pMD18-NS was calculated. Serial 10-fold dilutions of plasmid pMD18-NS, were diluted using EASY Dilution (TaKaRa, Dalian, China). All the serial diluted plasmids (ranging from 2.97×10^7 to 2.97×10^0 copies/µl), were stored at − 80 °C until use.

Real-time PCR assay

Real-time PCR amplification and detection were carried out on Mastercycler ep realplex (Eppendorf, Germany). The concentration of the primers, probe, and templates were optimized based on the obtained fluorescence and lowest threshold cycle (Ct). The optimized TaqMan-based PCR was prepared in a final volume of 25 µl containing 1 µl DNA template, 12.5 µl Premix Ex Taq (Probe qPCR, TaKaRa, Dalian, China), 0.5 µl of each primer (MDPV-qF and MDPV-qR, 10 µmol/l each), 2 µl probe (MDPV-qP, 5 µmol/l), and 8.5 µl Nuclease-free water to to adjust the reaction volume (total reaction volume of 25 µl). The mixed reactions was conducted in a single tube in a Mastercycler ep realplex by using the following thermoprofile: 40 cycles of 95 °C for 5 s, 58 °C for 10 s, and 72 °C for 15 s. 10-fold serial dilutions of plasmid pMD18-NS, containing different copy numbers of DNA (2.97×10^6 to

2.97×10^1 copies/µl) were conducted to generate the standard curve. All of the reactions were conducted in triplicate simultaneously. Analysis of each assay was conducted with CalQplex software (Mastercycler ep realplex, Eppendorf, Germany) according to the instruction manual. The software automatically uses the Ct values of serial dilutions of standards to calculate a standard curve, which shows the Ct values as a function of the amount of different copy numbers of DNA.

Sensitivity analysis

To evaluate the limit of detection (LOD) of the qPCR assay, 10-fold serial dilutions of plasmid DNA standard (ranging from 2.97×10^5 to 2.97×10^0 copies/µl) were prepared to determine the sensitivity. Each concentration was run in triplicate. Meanwhile, conventional PCR (cPCR) was performed with primers (NSF1 and NSR1) (NSF1, 5'- CAATGGGCTTTTACCAATATGC-3' and NSR1, 5'- ATTTTTCCCTCCTCCCACCA-3') and the same standard plasmid, in order to determine the LOD of cPCR assay. The cPCR reaction mixtures and thermal profile was described the same as previously reported [12]. PCR products were visualized following electrophoresis of 5 µl of each reaction in a 1.0% agarose gel according to instructions provided by the manufacturer. The LOD between the cPCR and qPCR assay were then compared.

Specificity analysis and reproducibility analysis

To determine the specificity of the qPCR assay, ten ng of extracted DNAs and cDNAs templates were used for the specificity analysis. The qPCR assay was carried out in triplicate to amplify a panel of duck-derived pathogens, i.e. H9N2 AIV, ATmV, DHAV-1, DHAV-3, MDRV, DAdV-A, DEV, GPV, N-GPV, E. coli., P.M., R.A. and S.S.. Nuclease-free water also used in the qPCR run to validate the specificity of qPCR assay as negative control. For reproducibility analysis of the qPCR assay, the 10-fold dilutions of pMD18-NS (concentration with 2.97×10^5, 2.97×10^3, and 2.97×10^1 copies/µl) were tested to evaluate the coefficient of variation (CV). For intra-assay variability of qPCR assay, triplicates of each dilution were analyzed, and the CVs were calculated according to the formula of the geometric mean Ct values deviation. For inter-assay variability, a coefficient of variation which expresses the standardized measure of dispersion with different time.

Table 1 Percentage nucleotide (nt) identity within and between MDPVs and GPVs

Viruses	Winthin MDPVs	Winthin GPVs	Between MDPVs and GPVs
Percentage	98.0–100%	93.3%--100%	80.8–83.4%

Table 2 Sequence variation in multiple sequences alignment

Primers	Sequence(5' → 3')[b1]	Frequency	GenBank accession numbers MDPVs[b2]
MDPV-qF	TACGAATGAACAAACCAA	80.0% (12/15)	/
	CACGAATGAACAAACCAA	13.33% (2/15)	KU844281, JF926697
	TACGAATGAAC**G**AACCAA	6.67% (1/15)	KC171936
MDPV-qR	CGCTCTTAATATCTCCTCTA	93.33% (14/15)	/
	TAGA**A**GAGATATTAAGAGCG	6.67% (1/15)	KT865605
MDPV-qP	TGAACGAGCGAATGAGCCTTCC	100% (15/15)	/

[b1]The variations are marked as Bold and underline
[b2]The GenBank accession numbers of MDPV strains used in this study are as follows: U22967, KU844282, KX000918, JF926698, KM093740, KY744743, KY069274, JF926695, KY511293, JF926696, X75093, KU844281, JF926697, KC171936, and KT865605. Only variations from GenBank accession numbers of MDPV strains have been marked

Clinical samples detection of MDPV

Detection of MDPV in field samples

In order to validate the qPCR assay, a total of 75 individuals Muscovy duckling-origin cloacal swabs samples (less than three-week) with diarrheal symptoms were collected in Fujian, Jiangxi, Guangdong, Jiangsu, and Zhejiang provinces, China. All of the suspensions were subjected to three freeze-thaw cycles and then centrifuged at 8000 rpm at 4 °C for 30 min. Viral DNAs were extracted from the harvested supernatants using EasyPure Micro Genomic DNA Kit (Transgen Bioteck, Beijing, China). Conventional PCR was preformed simultaneously.

Detection of MDPV in Muscovy duck embryos and newly hatched ducklings

Previous study indicates possible vertical transmission of N-GPV and suggests that N-GPV may be transmitted from breeder ducks to ducklings in *ovo* [23]. In order to determine the hypothesis whether MDPV could be possible vertically transmitted or not, 20 Muscovy duck embryos (15-day post fertilization) and 20 newly hatched Muscovy ducklings (1-day-old) were collected from the diseased farms where the virus (FJM5 strain) was discovered [8]. The liver of each embryo and newly hatched duckling was pooled and regarded as one sample. These samples were homogenized in phosphate-buffered saline (PBS) (20%, *w/v*). Viral DNAs were extracted from tissue homogenates of liver using EasyPure Micro Genomic DNA Kit (Transgen Bioteck, Beijing, China). Conventional PCR was preformed simultaneously. All of the cPCR-positive amplicons were harvested, T-A cloned, and then

sequenced to verify the results for Sanger seuquencing at Sangon (Shanghai, China) in both directions.

Results

NS gene analysis

We compared a total of 52 NS gene sequences (including 15 MDPVs and 37 GPVs) downloaded from the GenBank database. We found that within the MDPV cluster, there was a higher nucleotide identity (more or equal than 98.0%) than within the GPV cluster (more or equal than 93.3%). In addition, the NS gene homology between GPV cluster and MDPV cluster ranged from 80.8 to 83.4% (Table 1). The primers MDPV-qF and MDPV-qR, and the TaqMan probe MDPV-qP variation within MDPVs are listed in Table 2. Moreover, the 3'-terminal of MDPV-qP at position 1521–1524 was TTCC, while the sequence at this position in the 37 GPVs was either *GGAG* (54.05%, 20/37), *AGAG* (40.54%, 15/37), *AGAT* (2.70%, 1/37), or *GAAG* (2.70%, 1/37). This data demonstrated that sequences at positions 1521–1524 were significantly different between MDPVs and GPVs, which can be chosen for specific probe design.

Real-time PCR

The generated standard curve showed linearity over the 2.97×10^6 to 2.97×10^1 copies/µl range, with a slope of -3.358 and the Y-intercept was 34.93, efficiency of 0.99, which was conducted by software of Mastercycler ep realplex (Eppendorf, Germany) (Fig. 1). Using the viral DNAs, cDNAs, bacterial DNAs and Nuclease-free water,

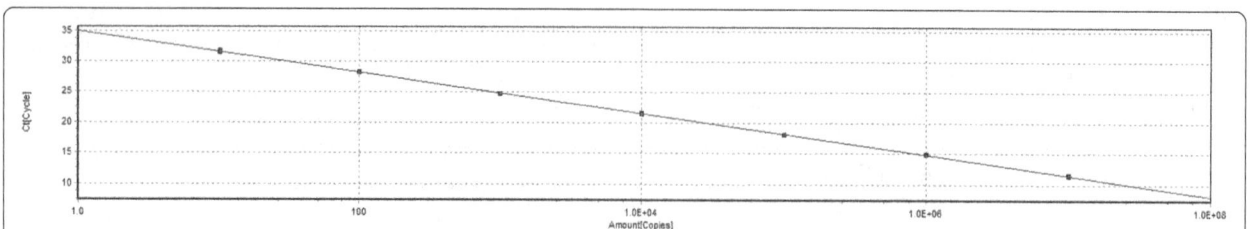

Fig. 1 Standard curve of TaqMan-based real-time PCR assay

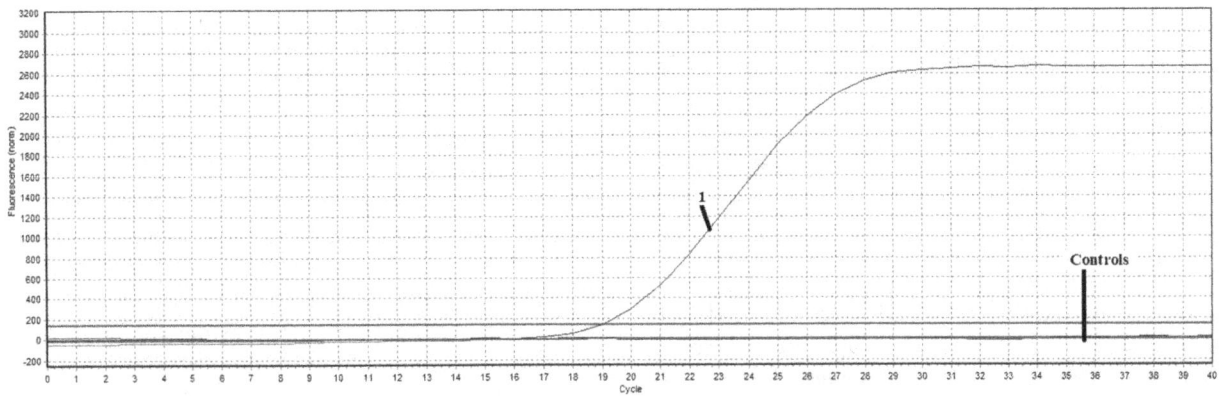

Fig. 2 Specificity of TaqMan-based real-time PCR assay. 1: MDPV; Controls: H9N2 AIV, ATmV, DHAV-1, DHAV-3, MDRV, DAdV-A, DEV, GPV, N-GPV, *E. coli.*, *P.M.*, *R.A.*, *S.S.* and Nuclease-free water. These controls were all found with no fluorescence signal

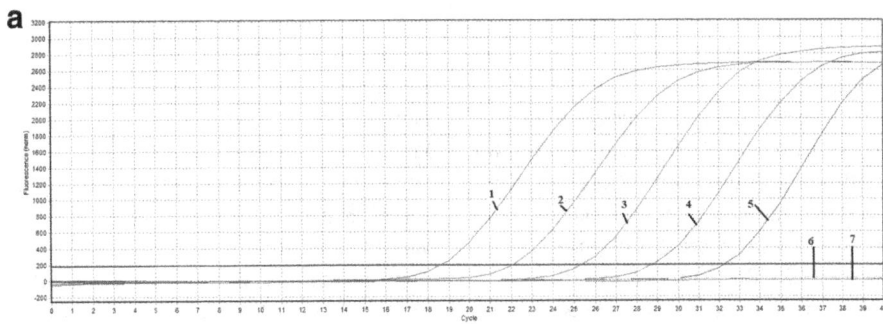

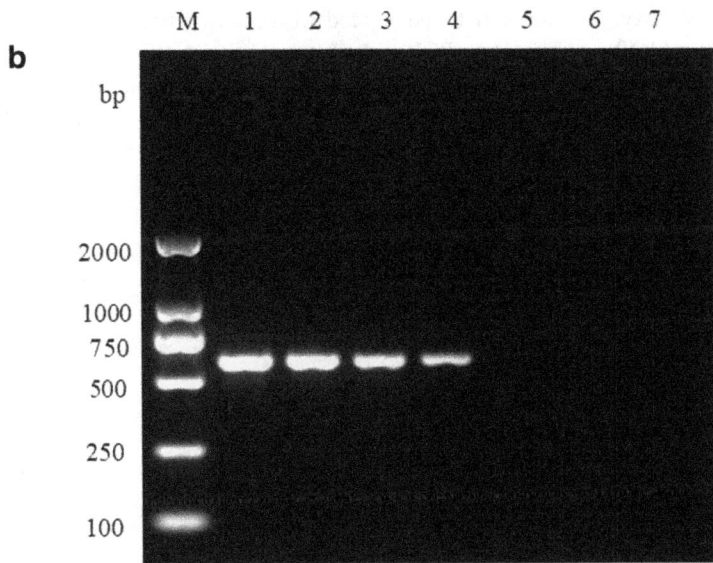

Fig. 3 a Sensitivity of TaqMan-based real-time PCR assay for MDPV detection. **b** Sensitivity of conventional PCR assay for MDPV detection. 1–6: a serial of ten-fold dilutions plasmid DNA (2.97×10^5 to 2.97×10^0 copies/μl); 7: negative control; M: DL2000 DNA Marker

Table 3 Intra- and inter-assay reproducibility for TaqMan-based PCR

Concentration of standard plasmid (copies/µl)	Intra-assay variability		Inter-assay variability	
	$\overline{X} \pm SD$	CV (%)	$\overline{X} \pm SD$	CV (%)
2.97×10^5	18.65 ± 0.11	0.59	18.72 ± 0.12	0.66
2.97×10^3	25.40 ± 0.11	0.44	25.46 ± 0.16	0.63
2.97×10^1	32.32 ± 0.46	1.43	32.48 ± 0.72	2.21

the results showed that only MDPV was detected by the TaqMan-based real-time PCR assay. No fluorescence signal was detected in control samples of H9N2 AIV, ATmV, DHAV-1, DHAV-3, MDRV, DAdV-A, DEV, GPV, N-GPV, *E. coli.*, *P.M.*, *R.A.*, *S.S.* and Nuclease-free water (Fig. 2), which indicated the excellent specificity of the qPCR assay. The LOD of the qPCR assay was 29.7 copies/µl (Fig. 3a). In contrast, conventional PCR had a detection limit of 2.97×10^2 copies/µl (Fig. 3b). The intra-assay CVs of pMD18-NS (concentration with 2.97×10^5, 2.97×10^3, and 2.97×10^1 copies/µl) were 0.59%, 0.44% and 1.43%. The inter-assay CVs were 0.66%, 0.63% and 2.21% (listed in Table 3), which indicate a high degree of repeatability.

Clinical samples application

Senventy-five cloacal swabs from Muscovy ducklings with diarrhea were evaluated of MDPV diagnosis by the TaqMan-based real-time PCR and conventional PCR assays. As summarized in Table 4, for 75 cloacal swabs evaluated samples, 10 (13.33%) were MDPV-positive by qPCR assay. However, cPCR results showed 7 were MDPV-positive. For embryonic samples, 4 of 20 (20.0%) were MDPV-positive by qPCR assay and 2 of 20 (10.0%) were MDPV-positive by cPCR assay. For newly hatched ducklings, 5 of 20 (25.0%) were MDPV-positive by qPCR assay and 4 of 20 (20.0%) were MDPV-positive by cPCR assay, respectively. The samples tested with MDPV-positive by cPCR assay, were also tested MDPV-positive by

qPCR assay. The viral DNA copy numbers in the positive samples when detecting MDPV DNAs were listed in Table 4. These findings provide evidence of possible vertical transmission of MDPV.

Sequences analysis

A total of 13 (7 from cloacal swabs, 2 from embryonic samples, and 4 from newly hatched ducklings) amplicons from cPCR positive were harvested, purified, T-A cloned and then sequenced. The cloned sequences shared ≥99.1% nucleotide identity with MDPV (strain FM). Moreover, the cloned sequences from cPCR positive samples could specifically (100%) match the primers (MDPV-qF and MDPV-qR) and probe (MDPV-qP) listed in Table 2.

Discussion

The current methods for the detection of MDPV, such as virus isolation, immunological-based assays, and electron microscopy, have proven to be laborious and time-consuming. Conventional PCR technology has been used for differentiation between MDPV and GPV, and the process includes restriction enzyme digestion, agarose gel electrophoresis, and DNA sequencing [12, 24]. Moreover, the PCR method for the specific detection of MDPV requires high precision primer design. Loop-mediated isothermal amplification (LAMP) [25] and an aptamer by label-free aptasensor [26] against MDPV for a highly sensitive, rapid visual detection was designed that targeted the VP3 gene of MDPVs. The VP3 protein is the most variable and abundant protein of MDPV. VP3 can induce neutralizing antibodies in GPV- or MDPV-infected waterfowl and confers protective cross-immunity in waterfowl parvoviruses [13, 27–30]. Moreover, VP3 has been associated with many genetic variations because of virus evolution selection pressure [6, 7, 31–33]. In addition, due to lack of information on the complete genomic sequence data of GPVs, especially MDPVs, false results can be obtained if primers are designed from unsuitable regions.

Table 4 Clinical samples tested for MDPV infection by conventional PCR and TaqMan-based PCR for MDPV detection

Samples	Region	Number of samples	Number of MDPV positive samples		Copy number for positive (copies/µL)	
			cPCR	qPCR	Both	Only
Cloacal Swabs	Fujian	15	2	4	7.45×10^4, 1.72×10^5	3.59×10^1, 8.24×10^1
	Jiangxi	15	1	1	4.45×10^4	–
	Guangdong	15	2	3	9.72×10^3, 2.45×10^4	1.74×10^2
	Jiangsu	15	1	1	8.61×10^4	–
	Zhejiang	15	1	1	1.07×10^4	–
Embryos	Fujian	20	2	4	2.89×10^3, 5.72×10^3	7.94×10^1, 1.18×10^2
Newly hatched Muscovy ducklings	Fujian	20	4	5	3.89×10^3, 6.02×10^3, 9.17×10^3, 2.21×10^4	9.47×10^1

cPCR means conventional PCR; qPCR means TaqMan-based real-time PCR; Both means the samples were tested both cPCR-positive and qPCR-positive; Only means the samples were tested only by qPCR-positive. "-" means no only qPCR sample available

The real-time PCR technology has advantages with its remarkable sensitivity, specificity, reproducibility, visualization results, time-saving benefits, high-throughput analysis and less-contamination potential compared to other diagnostic methods, that have been widely used for viral pathogenesis research and epidemiology surveillance [16–21]. Woźniakowski et al. developed a TaqMan-based real-time PCR method for GPVs and MDPVs in which the designed primers and TaqMan probe for real-time PCR were comple-mentary to GPV and MDPV inverted terminal repeats region (ITR) [34]. The calculation of results is often confounded due to the two ITR repeat regions at the 5′- and 3′-terminus in both the GPV and MDPV ge-nomes. Moreover, mutation and deletion have been found in ITR of both GPVs and MDPVs, which may cause high risk of failure for the specific detection of MDPV infection [35–37]. In this study, we compared a total of 52 NS gene sequences (including 15 MDPVs and 37 GPVs) retrieved from the GenBank. We found that within the MDPV cluster, the samples shared higher nucleotide identity (more or equal than 98.0%) than within GPV cluster (more or equal than 93.3%). In addition, between GPV cluster and MDPV cluster, the NS gene homology ranged from 80.8 to 83.4%. These data indicate that false results may be obtained if these primers are designed at the NS gene-specific regions. These NS genetic comparison data allowed us to identify the MDPV-specific suitable regions, which can be used to establish a TaqMan-based real-time PCR assay for the detection of MDPV with more precision.

In this study, we clearly developed and evaluated the applicability of TaqMan-based real-time PCR for the de-tection and quantification of MDPV. The data of the qPCR assay was performed over a wide dynamic range with no fluorescence signal can be observed from other duck-derived pathogens, and low intra- and inter-assay variation (less than 2.21%). Field samples from cloacal swabs detection and those used in previous studies of MDPV infection provided the evidence that MDPV can be horizontally transmitted. Furthermore, the positive fluorescence signals observed in embryos and newly hatched Muscovy ducklings support the view that MDPV could be possible vertically transmitted as well as N-GPV [23].

Conclusions

A specific and reliable TaqMan-based real-time PCR for the detection and quantification of MDPV was developed, because of significantly conserved region of NS genes between MDPVs and GPVs were selected for primers and probe design. The detection of the MDPV-positive in Muscovy duck embryos and newly hatched Muscovy ducklings, provide evidence of possible vertical transmission of MDPV from breeding Muscovy ducks to Muscovy ducklings.

Abbreviations

AIV: avian influenza virus;; ATmV: avian Tembusu virus; BLAST: Basic Local Alignment Search Tool; cPCR: conventional PCR; Ct: threshold cycle; CV: Coefficient of variation; DAdV-A: Duck adenovirus A; DEV: Duck enteritis virus; DHAV: Duck hepatitis virus; GPV: Goose parvovirus; ICTV: International Committee on Taxonomy of Viruses; ITR: Inverted terminal repeats; LAMP: Loop-mediated isothermal amplification; MDPV: Muscovy duck parvovirus; MDRV: Muscovy duck reovirus; N-GPV: Novel goose parvovirus; NS: Nonstructural; ORF: Open reading frame; *P.M.*: Pasteurellamultocida; qPCR: real-time quantitative PCR; *R.A.*: Rimerella anatipstifer; R²: correlation; *S.S.*: Salmonella spp

Funding

This work was funded by the Natural Science Foundation of China (31602068), China Agriculture Research System (CARS-42), Fujian Academy of Agriculture Science Innovative Research Team Project (STIT2017-3-10) and Young Talent Program Project (YC2015–12), and the Fujian Public Welfare Project (2018R1023–5). The funders had no role in study design, data collection and analysis, decision to publish, or preparation of the manuscript.

Authors' contributions

CW assembled the sequence data, designed primers and performed the experiments, analyzed the data and drafted the manuscript. CC, LC, HC, QF, SS, GF, and RL collected samples and helped in laboratory analysis. YH contributed to experimental design and supervised the study. All authors read, commented on and approved the final version of the manuscript.

Consent for publication

Not applicable.

Competing interests

The authors declare that they have no competing interests.

References

1. Zadori Z, Erdei J, Nagy J, Kisary J. Characteristics of the genome of goose parvovirus. Avian Pathol. 1994;23:359–64.
2. Zádori Z, Stefancsik R, Rauch T, Kisary J. Analysis of the complete nucleotide sequences of goose and muscovy duck parvoviruses indicates common ancestral origin with adeno-associated virus 2. Virology. 1995;212:562–73.
3. Le Gall-Reculé G, Jestin V. Biochemical and genomic characterization of muscovy duck parvovirus. Arch Virol. 1994;139:121–31.
4. Chen H, Dou Y, Tang Y, Zhang Z, Zheng X, Niu X, Yang J, Yu X, Diao Y. Isolation and genomic characterization of a duck-origin GPV-related parvovirus from Cherry Valley ducklings in China. PLoS One. 2015;10:e0140284.
5. Liu H, Wang H, Tian X, Zhang S, Zhou X, Qi K, Pan L. Complete genome sequence of goose parvovirus Y strain isolated from Muscovy ducks in China. Virus Genes. 2014;48:199–202.
6. Zhu Y, Zhou Z, Huang Y, Yu R, Dong S, Li Z, Zhang Y. Identification of a recombinant Muscovy duck parvovirus (MDPV) in shanghai, China. Vet Microbiol. 2014;174:560–4.
7. Wang J, Ling J, Wang Z, Huang Y, Zhu J, Zhu G. Molecular characterization of a novel Muscovy duck parvovirus isolate: evidence of recombination between classical MDPV and goose parvovirus strains. BMC Vet Res. 2017;13:327.
8. Wan C, Chen H, Fu Q, Fu G, Cheng L, Shi S, Huang Y, Hu K. Genomic characterization of goose parvovirus and Muscovy duck parvovirus co-infection in Fujian. China Kafkas Univ Vet Fak Derg. 2015;21:923–8.
9. Shao H, Lv Y, Ye J, Qian K, Jin W, Qin A. Isolation of a goose parvovirus from swan and its molecular characteristics. Acta Virol. 2014;58:194–8.

10. Wan C, Huang Y, Shi S, Fu G, Cheng L, Chen H, Fu Q, Chen C, Hu K. Complete genome sequence of goose parvovirus isolated from Anser cygnoides in China. Kafkas Univ Vet Fak Derg. 2016;22:155–8.

11. Lin S, Yu X, Chen B, Chen B. The diagnosis of a new muscovy duck virus infection. Chin J Prevent Vet Med. 1991;2:27–8. (in Chinese)

12. Wan C, Chen H, Fu Q, Shi S, Fu G, Cheng L, Chen C, Huang Y, Hu K. Development of a restriction length polymorphism combined with direct PCR technique to differentiate goose and Muscovy duck parvoviruses. J Vet Med Sci. 2016;78:855–8.

13. Zhang Y, Li Y, Liu M, Zhang D, Guo D, Liu C, Zhi H, Wang X, Li G, Li N, Liu S, Xiang W, Tong G. Development and evaluation of a VP3-ELISA for the detection of goose and Muscovy duck parvovirus antibodies. J Virol Methods. 2010;163:405–9.

14. Li M, Yu TF. Immunologic cross-reactivity between Muscovy duck parvovirus and goose parvovirus on the basis of epitope prediction. Braz J Microbiol. 2013;44:519–21.

15. Yu TF, Li M, Yan B, Shao SL, Fan XD, Wang J, Wang DN. Identification of antigenic domains in the non-structural protein of Muscovy duck parvovirus. Arch Virol. 2016;161:2269–72.

16. Niesters HG. Quantitation of viral load using real-time amplification techniques. Methods. 2001;25:419–29.

17. Mackay IM, Arden KE, Nitsche A. Real-time PCR in virology. Nucleic Acids Res. 2002;30:1292–305.

18. Niu X, Chen H, Yang J, Yu X, Ti J, Wang A, Diao Y. Development of a TaqMan-based real-time PCR assay for the detection of novel GPV. J Virol Methods. 2016;237:32–7.

19. Wang J, Wang J, Cui Y, Nan H, Yuan W. Development of a taqman-based real-time PCR assay for the rapid and specific detection of novel duck-origin goose parvovirus. Mol Cell Probes. 2017;34:56–8.

20. Wan C, Cheng L, Fu G, Chen C, Liu R, Shi S, Chen H, Fu Q, Huang Y. Rapid detection of goose hemorrhagic polyomavirus using TaqMan quantitative real-time PCR. Mol Cell Probes. 2018;39:61–4.

21. Wan C, Chen C, Cheng L, Guanghua F, Shi S, Liu R, Chen H, Fu Q, Huang Y. Development of a TaqMan-based real-time PCR for detecting duck adenovirus 3. J Virol Methods. 2018; https://doi.org/10.1016/j.jviromet.2018.08.011.

22. Yun JJ, Heisler LE, Hwang II, Wilkins O, Lau SK, Hyrcza M, Jayabalasingham B, Jin J, McLaurin J, Tsao MS, Der SD. Genomic DNA functions as a universal external standard in quantitative real-time PCR. Nucleic Acids Res. 2006;34:e85.

23. Chen H, Tang Y, Dou Y, Zheng X, Diao Y. Evidence for vertical transmission of novel duck-origin goose parvovirus-related parvovirus. Transbound Emerg Dis. 2016;63:243–7.

24. Sirivan P, Obayashi M, Nakamura M, Tantaswasdi U, Takehara K. Detection of goose and Muscovy duck parvoviruses using polymerase chain reaction-restriction enzyme fragment length polymorphism analysis. Avian Dis. 1998;42:133–9.

25. Ji J, Xie QM, Chen CY, Bai SW, Zou LS, Zuo KJ, Cao YC, Xue CY, Ma JY, Bi YZ. Molecular detection of Muscovy duck parvovirus by loop-mediated isothermal amplification assay. Poult Sci. 2010;89:477–83.

26. Lu T, Ma Q, Yan W, Wang Y, Zhang Y, Zhao L, Chen H. Selection of an aptamer against Muscovy duck parvovirus for highly sensitive rapid visual detection by label-free aptasensor. Talanta. 2018;176:214–20.

27. Le Gall-Reculé G, Jestin V, Chagnaud P, Blanchard P, Jestin A. Expression of muscovy duck parvovirus capsid proteins (VP2 and VP3) in a baculovirus expression system and demonstration of immunity induced by the recombinant proteins. J Gen Virol. 1996;77:2159–63.

28. Yin X, Zhang S, Gao Y, Li J, Tan S, Liu H, Wu X, Chen Y, Liu M, Zhang Y. Characterization of monoclonal antibodies against waterfowl parvoviruses VP3 protein. Virol J. 2012;9:288.

29. Tarasiuk K, Woźniakowski G, Holec-Gąsior L. Expression of goose parvovirus whole VP3 protein and its epitopes in Escherichia coli cells. Pol J Vet Sci. 2015;18:879–80.

30. Li C, Liu H, Li J, Liu D, Meng R, Zhang Q, Shaozhou W, Bai X, Zhang T, Liu M, Zhang Y. A conserved epitope mapped with a monoclonal antibody against the VP3 protein of goose parvovirus by using peptide screening and phage display approaches. PLoS One. 2016;11:e0147361.

31. Chu CY, Pan MJ, Cheng JT. Genetic variation of the nucleocapsid genes of waterfowl parvovirus. J Vet Med Sci. 2001;63:1165–70.

32. Wang S, Cheng X, Chen S, Zhu X, Chen S, Lin F, Li Z. Genetic characterization of a potentially novel goose parvovirus circulating in Muscovy duck flocks in Fujian Province, China. J Vet Med Sci. 2013;75:1127–30.

33. Wang S, Cheng X, Chen S, Lin F, Chen S, Zhu X, Wang J. Evidence for natural recombination in the capsid gene VP2 of Taiwanese goose parvovirus. Arch Virol. 2015;160:2111–5.

34. Woźniakowski G, Samorek-Salamonowicz E, Kozdruń W. Quantitative analysis of waterfowl parvoviruses in geese and Muscovy ducks by real-time polymerase chain reaction: correlation between age, clinical symptoms and DNA copy number of waterfowl parvoviruses. BMC Vet Res. 2012;8:29.

35. Shien JH, Wang YS, Chen CH, Shieh HK, Hu CC, Chang PC. Identification of sequence changes in live attenuated goose parvovirus vaccine strains developed in Asia and Europe. Avian Pathol. 2008;37:499–505.

36. Wang J, Duan J, Zhu L, Jiang Z, Zhu G. Sequencing and generation of an infectious clone of the pathogenic goose parvovirus strain LH. Arch Virol. 2015;160:711–8.

37. Wang J, Huang Y, Zhou M, Zhu G. Analysis of the genome sequence of the pathogenic Muscovy duck parvovirus strain YY reveals a 14-nucleotide-pair deletion in the inverted terminal repeats. Arch Virol. 2016;161:2589–94.

Molecular detection of enteric viruses and the genetic characterization of porcine astroviruses and sapoviruses in domestic pigs from Slovakian farms

Slavomira Salamunova, Anna Jackova, Rene Mandelik, Jaroslav Novotny, Michaela Vlasakova and Stefan Vilcek*[iD]

Abstract

Background: Surveillance and characterization of pig enteric viruses such as transmissible gastroenteritis virus (TGEV), porcine epidemic diarrhea virus (PEDV), rotavirus, astrovirus (PAstV), sapovirus (PSaV), kobuvirus and other agents is essential to evaluate the risks to animal health and determination of economic impacts on pig farming. This study reports the detection and genetic characterization of PAstV, PSaV in healthy and diarrheic domestic pigs and PEDV and TGEV in diarrheic pigs of different age groups.

Results: The presence of PAstV and PSaV was studied in 411 rectal swabs collected from healthy ($n = 251$) and diarrheic ($n = 160$) pigs of different age categories: suckling ($n = 143$), weaned ($n = 147$) and fattening ($n = 121$) animals on farms in Slovakia. The presence of TGEV and PEDV was investigated in the diarrheic pigs ($n = 160$). A high presence of PAstV infections was detected in both healthy (94.4%) and diarrheic (91.3%) pigs. PSaV was detected less often, but also equally in clinically healthy (8.4%) and diarrheic (10%) pigs. Neither TGEV nor PEDV was detected in any diarrheic sample. The phylogenetic analysis of a part of the RdRp region revealed the presence of all five lineages of PAstV in Slovakia (PAstV-1 – PAstV-5), with the most frequent lineages being PAstV-2 and PAstV-4. Analysis of partial capsid genome sequences of the PSaVs indicated that virus strains belonged to genogroup GIII. Most of the PSaV sequences from Slovakia clustered with sequences originating from neighbouring countries.

Conclusions: Due to no significant difference between healthy and diarrheic pigs testing of the presence of PAstV and PSaV provides no diagnostic value. Genetic diversity of PAstV was very high as all five lineages were identified in pig farms in Slovakia. PSaV strains were genetically related to the strains circulating in Central European region.

Keywords: Porcine astrovirus, Porcine sapovirus, Diarrhea, Pig, Phylogenetic analysis

Background

Viral gastroenteritis is one of the serious diseases in pigs with high morbidity observed worldwide, causing significant financial losses. Therefore, surveillance and characterization of pig enteric viruses such as transmissible gastroenteritis virus (TGEV), porcine epidemic diarrhea virus (PEDV), rotavirus, astrovirus, sapovirus, kobuvirus and other agents is essential to evaluate the possible risks to animal health as well as for the

epidemiological analysis and determination of economic impacts on pig farming.

Many enteric RNA viruses have been found in swine herds including those from genera *Mamastrovirus* and *Sapovirus*. The astroviruses from genus *Mamastrovirus*, family *Astroviridae* affect a wide variety of mammalian species [1]. They consist of small (28–30 nm), non-enveloped, single-stranded, positive-sense RNA molecule of 6.4–7.7 kb in size. Their genome is arranged in three open reading frames (ORF). The ORF1a and ORF1b at the 5'end encode non-structural proteins and ORF2 at the 3' end encodes the structural proteins [1].

* Correspondence: vilcek@uvm.sk
University of Veterinary Medicine and Pharmacy, Komenskeho 73, 040 00 Kosice, Slovakia

Porcine astrovirus (PAstV) isolated from pigs worldwide [2–4] includes five highly genetic variable lineages (PAstV-1 to PAstV-5) [4, 5]. In Europe, all five lineages have been detected [6], with the most common lineage being PAstV-4. PAstV has also been detected in several Central European countries [7, 8].

Genus *Sapovirus* belongs within the family *Caliciviridae*. Similar to astrovirus, it is a non-enveloped virus, containing a single-stranded, positive-sense RNA genome of 7.4–8.3 kb. The virus genome is organized into two or three ORFs, which encode both structural and non-structural proteins, including the RNA-dependent RNA polymerase (RdRp) and the major capsid protein (VP1) [1]. Due to great genetic variability and recombination capacity, the genus is divided into fifteen different genogroups based on analysis of both capsid and RdRp sequences [9, 10]. Sapoviruses infect a wide variety of hosts including humans, pigs, mink, canines and bats [11–14]. The porcine sapovirus (PSaV), formerly known as porcine enteric calicivirus (PEC), is divided into GIII, GV – GXI genogroups, but GIII is considered as the prototypic genogroup with the only cultivable Cowden strain [9].

While representatives of the genera *Mamastrovirus* and *Sapovirus* were mainly associated with gastroenteritis in infants and young children [1, 15], the epidemiology of these viruses in animals especially in pigs is not so well understood and is intensively studied. The virus infections are associated with enteric diseases, but their role in diarrhea or gastroenteritis is still unknown due their detection in healthy as well as in diarrheic pigs [7, 16].

TGEV and PEDV, however, are considered to be important causative agents of diarrhea in pigs. TGEV and PEDV are members of the genus *Alphacoronavirus* of the family *Coronaviridae* with enveloped virions, containing longer positive single-stranded RNA genomes approximately 28 kb in length [1]. These viruses are causative agents of acute diarrhea in all age categories and can cause high mortality in neonatal piglets, resulting in significant economic losses [17].

In Slovakia, there is limited information on the occurrence of enteric viruses in pig farms so far. We have selected for our research project three important enteric viruses, namely rotavirus A (RVA), TGEV and PEDV. In addition we were also interested in porcine kobuvirus 1 – PKV-1, PAstV and PSaV, which are believed to be enteric viruses too. Recently we have analysed a broad collection of pig rectal swabs for the detection of RVA and PKV-1 [18]. This study on the same samples reports the molecular detection and partial genetic characterization of PAstV, PSaV in healthy and diarrheic domestic pigs and PEDV and TGEV in diarrheic domestic pigs of different age groups.

Results

Molecular detection of TGEV and PEDV

Neither TGEV nor PEDV was detected in any of the rectal swabs collected from 160 diarrheic pigs.

Molecular detection of porcine astrovirus

Results of the PAstV detection in investigated pigs are presented on Fig. 1.

PAstV RNA was detected in all 17 investigated farms with an overall prevalence 93.2% (383/411).

Among healthy pigs 94.4% (237/251) were found to be infected by PAstV. The most affected group across age categories was the fattening pigs where nearly all animals

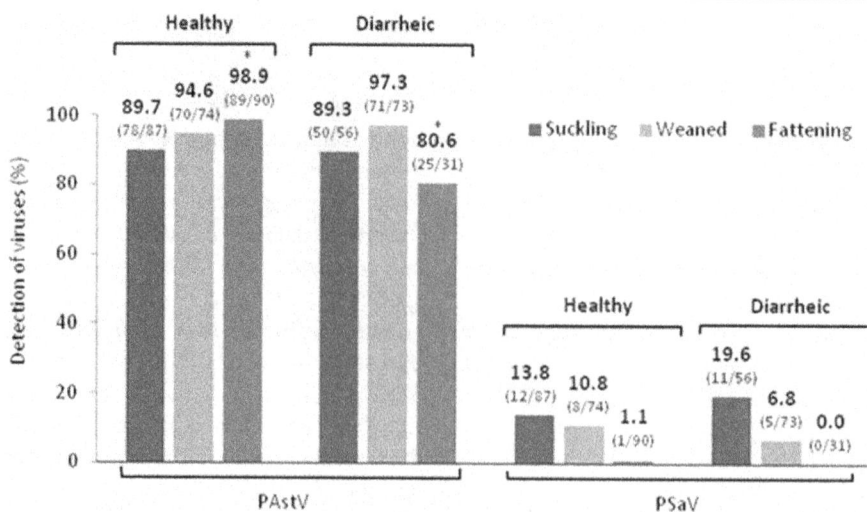

Fig. 1 PAstV and PSaV in healthy and diarrheic pigs of three age categories. Brackets indicate positive/total number of samples analysed. The star indicates statistically significant ($p < 0.01$) value

were virus positive (98.9%; 89/90) followed by weaned pigs (94.6%; 70/74), and suckling piglets (89.7%; 78/87).

In diarrheic animals, 91.3% (146/160) pigs were infected by PAstV. The most affected group across age categories were weaned pigs (97.3%; 71/73), followed by suckling piglets (89.3%; 50/56) and fattening pigs (80.6%; 25/31).

Statistical analysis of individual age categories revealed high significant differences of astrovirus infection between diarrheic fattening (80.6%) pigs and healthy fattening (98.6%) pigs ($p < 0.01$, $\chi^2 = 14.08$) with even higher prevalence in healthy animals. No significant differences were observed between healthy and diarrheic in the other two age categories.

Molecular detection of porcine sapovirus

The values of detection of PSaV in pigs are presented on Fig. 1.

Only 7 of the 17 farms investigated were positive for PSaV RNA. Of all investigated samples, 9% (37/411) were virus positive.

Approximately the same number of animals were infected by virus in the group of healthy pigs (8.4% - 21/251) as in diarrheic (10% - 16/160) animals. In both, healthy and diarrheic animals, the most affected group across age categories were suckling piglets (13.8% - 12/87 healthy and 19.6% - 11/56 diarrheic pigs), followed by weaned pigs (10.8% - 8/74 and 6.8% - 5/73). None of the diarrheic pigs and only one positive sample (1.1% - 1/90) of PSaV in healthy pigs was detected in the fattening age category. All PSaV positive pigs were also infected with PAstV.

There were no statistically significant differences between healthy and diarrheic groups of pigs in any of the three age categories mentioned above.

Sequence and phylogenetic analysis of partial RdRp region of PAstV

Twenty five amplicons from PAstV positive samples coming from 13 farms were selected for nucleotide sequencing. The phylogenetic analysis of RdRp region fragments is shown in Fig. 2.

Five lineages PAstV-1 to PAstV-5 were detected in investigated pigs from Slovakia. The lineages PAstV-2 ($n = 12$) and PAstV-4 ($n = 10$) were the most represented. The remaining tree lineages were each represented by one sequence. The nucleotide sequence identities between Slovakian PAstV lineages were 52.0–65.4%. Variability of Slovakian sequences inside the lineage PAstV-2 was in the range 74.9–100%, and variability inside the lineage PAstV-4 was 84.6–100%.

All sequences from Slovakian samples clustered together with viral sequences from European countries, particularly Croatia, but also with sequences from North America and China. The clustering of Slovakian sequences

with those of neighbouring countries was sporadic. Farm-specific grouping was observed with the most of PAstV sequences. However, isolates circulating on one farm belonging to 2–3 different lineages were confirmed on three investigated farms (Fig. 2, farms ME, MO and RI).

Sequence and phylogenetic analysis of partial capsid protein of PSaV

Twelve PSaV PCR amplicons were sequenced, which were selected to include the majority of infected farms, different health status and age categories of pigs. The phylogenetic analysis of 667 bp fragment (capsid region) is shown in Fig. 3.

All of the investigated viral sequences belong to genogroup GIII, showing nucleotide sequence identities in the range 78.7–100%. On the phylogenetic tree, all sequences from Slovakian pig samples clustered together with viral sequences from neighbouring countries, such as Czech Republic and Hungary, except two strains TOP C17 and TOP C27 which were clustered together with a sequence from Korea. PSaV sequences were clustered according to the farm origin, except RK O3 which was separate from sequences originating from the same farm.

Discussion

In this work, we have determined for the first time the presence of the PAstV and PSaV infections in pigs from Slovakia. PAstV has been detected widely in rectal swab samples of healthy and diarrheic pigs of all age categories as demonstrated by a greater than 90% prevalence of the virus. On the other hand, the incidence of PSaV has been found to be more sporadic (8–10%) in pig samples investigated independently if they originated from healthy or diarrheic pigs. Our results indicate that there is no strong correlation between detection of both viruses in enteric samples and diarrhea. This observation has been quite surprising as it is believed that porcine astrovirus and porcine sapovirus belong to the common agents of pig gastrointestinal infections, which are usually associated with diarrheic diseases [15, 19].

In our recent work [18] we tested the same collection of pig rectal swabs as used in the present work for the presence of porcine kobuvirus 1 (PKV-1) and RVA. The presencee of PKV-1 as a newly emerged virus, which role in gastroenteritis is still unknown, has been similar in both, healthy and diarrheic pigs, as has been observed for PAstV and PSaV in this work. However, the detection of RVA in suckling piglets, but not in weaned and fattening pigs, was significantly associated with diarrhea, confirming that this virus is a cause of gastroenteritis. In this context it is difficult to conclude, at least from our findings, that PAstV, PSaV and PKV-1 are typical enteric viruses associated with diarrhea which contrasts with conclusions by others [4, 19, 20]. This inconsistency

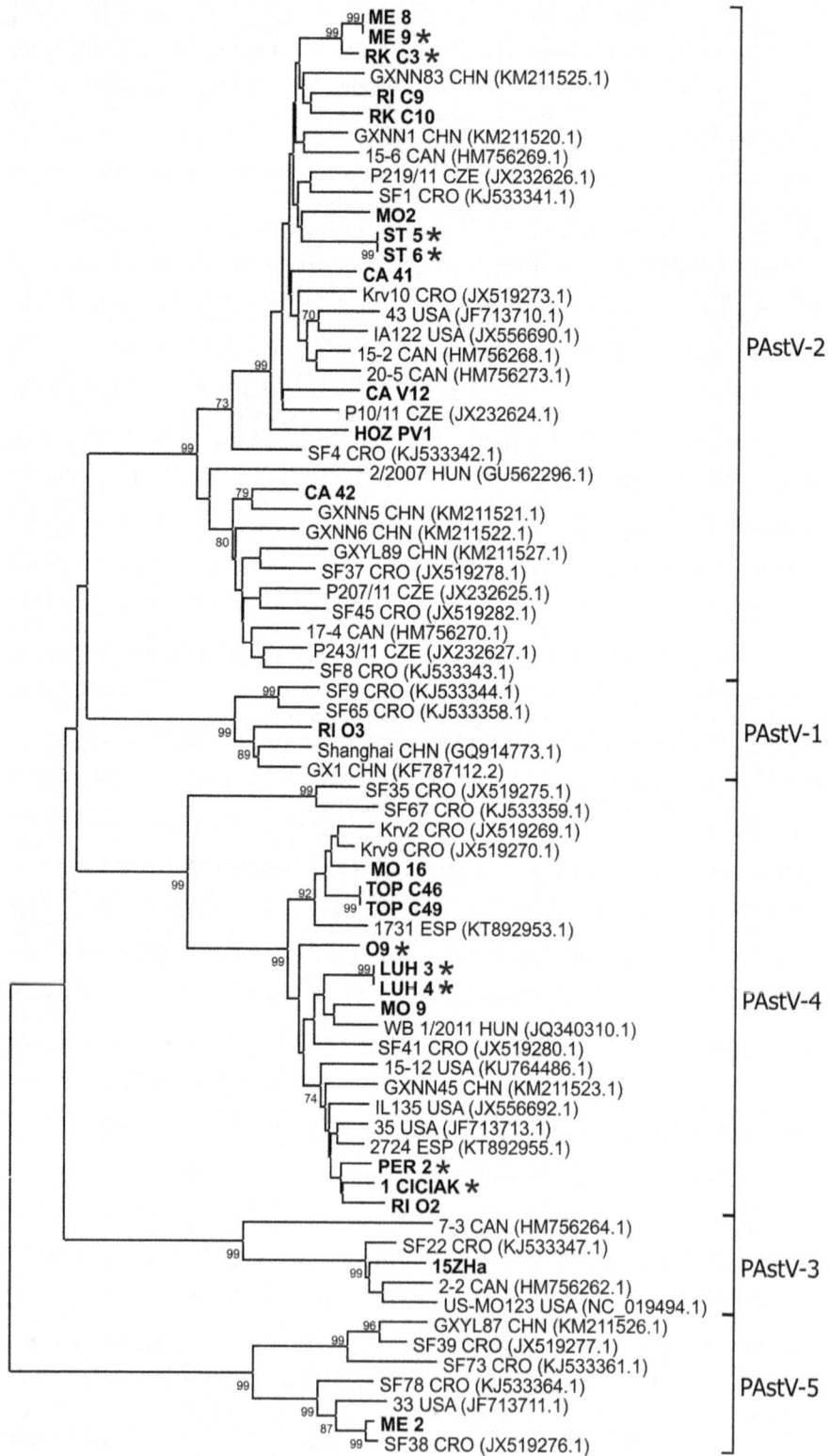

Fig. 2 (See legend on next page.)

(See figure on previous page.)

Fig. 2 Phylogenetic tree of PAstV. The tree was constructed on the base of 355 to 382 bp (depend on PAstV lineage) nucleotide sequences of RdRp of the PAstV and selected sequences from GenBank (labelled with accession number and country of origin). Sequences from Slovakian samples are in bolt, from diarrheic pigs are marked with stars. The tree was constructed using the neighbour-joining method implemented in MEGA6 software. Scale bar represents the number of nucleotide substitutions per site

indicates that viruses could express their pathogenicity in only a small percentage of cases as has been already hypothesized by Luo et al. [21] due to co-infection with other gastrointestinal viruses or due to immunodeficiency of animals. However, our data of the co-infection of viruses do not support this hypothesis. In two studies (this work and [18]) we did not detect strong differences between healthy and diarrheic animals with co-infection of PAstV and PSaV or in the case of rotavirus A co-infection with both PAstV and PSaV. Other studies also indicated that PAstV as a ubiquitous virus has been often found in co-infections with various viruses [7, 8]. We further speculate that the development of diarrhea in pigs is for some viruses a result of interaction the entire virome or microbiome with environment in the intestinal tract which may influence the course of viral infection. It is quite surprising that infection of gnotobiotic pigs, in which the virome is different from the virome in pigs on farms, with PSaV resulted in the development of diarrhea [22].

Despite the worldwide spread of PAst and PSa viruses in pig populations, their occurrence is significantly different in European countries, and range from approximately 20% in Germany up to 100% in Austria for PAstV [8, 23], and from 1.6% in Hungary up to 44% in Slovenia for PSaV [24]. The heterogeneous distribution of both viruses was observed also in other countries, such as USA, Canada, Korea or China [5, 16, 25, 26]. Nevertheless, in most of these studies no direct association between diarrhea and the presence of PAstV or PSaV has been observed.

In our experiments the detection of PSaV was significantly lower than PAstV, similar to findings in neighbouring countries e.g. Czech Republic or Hungary [7, 24]. In addition, PSaV together with PKV-1 and RVA [18] affected mainly suckling piglets, while in the older pigs only a few positive samples were detected in both healthy and diarrheic pigs. Some studies suggested that the piglets are protected against infection by maternal antibodies [7, 24] but our results are not in accordance with this opinion. Whether viral infections of suckling piglets could be mostly due to vertical transmission from sows will be examined in our future experiments.

The search for TGEV and PEDV has been done in diarrheic animals only, not in healthy pigs, because we believe that viral infection would be manifested with strong clinical signs (diarrhea, vomiting) which would not be observed, of course, in healthy animals. Neither virus was found in any clinical sample. However, TGEV has been already detected in neighbouring countries. TGEV is present in Hungary and causes sporadic outbreaks in pig herds [27]. Antibodies against TGEV have also been found in wild boars in Czech Republic [28]. PEDV has been detected in many European countries, including neighbouring Austria where virus was probably introduced by purchasing piglets from a German source [29].

Recently, a novel chimeric swine enteric coronavirus (SeCoV) has been identified in diseased pigs in Slovakia with clinical signs as diarrhea, vomiting and severe dehydration in young piglets resulting in a high level of mortality. Genetic analyses revealed that most of the genome of this virus was derived from TGEV, but the S-gene originated from PEDV [30]. The chimeric SeCoV detected in Slovakia in early 2015 was highly related, but not identical, to SeCoV isolates found in Italy in 2009–2012 [31] and later in Germany [32]. The isolates from Slovakia could be detected by RT-PCR with primers selected from the S gene. We also used primers selected from S gene [33] in our RT-PCR assay. Since all samples were negative for PEDV, no SeCoV isolate(s) circulated in any farms investigated in this study.

Phylogenetic analysis of the partial RdRp region of our selected PAstV positive swabs revealed a high genetic diversity among them including all known five viral lineages in pigs in Slovakia. The comparison with available sequences from GenBank database revealed that the majority of PAstV sequences belong to the lineages PAstV-2 and PAstV-4, which were also the most often detected in neighbouring countries e.g. Czech Republic and Hungary [7, 8], but also in North America [4, 21]. In addition we detected another three rare lineages PAstV-1, PAstV-3 and PAstV-5; each of them was represented by one sequence. Until now, all five lineages were detected inside Europe only in Croatia [6].

Despite the majority of PAstV sequences being farm specific, three additional different lineages - PAstV-1, PAstV-2 and PAstV-4 were found on one farm (Fig. 2, farm named RI). We did not detect more than one PAstV strain in a single pig as it was reported by Luo et al. [21]. In some sequence chromatograms, however, several smaller nucleotide peaks on the bottom of higher peaks were observed which could be the result of the presence of a minor viral isolate in the pig sample.

All 12 PSaV sequences analysed in this study fall in the phylogenetic tree into genogroup GIII, which is the

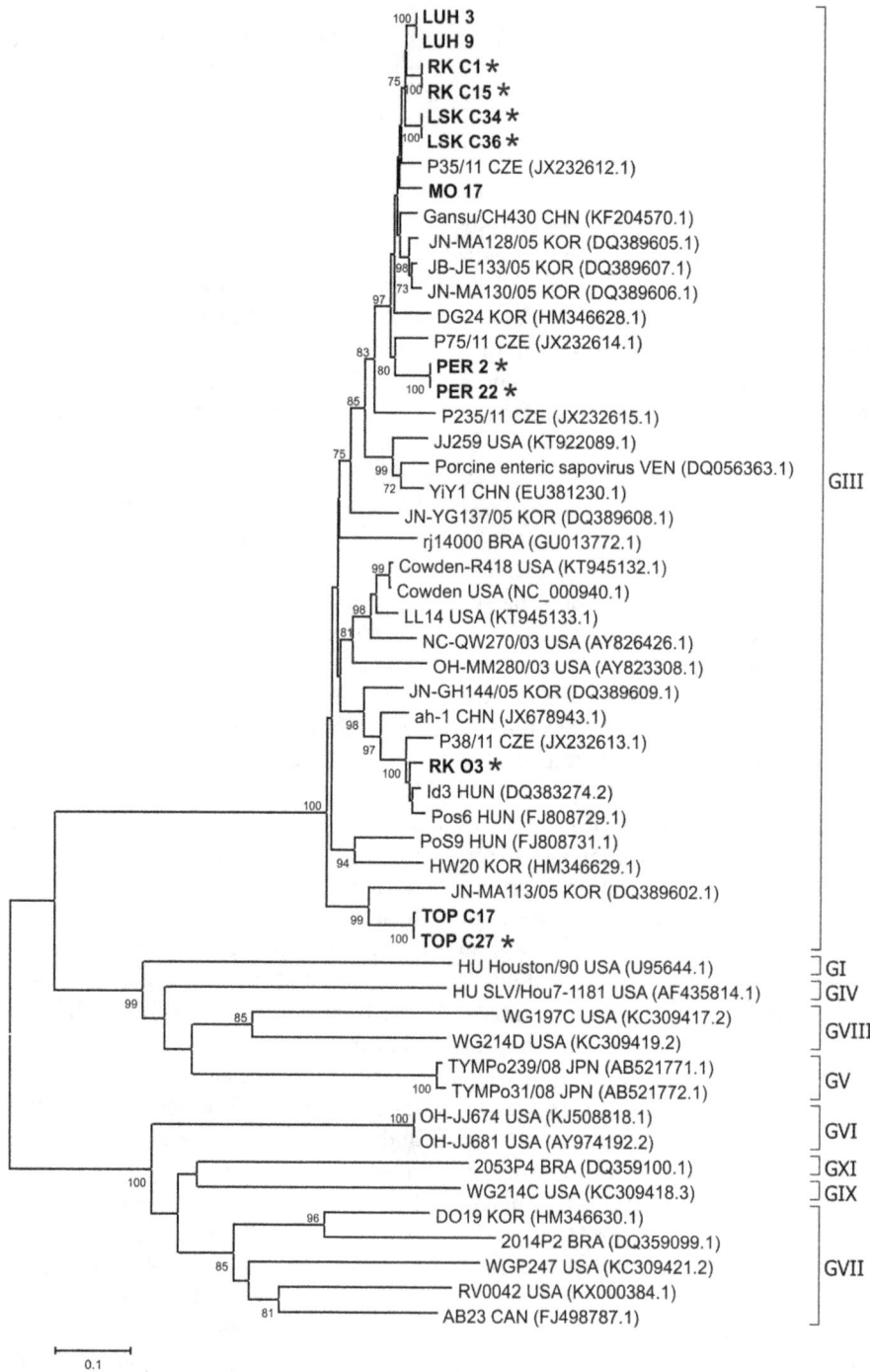

Fig. 3 Phylogenetic tree of PSaV. The tree was constructed on the base of 667 bp nucleotide sequences of capsid PSaV and selected sequences from GenBank (labelled with accession number and country of origin). Two references human strains from GenBank are also included (HU). Sequences from Slovakian samples are in bolt, from diarrheic pigs are marked with stars. The tree was constructed using the neighbour-joining method implemented in MEGA6 software. Scale bar represents the number of nucleotide substitutions per site

most prevalent genogroup worldwide [24]. The phylogenetic tree indicated three different genotypes inside GIII (Fig. 3) as describe by Jeong et al. [34] and Dufkova et al. [7]. The RK O3 isolate showed a phylogenetic relatedness with Czech strain P38/11/CZ which was proposed as "P38/11-like" genotype [7]. This isolate was different from other isolates coming from the same farm and represented the only exception when viral isolates from the same farm did not belong to the same cluster.

Generally, most of our PSaV sequences clustered together with sequences originating from PSaV of neighbouring countries. This phenomenon indicates the circulation of related viral isolates in Central Europe. None of the Slovakian isolates has been found to be related to human sapovirus. Thus, at least at present PSaV has no zoonotic risk for human populations.

Conclusion

This work documented the molecular detection and diversity of enteric viral agents in suckling, weaned and fattening pigs on farms in Slovakia. PAstV was found as a dominant virus species with high presence (80–99%) in investigated farms, but the presence was not depending on the health status of pigs. On the other side PSaV was found in a small percentage (around 9%) of both, healthy and diarrheic animals with higher occurrence in suckling piglets. The equal presence of both viruses in healthy and diarrheic pigs does not clearly clarify their role in gastrointestinal diseases and their detection has no diagnostic value. In addition, the phylogenetic analysis revealed all five genetic lineages of PAstV sequences. No TGEV and PEDV have been found in diarrheic pigs of different ages.

Methods

Farms, sample collection and preparation

The samples were collected from 17 pig farms located in Slovakia between years 2013 and 2016. The size of farms varied between 2 and 400 pigs (11 small farms) and 1.000–6.000 pigs (6 large farms). There were not the farms with all age categories of pigs because most Slovakian farms are specialized for breeding of some age categories of animals. The selection of farms was based on the information of problems with health status of pigs as inappetence, vomiting, loss of weight, growth disorders, wasting and high morbidity. Our attention was especially focused on animals with strong diarrhea.

A total of 411 rectal swabs were used in this study. The pigs belonged to three different age stages: suckling piglets (< 28 days, $n = 143$), weaned pigs (28–70 days, $n = 147$) and fattening pigs (> 70 days, $n = 121$). Rectal swabs were collected from both healthy pigs ($n = 251$) and diarrheic pigs ($n = 160$).

The swabs collected from each animal were placed into 1 ml of 0.01 mol/l PBS (Merck Milipore Corp., USA) for 30 min, and subsequently vortexed at 2000 rev. min^{-1} for 3 min and centrifuged at 12000 rpm for 5 min. After the procedure, if samples were not directly used for RNA isolation, they were stored at − 80 °C.

RNA isolation

Total RNA was isolated from 200 μl of suspension using TRIzol Reagent (Life Technologies, USA) according to the manufacturer's instructions, eluted in 20 μl of molecular grade water (Merck, GmbH, Germany), and used immediately or stored at − 80 °C.

RT-PCR and nested-PCR

Reverse transcription (RT) and polymerase chain reaction (PCR) were performed separately. The synthesis of cDNA was carried out in a 20 μl reaction mixture using RevertAid Premium reverse transcriptase (ThermoScienfitic, USA) containing 5 μl of isolated RNA. The cDNA synthesis was performed according manufacturer's instructions.

PCR was carried out in 50 μl reaction volume consisting of 1xThermoPol reaction buffer (New England Biolabs, USA), 0.2 mM dNTPs (ThermoScienfitic, USA), 0.3 μM of each primer, 0.5 U Taq DNA polymerase (New England Biolabs, USA), 4 μl cDNA and molecular grade water. Primers used for PCR detection of TGEV, PEDV, PSaV and PAstV are shown in Table 1. A single PCR was carried out for the detection of TGEV, PEDV and PSaV. A semi-nested PCR was used for PAstV detection. PCR primers panAV-F11 (forward), panAV-F12 (forward) and panAV-R1

Table 1 Sequence of primers used for RT-PCR and sequencing

Virus	Primer	Sequence (5'-3')	Region	Size (bp)	References
PAstV	panAV-F11	GARTTYGATTGGRCKCGKTAYGA	RdRp	420–400	[36]
	panAV-F12	GARTTYGATTGGRCKAGGTAYGA			
	panAV-F21	CGKTAYGATGGKACKATHCC			
	panAV-F22	AGGTAYGATGGKACKATHCC			
	panAV-R1	GGYTTKACCCACATICCRAA			
PSaV	PECVcapsidF	CTCATCAACCCTTTTGAAAC	Capsid protein	757	[26]
	PECVcapsidR	AAAGCATGATGTTGTTAGGC			
TGEV	T1	GTGGTTTTGGTYRTAʌATGC	Spike protein	858	[33]
	T2	CACTAACCAACGTGGARCTA			
PEDV	P1	TTCTGAGTCACGAACAGCCA	Spike protein	650	
	P2	CATATGCAGCCTGCTCTGAA			

(reverse) were used in an initial RT-PCR, while primers panAV-F21 (forward), panAV-F22 (forward) and panAV-R1 were used in the semi-nested PCR (Table 1).

The PCR program used for detection of all investigated viruses had the following thermal profile: initial denaturation at 95 °C for 1 min, followed by 35–37 cycles with denaturation at 95 °C for 30 s, annealing at 50–53 °C for 1 min, extension at 68 °C for 1 min and final extension at 68 °C for 5 min.

The size of PCR products was analyzed by 2% agarose gel electrophoresis and visualized by Gel Doc EZ imager (Bio-Rad Laboratories, USA) after staining with GelRed™ (Biotum, USA).

While tests for PAstV and PSaV were done on samples collected from healthy and diarrheic pigs ($n = 411$), tests for TGEV and PEDV were done only on samples from diarrheic samples ($n = 160$).

Sequencing of DNA and phylogenetic analysis

PCR amplicons ($n = 25$) from PAstV positive samples from 13 farms and PCR amplicons ($n = 12$) obtained from PSaV positive samples originating from 6 farms were used for sequencing and phylogenetic analysis. The selection of PCR amplicons was carried out on the basis of their quality and to cover healthy and diarrheic pigs from the entire geographic area. The purification of PCR amplicons and sequencing in both directions using Sanger's method employing fluorescently labelled ddNTPs was carried out by the commercial company Microsynth Austria GmbH (Austria). The chromatograms were checked and edited by the computer programme SeqMan in our laboratory. The nucleotide sequences were deposited into GenBank under accession numbers MG051040-MG051076.

The nucleotide sequences were aligned by the MegAlign program (Lasergene, DNASTAR, Inc., USA). The phylogenetic trees were constructed by the neighbour-joining method using the Kimura-2 parameter using program MEGA6 [35].

Statistical analysis

Statistical analyses of data were performed by chi-square (χ^2) test with confidence limits of 95%, $P < 0.05$ or 99%, $P < 0.01$ using GraphPad Prism 5 for Windows (GraphPad Software, USA).

Abbreviations

PAstV: Porcine astrovirus; PEDV: Porcine epidemic diarrhea virus; PKV-1: Porcine kobuvirus 1; PSaV: Porcine sapovirus; RdRp: RNA-dependent RNA polymerase; TGEV: Transmissible gastroenteritis virus

Acknowledgements

We would like to thank Peter Nettleton, Edinburgh for critical reading of manuscript and correction of English grammar.

Funding

This work was supported by projects APVV-15-0415 from The Slovak Research and Development Agency and Medical University Park in Košice (MediPark, Košice) ITMS: 26220220185 supported by Operational Programme Research and Development (OP VaV-2012/2.2/08-RO) (Contract No. OPVaV/12/2013).

Authors' contributions

SS participated in the molecular genetic studies, performed statistical analyses and drafted the manuscript. AJ and MV participated on molecular genetic studies and evaluated the results. RM and JN were involved in epidemiological studies and collection of samples on the farms. SV as a project leader conceived the study, participated in the design and coordination of experiments and helped to draft the manuscript. All authors read and approved the final manuscript.

Ethics approval

None of the pigs were killed with the purpose to fulfill the objectives of the present study. The collection of rectal swabs was carried out with verbal agreement of the pig farmers. The protocol for collection of clinical samples followed the guidelines stated in the Guide for the Care and Use of Animals (protocol number 3323/16–221/3) which was approved by the State Veterinary and Food Administration of the Slovak Republic and by Ethics Commission of the University of Veterinary Medicine and Pharmacy in Kosice, Slovakia.

Consent for publication

Not applicable.

Competing interests

The authors declare that they have no competing interests.

References

1. King AMQ, Adams MJ, Carstens EB, Lefkowitz EJ, editors. Virus taxonomy: classification and nomenclature of viruses (ninth report of the international committee on the taxonomy of viruses). New York: Elsevier; 2012.
2. Di Bartolo I, Tofani S, Angeloni G, Ponterio E, Ostanello F, Ruggeri FM. Detection and characterization of porcine caliciviruses in Italy. Arch Virol. 2014;159(9):2479–84.
3. Lee MH, Jeoung HY, Park HR, Lim JA, Song JY, An DJ. Phylogenetic analysis of porcine astrovirus in domestic pigs and wild boars in South Korea. Virus Genes. 2013;46(1):175–81.
4. Mor SK, Chander Y, Marthaler D, Patnayak DP, Goyal SM. Detection and molecular characterization of porcine astrovirus strains associated with swine diarrhea. J Vet Diagn Investig. 2012;24(6):1064–7.
5. Laurin MA, Dastor M, L'homme Y. Detection and genetic characterization of a novel pig astrovirus: relationship to other astroviruses. Arch Virol. 2011; 156(11):2095–9.
6. Brnić D, Jemeršić L, Keros T, Prpić J. High prevalence and genetic heterogeneity of porcine astroviruses in domestic pigs. Vet J. 2014;202(2): 390–2.
7. Dufkova L, Scigalkova I, Moutelikova R, Malenovska H, Prodelalova J. Genetic diversity of porcine sapoviruses, kobuviruses, and astroviruses in asymptomatic pigs: an emerging new sapovirus GIII genotype. Arch Virol. 2013;158(3):549–58.
8. Zhou W, Ullman K, Chowdry V, Reining M, Benyeda Z, Baule C, Juremalm M, Wallgren P, Schwarz L, Zhou E, Pedrero SP, Henning-Pauka I, Segales J, Liu L. Molecular investigation on the prevalence and viral load of enteric viruses in pigs from five European countries. Vet Microbiol. 2016;182:75–81.
9. Oka T, Lu Z, Phan T, Delwart EL, Saif L, Wang Q. Genetic characterization and classification of human and animal sapoviruses. PLoS One. 2016;11(5): e0156373. https://doi.org/10.1371/journal.pone.0156373.
10. Scheuer AK, Oka T, Hoet AE, Gebreyes WA, Molla BZ, Saif LJ, Wang QH. Prevalence of porcine noroviruses, molecular characterization of emerging porcine sapoviruses from finisher swine in the United States, and unified classification scheme for sapoviruses. J Clin Microbiol. 2013; 51(7):2344–53.
11. Kapoor A, Li L, Victoria J, Oderinde B, Mason C, Pandey P, Zaidi SZ, Delwart E. Multiple novel astrovirus species in human stool. J Gen Virol. 2009;90(Pt 12):2965–72.

Molecular detection of enteric viruses and the genetic characterization of porcine astroviruses...

199

12. Saif LJ, Bohl EH, Theil KW, Cross RF, House JA. Rotavirus-like, calicivirus-like, and 23-nm virus-like particles associated with diarrhea in young pigs. J Clin Microbiol. 1980;12(1):105–11.

13. Li L, Pesavento PA, Shan T, Leutenegger CM, Wang C, Delwart E. Viruses in diarrhoeic dogs include novel kobuviruses and sapoviruses. J Gen Virol. 2011;92(Pt 11):2534–41.

14. Tse H, Chan WM, Li KS, Lau SK, Woo PC, Yuen KY. Discovery and genomic characterization of a novel bat sapovirus with unusual genomic features and phylogenetic position. PLoS One. 2012;7(4):e34987. https://doi.org/10.1371/journal.pone.0034987.

15. De Benedictis P, Schultz-Cherry S, Burnham A, Cattoli G. Astrovirus infections in humans and animals-molecular biology, genetic diversity, and interspecies transmissions. Infect Genet Evol. 2011;11(7):1529–44.

16. Zhang B, Tang C, Yue H, Ren Y, Song Z. Viral metagenomics analysis demonstrates the diversity of viral flora in piglet diarrhoeic faeces in China. J Gen Virol. 2014;95(Pt 7):1603–11.

17. Sun RQ, Cai RJ, Chen YQ, Liang PS, Chen DK, Song CX. Outbreak of porcine epidemic diarrhea in suckling piglets, China. Emerg Infect Dis. 2012;18(1):161–3.

18. Jackova A, Sliz I, Mandelik R, Salamunova S, Novotny J, Kolesarova M, Vlasakova M, Vilcek S. Porcine kobuvirus 1 in healthy and diarrheic pigs: genetic detection and characterization of virus and co-infection with rotavirus A. Infect Genet Evol. 2017;49:73–7.

19. Zhang W, Shen Q, Hua X, Cui L, Yang S. The first Chinese porcine sapovirus strain that contributed to an outbreak of gastroenteritis in piglets. J Virol. 2008;82(16):8239–40.

20. Park SJ, Kim HK, Moon HJ, Song DS, Rho SM, Han JY, Nguyen VG, Park BK. Molecular detection of porcine kobuviruses in pigs in Korea and their association with diarrhea. Arch Virol. 2010;155(11):1803–11.

21. Luo Z, Roi S, Dastor M, Gallice E, Laurin MA, L'Homme Y. Multiple novel and prevalent astroviruses in pigs. Vet Microbiol. 2011;149(3–4):316–23.

22. Guo M, Hayes J, Cho KO, Parwani AV, Lucas LM, Saif LJ. Comparative pathogenesis of tissue culture-adapted and wild-type Cowden porcine enteric calicivirus (PEC) in gnotobiotic pigs and induction of diarrhea by intravenous inoculation of wild-type PEC. J Virol. 2001;75(19):9239–51.

23. Machnowska P, Ellerbroek L, Johne R. Detection and characterization of potentially zoonotic viruses in faeces of pigs at slaughter in Germany. Vet Microbiol. 2014;168(1):60–8.

24. Reuter G, Zimsek-Mijovski J, Poljsak-Prijatelj M, Di Bartolo I, Ruggeri FM, Kantala T, Maunula L, Kiss I, Kecskeme'ti S, Halaihel N, Buesa J, Johnsen C, Hjulsager CK, Larsen LE, Koopmans M, Böttiger B. Incidence, diversity, and molecular epidemiology of sapoviruses in swine across Europe. J Clin Microbiol. 2010;48(2):363–8.

25. Wang Q-H, Souza M, Funk JA, Zhang W, Saif LJ. Prevalence of noroviruses and sapoviruses in swine of various ages determined by reverse transcription-PCR and microwell hybridization assays. J Clin Microbiol. 2006;44(6):2057–62.

26. Kim HJ, Cho HS, Cho KO, Park NY. Detection and molecular characterization of porcine enteric calicivirus in Korea, genetically related to sapoviruses. J Vet Med B Infect Dis Vet Public Health. 2006;53(4):155–9.

27. Lorincz M, Biksi I, Andersson S, Cságola A, Tuboly T. Sporadic re-emergence of enzootic porcine transmissible gastroenteritis in Hungary. Acta Vet Hung. 2014;62(1):125–33.

28. Sedlak K, Bartova E, Machova J. Antibodies to selected viral disease agents in wild boars from the Czech Republic. J Wildl Dis. 2008;44(3):777–80.

29. Steinrigl A, Fernández SR, Stoiber F, Pikalo J, Sattler T, Schmoll F. First detection, clinical presentation and phylogenetic characterization of porcine epidemic diarrhea virus in Austria. BMC Vet Res. 2015;11:310. https://doi.org/10.1186/s12917-015-0624-1.

30. Belsham GJ, Rasmussen TB, Normann P, Vaclavek P, Strandbygaard B, Bøtner A. Characterization of a novel chimeric swine enteric coronavirus from diseased pigs in Central Eastern Europe in 2016. Transbound Emerg Dis. 2016;63(6):595–601.

31. Boniotti MB, Papetti A, Lavazza A, Alborali G, Sozzi E, Chiapponi C, Faccini S, Bonilauri P, Cordioli P, Marthaler D. Porcine epidemic diarrhea virus and discovery of a recombinant swine enteric coronavirus, Italy. Emerg Infect Dis. 2016;22(1).83–7.

32. Akimkin V, Beer M, Blome S, Hanke D, Höper D, Jenckel M, Pohlmann A. New chimeric porcine coronavirus in swine feces, Germany, 2012. Emerg Infect Dis. 2016;22(1):1314–5.

33. Kim SY, Song DS, Park BK. Differential detection of transmissible gastroenteritis virus and porcine epidemic diarrhea virus by duplex RT-PCR. J Vet Diagn Investig. 2001;13(6):516–20.

34. Jeong C, Park SI, Park SH, Kim HH, Park SJ, Jeong JH, Choy HE, Saif LJ, Kim SK, Kang MI, Hyun BH, Cho KO. Genetic diversity of porcine sapoviruses. Vet Microbiol. 2007;122(3–4):246–57.

35. Tamura K, Stecher G, Peterson D, Filipski A, Kumar S. MEGA6: molecular evolutionary genetics analysis version 6.0. Mol Biol Evol. 2013;30(12):2725–9.

36. Chu DK, Poon LL, Guan Y, Peiris JS. Novel astroviruses in insectivorous bats. J Virol. 2008;82(18):9107–14.

Permissions

The contributors of this book come from diverse backgrounds, making this book a truly international effort. This book will bring forth new frontiers with its revolutionizing research information and detailed analysis of the nascent developments around the world.

We would like to thank all the contributing authors for lending their expertise to make the book truly unique. They have played a crucial role in the development of this book. Without their invaluable contributions this book wouldn't have been possible. They have made vital efforts to compile up to date information on the varied aspects of this subject to make this book a valuable addition to the collection of many professionals and students.

This book was conceptualized with the vision of imparting up-to-date information and advanced data in this field. To ensure the same, a matchless editorial board was set up. Every individual on the board went through rigorous rounds of assessment to prove their worth. After which they invested a large part of their time researching and compiling the most relevant data for our readers.

The editorial board has been involved in producing this book since its inception. They have spent rigorous hours researching and exploring the diverse topics which have resulted in the successful publishing of this book. They have passed on their knowledge of decades through this book. To expedite this challenging task, the publisher supported the team at every step. A small team of assistant editors was also appointed to further simplify the editing procedure and attain best results for the readers.

Apart from the editorial board, the designing team has also invested a significant amount of their time in understanding the subject and creating the most relevant covers. They scrutinized every image to scout for the most suitable representation of the subject and create an appropriate cover for the book.

The publishing team has been an ardent support to the editorial, designing and production team. Their endless efforts to recruit the best for this project, has resulted in the accomplishment of this book. They are a veteran in the field of academics and their pool of knowledge is as vast as their experience in printing. Their expertise and guidance has proved useful at every step. Their uncompromising quality standards have made this book an exceptional effort. Their encouragement from time to time has been an inspiration for everyone.

The publisher and the editorial board hope that this book will prove to be a valuable piece of knowledge for researchers, students, practitioners and scholars across the globe.

List of Contributors

Suya Cao and Yulong Wang
Department of Wildlife Medicine, Wildlife Resources Faculty, Northeast Forestry University, Harbin 150040, China

Jian Ma
State Key Laboratory of Veterinary Biotechnology, Harbin Veterinary Research Institute, the Chinese Academy of Agriculture Sciences, Harbin 150001, China

Cheng Cheng and Wendong Ju
Heilongjiang International Travel Healthcare Center, Harbin 150001, China

Hongwei Zhu, Qingrong Huang, Jianlong Zhang, Linlin Jiang, Xin Yu and Xingxiao Zhang
School of Life Sciences, Ludong University, No. 186 Hongqi Middle Rd., Zhifu District, Yantai 264025, China

Xiaoliang Hu
Harbin Veterinary Research Institute, Chinese Academy of Agricultural Sciences, Harbin 150069, China

Shipeng Cheng and Wenhui Chu
Institute of Special Economic Animal and Plant Sciences, Chinese Academy of Agricultural Sciences, No. 4899 Juye St., Jingyue District, Changchun 130112, China

Ruiqiao Li, Kangkang Guo, Caihong Liu, Jing Wang, Dan Tan, Xueying Han, Chao Tang, Yanming Zhang and Jingyu Wang
College of Veterinary Medicine, Northwest A&F University, Yangling, Shannxi 712100, China

Zhixun Zhao, Guohua Wu, Xinmin Yan, Xueliang Zhu, Jian Li, Haixia Zhu, Zhidong Zhang and Qiang Zhang
Key Laboratory of Animal virology of the Ministry of Agriculture, State Key Laboratory of Veterinary Etiological Biology, Lanzhou Veterinary Research Institute, CAAS, Lanzhou, People's Republic of China

Urska Jamnikar-Ciglenecki and Andrej Kirbis
Institute of Food safety, Feed and Environment, Veterinary Faculty, University of Ljubljana, Gerbičeva 60, 1000 Ljubljana, Slovenia

Urska Kuhar
Institute of Microbiology and Parasitology, Veterinary Faculty, University of Ljubljana, Gerbičeva 60, 1000 Ljubljana, Slovenia

Andrej Steyer
Institute of Microbiology and Immunology, Faculty of Medicine, University of Ljubljana, Zaloška 4, 1000 Ljubljana, Slovenia

Han-Siang Hsu and Ting-Han Lin
Department of Veterinary Medicine, College of Veterinary Medicine, National Pingtung University of Science and Technology, Neipu, Pingtung, Taiwan

Lee-Shuan Lin and Cheng-Shu Chung
Department of Veterinary Medicine, College of Veterinary Medicine, National Pingtung University of Science and Technology, Neipu, Pingtung, Taiwan
Veterinary Hospital, College of Veterinary Medicine, National Pingtung University of Science and Technology, Neipu, Pingtung, Taiwan

Ming-Tang Chiou and Chao-Nan Lin
Department of Veterinary Medicine, College of Veterinary Medicine, National Pingtung University of Science and Technology, Neipu, Pingtung, Taiwan
Animal Disease Diagnostic Center, College of Veterinary Medicine, National Pingtung University of Science and Technology, Neipu, Pingtung, Taiwan

Hung-Yi Wu
Graduate Institute of Veterinary Pathobiology, College of Veterinary Medicine, National Chung-Hsing University, South Dist, Taichung, Taiwan

Antonio Rivero-Juarez, María A. Risalde, Mario Frias, Pedro Lopez-Lopez and Angela Camacho
Infectious Diseases Unit. Instituto Maimonides de Investigación Biomédica de Córdoba (IMIBIC), Hospital Universitario Reina Sofía de Córdoba. Universidad de Córdoba, 2° Floor, Avenida Menendez Pidal s/n, 14004 Córdoba, Spain

Antonio Rivero
Infectious Diseases Unit. Instituto Maimonides de Investigación Biomédica de Córdoba (IMIBIC), Hospital Universitario Reina Sofía de Córdoba. Universidad de Córdoba, 2° Floor, Avenida Menendez Pidal s/n, 14004 Córdoba, Spain
Unidad de Enfermedades Infecciosas. Hospital Provincial, Complejo Hospitalario reina Sofía de Córdoba, Avenida Menendez Pidal s/n, 14006 Cordoba, Spain

Ignacio García-Bocanegra, Saul Jimenez-Ruiz and David Cano-Terriza
Animal Health Department. Veterinary Science College, Universidad de Córdoba, 14014 Cordoba, Spain

Jose C. Gomez-Villamandos
Animal Pathology Department. Veterinary Science College, Universidad de Córdoba, Cordoba, Spain

Brigitte Böhm, Benjamin Schade, Benjamin Bauer and Jens Böttcher
Bavarian Animal Health Service, Senator-Gerauer-Straße 23, 85586 Poing, Germany

Bernd Hoffmann, Donata Hoffmann and Martin Beer
Institute of Diagnostic Virology, Friedrich-Loeffler-Institute, Südufer 10, 17493 Greifswald-Insel Riems, Germany

Ute Ziegler
Institute of Novel and Emerging Infectious Diseases, Friedrich-Loeffler-Institute, Südufer 10, 17493 Greifswald-Insel Riems, Germany

Christine Klaus
Institute of bacterial Zoonoses and Infections, Friedrich-Loeffler-Institute, Naumburger Straße 96 a, 07743 Jena, Germany

Herbert Weissenböck
Institute of Pathology and Forensic Veterinary Medicine, University of Veterinary Medicine, Veterinärplatz 1, 1210 Vienna, Austria

Xue Lian, Xin Ming, Jiarong Xu, Yong-Sam Jung and Yingjuan Qian
MOE Joint International Research Laboratory of Animal Health and Food Safety, College of Veterinary Medicine, Nanjing Agricultural University, Nanjing, China

Wangkun Cheng
Nanjing Hongshan Forest Zoo, Nanjing, China

Xunhai Zhang
Anhui Provincial Key Laboratory for Control and Monitoring of Poultry Diseases, Anhui Science and Technology University, Fengyang, China

Hongjun Chen and Chan Ding
Shanghai Veterinary Research Institute, Chinese Academy of Agricultural Sciences, Shanghai, China

Peck Toung Ooi and Zeenathul Nazariah Binti Allaudin
Department of Clinical Studies, Faculty of Veterinary Medicine, Universiti Putra Malaysia, UPM, Serdang, Selangor 43400, Malaysia

Seetha Jaganathan King
Department of Clinical Studies, Faculty of Veterinary Medicine, Universiti Putra Malaysia, UPM, Serdang, Selangor 43400, Malaysia
Asia-Pacific Special Nutrients Sdn. Bhd, Lot 18B, Jalan 241, Section 51A, Petaling Jaya, Selangor 46100, Malaysia

Lai Yee Phang
Department of Biotechnology, Faculty of Biotechnology & Molecular Science, Universiti Putra Malaysia, UPM, Serdang, Selangor 43400, Malaysia

Shiao Pau How
Vet Food Agro Diagnostic Sdn. Bhd, Lot 18B, Jalan 241, Section 51A, Petaling Jaya, Selangor 46100, Malaysia

Chiou Yan Tee, Lai Siong Yip, Pow Yoon Choo, Ban Keong Lim and Wei Hoong Loh
Vet Food Agro Diagnostic (M) Sdn. Bhd, Lot 18B, Jalan 241, Section 51A, Petaling Jaya, Selangor 46100, Malaysia

Xibao Shi
College of Life Sciences, Henan Normal University, Xinxiang 453007, China
Key Laboratory of Animal Immunology of the Ministry of Agriculture, Henan Provincial Key Laboratory of Animal Immunology, Henan Academy of Agricultural Sciences, Zhengzhou, Henan 450002, China

Xiaozhuan Zhang
College of Life Sciences, Henan Normal University, Xinxiang 453007, China
College of Veterinary Medicine and Animal Science, Henan Agricultural University, Zhengzhou, Henan 450002, China

Yongzhe Chang and Ruiguang Deng
Key Laboratory of Animal Immunology of the Ministry of Agriculture, HenanProvincial Key Laboratory of Animal Immunology, Henan Academy of Agricultural Sciences, Zhengzhou, Henan 450002, China

Gaiping Zhang
Key Laboratory of Animal Immunology of the Ministry of Agriculture, HenanProvincial Key Laboratory of Animal Immunology, Henan Academy of Agricultural Sciences, Zhengzhou, Henan 450002, China
College of Veterinary Medicine and Animal Science, Henan Agricultural University, Zhengzhou, Henan 450002, China

Aiping Wang
Department of Bioengineering, Zhengzhou University, Zhengzhou, Henan 450000, 450002, China

Bo Jiang
Office of Science & Technology, Chongqing Police College, Chongqing 401331, China

Xiaoyue Chen
Research Center of Avian Disease, College of Veterinary, Medicine of Sichuan Agricultural University, Wenjiang District, Chengdu 611130, Sichuan Province, China
Key Laboratory of Animal Disease and Human Health of Sichuan Province, Wenjiang District, Chengdu 611130, Sichuan Province, China

Renyong Jia, Mingshu Wang, Shun Chen, Mafeng Liu, Dekang Zhu, Xinxin Zhao, Kunfeng Sun, Qiao Yang, Ying Wu and Anchun Cheng
Research Center of Avian Disease, College of Veterinary, Medicine of Sichuan Agricultural University, Wenjiang District, Chengdu 611130, Sichuan Province, China
Key Laboratory of Animal Disease and Human Health of Sichuan Province, Wenjiang District, Chengdu 611130, Sichuan Province, China

Institute of Preventive Veterinary Medicine, Sichuan Agricultural University, Wenjiang District, Chengdu 611130, Sichuan Province, China

Jiakun Zhou and Xianglong Wu
Research Center of Avian Disease, College of Veterinary, Medicine of Sichuan Agricultural University, Wenjiang District, Chengdu 611130, Sichuan Province, China
Institute of Preventive Veterinary Medicine, Sichuan Agricultural University, Wenjiang District, Chengdu 611130, Sichuan Province, China

Zhongqiong Yin
Key Laboratory of Animal Disease and Human Health of Sichuan Province, Wenjiang District, Chengdu 611130, Sichuan Province, China

Jue Wang
BGI Genomics Co,shenzhen Ltd, Shenzhen 518083, Guangdong Province, China

Hiroshi Bannai, Manabu Nemoto, Hidekazu Niwa, Koji Tsujimura, Takashi Yamanaka and Takashi Kondo
Equine Research Institute, Japan Racing Association, 1400-4 Shiba, Shimotsuke, Tochigi 329-0412, Japan

Satoshi Murakami
Thermo Fisher Scientific, Life Technologies Japan Ltd, 4-2-8 Shibaura, Minato-ku, Tokyo, Japan

Tomoko Kato, Tohru Yanase and Hiroaki Shirafuji
Kyushu Research Station, National Institute of Animal Health, NARO, 2702 Chuzan, Kagoshima 891-0105, Japan

Moemi Suzuki, Yoshito Katagiri and Katsunori Takayoshi
Okinawa Prefectural Institute of Animal Health, 1-24-29 Kohagura, Naha, Okinawa 900-0024, Japan

Kazufumi Ikemiyagi
Yaeyama Livestock Hygiene Service Center, 1-2 Miyara, Ishigaki, Okinawa 907-0022, Japan

Seiichi Ohashi
Viral Disease and Epidemiology Research Division, National Institute of Animal Health, NARO, 3-1-5 Kannondai, Tsukuba, Ibaraki 305-0856, Japan

Kazuo Yoshida and Makoto Yamakawa
Exotic Disease Research Station, National Institute of Animal Health, 6-20-1 Josuihoncho, Kodaira, Tokyo 187-0222, Japan

Tomoyuki Tsuda
National Institute of Animal Health, NARO, 3-1-5 Kannondai, Tsukuba, Ibaraki 305-0856, Japan

Lingna Zhao, Wei Liu, Lin Liang and Gang Li
State Key Laboratory of Animal Nutrition, Institute of Animal Science, Chinese Academy of Agricultural Sciences, Beijing 100193, People's Republic of China

Makay Zheney
State Key Laboratory of Animal Nutrition, Institute of Animal Science, Chinese Academy of Agricultural Sciences, Beijing 100193, People's Republic of China
Faculty of Veterinary, Kazakh National Agrarian University, Almaty 050013, Republic of Kazakhstan

Zhambul Kaziyev and Gulmira Kassenova
Faculty of Veterinary, Kazakh National Agrarian University, Almaty 050013, Republic of Kazakhstan

Mathias Ackermann and Kurt Tobler
Institute of Virology. Vetsuisse Faculty, University of Zurich Winterthurerstrasse 266a, 8057 Zurich, Switzerland

Marco Geisseler
Institute of Virology. Vetsuisse Faculty, University of Zurich Winterthurerstrasse 266a, 8057 Zurich, Switzerland
Dermatology Department, Clinic for Small Animal Internal Medicine, Vetsuisse Faculty, University of Zurich, Winterthurerstrasse 260, 8057 Zurich, Switzerland
Amt für Landwirtschaft und Natur des Kantons Bern, Veterinärdienst, Herrengasse 1, 3011 Berne, Switzerland

Christian E. Lange
Institute of Virology. Vetsuisse Faculty, University of Zurich Winterthurerstrasse 266a, 8057 Zurich, Switzerland
Dermatology Department, Clinic for Small Animal Internal Medicine, Vetsuisse Faculty, University of Zurich, Winterthurerstrasse 260, 8057 Zurich, Switzerland
Department of Microbiology and Immunobiology, Harvard Medical School, 77 Avenue Louis Pasteur, Boston, MA 02115, USA

Claude Favrot and Nina Fischer
Dermatology Department, Clinic for Small Animal Internal Medicine, Vetsuisse Faculty, University of Zurich, Winterthurerstrasse 260, 8057 Zurich, Switzerland

Cun Liu, Xuehua Wei, Yue Zhang and Fulin Tian
Shandong Provincial Center for Animal Disease Control and Prevention, Ji'nan 250022, Shandong, China

Yanhan Liu and Ye Tian
College of Veterinary Medicine, China Agricultural University, Beijing 100193, China

Lisa Dähnert, Martin Eiden, Christine Fast and Martin H. Groschup
Institute of Novel and Emerging Infectious Diseases, Friedrich-Loeffler-Institut, Südufer 10, 17493 Greifswald, Insel Riems, Germany

Josephine Schlosser
Department of Veterinary Medicine, Institute of Immunology, Freie Universität Berlin, Robert-von-Ostertag-Straße 7-13, 14163 Berlin, Germany

Charlotte Schröder and Elke Lange
Department of Experimental Animal Facilities and Biorisk Management, Friedrich-Loeffler-Institut, 17493 Greifswald, Insel Riems, Germany

Albrecht Gröner
PathoGuard Consult, Fasanenweg 6, 64342, Seeheim-Jugenheim, Germany

Wolfram Schäfer
CSL Behring Biotherapies for Life, 35002 Marburg, Germany

Suquan Song
MOE Joint International Research Laboratory of Animal Health and Food Safety, Institute of Immunology and College of Veterinary Medicine, Nanjing Agricultural University, Nanjing 210095, People's Republic of China

Jianhua Hu, Jing Lei, Zhiyu Shi, Qian Xiao, Zhenwei Bi, Lu Yao, Yuan Li, Yuqing Chen, An Fang and Hui Li
MOE Joint International Research Laboratory of Animal Health and Food Safety, Institute of Immunology and College of Veterinary Medicine, Nanjing Agricultural University, Nanjing 210095, People's Republic of China
Jiangsu Engineering Laboratory of Animal Immunology, Institute of Immunology and College of Veterinary Medicine, Nanjing Agricultural University, Nanjing 210095, People's Republic of China

Liping Yan
MOE Joint International Research Laboratory of Animal Health and Food Safety, Institute of Immunology and College of Veterinary Medicine, Nanjing Agricultural University, Nanjing 210095, People's Republic of China
Jiangsu Engineering Laboratory of Animal Immunology, Institute of Immunology and College of Veterinary Medicine, Nanjing Agricultural University, Nanjing 210095, People's Republic of China
Jiangsu Detection Center of Terrestrial Wildlife Disease, Institute of Immunology and College of Veterinary Medicine, Nanjing Agricultural University, Nanjing 210095, People's Republic of China

Jiyong Zhou
MOE Joint International Research Laboratory of Animal Health and Food Safety, Institute of Immunology and College of Veterinary Medicine, Nanjing Agricultural University, Nanjing 210095, People's Republic of China
Jiangsu Engineering Laboratory of Animal Immunology, Institute of Immunology and College of Veterinary Medicine, Nanjing Agricultural University, Nanjing 210095, People's Republic of China
Jiangsu Detection Center of Terrestrial Wildlife Disease, Institute of Immunology and College of Veterinary Medicine, Nanjing Agricultural University, Nanjing 210095, People's Republic of China

Key Laboratory of Animal Virology, Ministry of Agriculture, Zhejiang University, Hangzhou 310058, People's Republic of China
Collaborative Innovation Center for Diagnosis and Treatment of Infectious Diseases, The First Affiliated Hospital, Zhejiang University, Hangzhou 310058, People's Republic of China

Min Liao
Key Laboratory of Animal Virology, Ministry of Agriculture, Zhejiang University, Hangzhou 310058, People's Republic of China
Collaborative Innovation Center for Diagnosis and Treatment of Infectious Diseases, The First Affiliated Hospital, Zhejiang University, Hangzhou 310058, People's Republic of China

Chunhe Wan, Cuiteng Chen, Longfei Cheng, Hongmei Chen, Qiuling Fu, Shaohua Shi, Guanghua Fu, Rongchang Liu and Yu Huang
Fujian Provincial Key Laboratory for Avian Diseases Control and Prevention, Fujian Animal Diseases Control Technology Development Center, Institute of Animal Husbandry and Veterinary Medicine of Fujian Academy of Agricultural Sciences, Xifeng Road No.100, Jiantian village, Jin'an district, Fuzhou 350013, China

Slavomira Salamunova, Anna Jackova, Rene Mandelik, Jaroslav Novotny, Michaela Vlasakova and Stefan Vilcek
University of Veterinary Medicine and Pharmacy, Komenskeho 73, 040 00 Kosice, Slovakia

Index